Forensic DNA Profiling Protocols

METHODS IN MOLECULAR BIOLOGY™

John M. Walker, SERIES EDITOR

METHODS IN MOLECULAR BIOLOGY™

Forensic DNA Profiling Protocols

Edited by

Patrick J. Lincoln

University of London, UK

and

Jim Thomson

LGC, Teddington, UK

Humana Press ✳ Totowa, New Jersey

This publication is printed on acid-free paper. ∞
ANSI Z39.48-1984 (American Standards Institute)
Permanence of Paper for Printed Library Materials.

Cover illustration: Fig. 6 in Chapter 24, "The Use of Capillary Electrophoresis in Genotyping STR Loci," by John M. Butler.

Cover design by Jill Nogrady.

For additional copies, pricing for bulk purchases, and/or information about other Humana titles, contact Humana at the above address or at any of the following numbers: Tel.: 973-256-1699; Fax: 973-256-8341; E-mail: humana@mindspring.com; or visit our Website: http://humanapress.com

Printed in the United States of America. 10 9 8 7 6 5 4 3 2 1

Library of Congress Cataloging in Publication Data

Main entry under title:

Methods in molecular biology™.

Forensic DNA profiling protocols/edited by Patrick J. Lincoln and Jim Thomson.
    p.   cm.—(Methods in molecular biology; 98)
    Includes index.
    ISBN 0-89603-443-7 (alk. paper)
    1. Forensic genetics—Laboratory manuals. 2. DNA fingerprinting—Laboratory manuals. 3. Polymerase chain reaction—Laboratory manuals. 4. DNA—Analysis—Laboratory manuals. I. Lincoln, Patrick J. II. Thomson, Jim, 1963– . III. Series: Methods in molecular biology (Totowa, NJ); 98.
    [DNLM: 1. DNA Fingerprinting—methods—laboratory manuals. 2. Forensic Medicine—laboratory medicine. 3. DNA—chemistry—laboratory manuals. W1 Me9616J v. 98 1998/W 625 D629 1998]
RA1057.5.D63 1998
614'.1—DC21
DNLM/DLC
for Library of Congress

        97-30911
        CIP

Preface

It is now more than ten years since Dr. Alec Jeffreys (now Professor Sir Alec Jeffreys, FRS) reported in *Nature* that the investigation of certain minisatellite regions in the human genome could produce what he termed DNA fingerprints and provide useful information in the fields of paternity testing and forensic analysis. Since that time we have witnessed a revolution in the field of forensic identification. A total change of technology, from serological or electrophoretic analysis of protein polymorphisms to direct investigation of the underlying DNA polymorphisms has occurred in a short space of time. In addition, the evolution and development of the DNA systems themselves has been rapid and spectacular.

In the last decade we have progressed from the multilocus DNA fingerprints, through single locus systems based on the same Southern blot RFLP technology, to a host of systems based on the PCR technique. These include Allele Specific Oligonucleotide (ASO)-primed systems detected by dot blots, the "binary" genotypes produced by mapping variations within VNTR repeats demonstrated by minisatellite variant repeat (MVR) analysis, and yet other fragment-length polymorphisms in the form of Short Tandem Repeat (STR) loci. Hand in hand with the increasing range of systems available has been the development of new instrumentation to facilitate their analysis and allow us to explore the possibilities of high volume testing in the form of mass screening and offender databases.

This explosion of new technologies with their vast potential has brought with it the need for forensic scientists to equip themselves and their laboratories with a whole array of new expertise. With the high discriminating power of the DNA systems has come high potential in evidentiary terms, high profile status for many investigations, and not least, a high degree of professional scrutiny of evidence produced by such technology.

Although there has been a two way exchange of learning between the forensic scientists taking molecular biology into the forensic arena and molecular biologists learning about the application of science to medicolegal problems, it is to those forensic scientists faced with having to apply new technologies that the chapters of this book are primarily addressed. Because of the speed with which DNA analysis has progressed in the field, some of the technologies described in *Forensic DNA Profiling Protocols* may be quite familiar to many workers and no longer in routine use in their laboratories; we expect that others will still benefit from their inclusion.

The present book provides protocols for the major methods of DNA analysis that have been introduced for identity testing in forensic laboratories. There are chapters addressing the recovery of DNA from a wide range of sample types, including blood, semen, skeletal remains, and saliva-stained material, such as cigaret butts. Analysis of the DNA recovered from such materials using single and multilocus VNTR probes and using all the PCR systems mentioned previously are covered in the chapters that follow. A valuable introduction to PCR primer design is also included. Several chapters are dedicated to the now widely used STR systems, including details of both manual and automated detection systems. An important chapter on the interpretation and statistical evaluation of the results of many of these DNA systems will serve as an introduction to an area of discussion and debate largely outside the scope of this book.

Finally, there are a number of chapters dealing with other methodologies such as repeat unit mapping (MVR), capillary electrophoresis direct blotting, and solid-phase minisequencing. Sex determination, mitochondrial DNA analysis, and species identification are also given chapters.

As has been a feature of the *Methods in Molecular Biology* series, this is primarily a practical guide with extensive Notes sections providing tips and useful hints on how to get the best out of a given method. These, we believe, are a valuable feature of the series and will help many scientists overcome the inevitable stumbling blocks that come with any new technique.

We are indebted to Professor John Walker, the series editor, for his patience and helpful encouragement through the gestation of this volume, but most of all we owe our thanks to all the contributing authors who so expertly provided manuscripts to our specifications and without whom there would be no book.

Patrick J. Lincoln
Jim Thomson

Contents

Contributors

ATSUSHI AKANE • *Department of Legal Medicine, Kansai Medical University, Moriguchi, Japan*

ROBERT C. ALLEN • *Department of Pathology and Laboratory Medicine, Medical University of South Carolina, Charleston, SC*

JULIA ANDERSEN • *Forensic Science Service, Metropolitan Laboratory, London, UK*

MICHAEL L. BAIRD • *Lifecodes Corporation, Stamford, CT*

RITA BARALLON • *LGC, Teddington, UK*

BERND BRINKMANN • *Institut für Rechtsmedizin, Universität Münster, Germany*

JOHN F. Y. BROOKFIELD • *Department of Genetics, University of Nottingham, UK*

BRUCE BUDOWLE • *Forensic Science Research and Training Center, FBI Academy, FBI Laboratory, Quantico, VA*

JOHN M. BUTLER • *DNA Technologies Group, National Institute of Standards and Technology, Gaithersburg, MD*

ANGEL CARRACEDO • *Instituto de Medicina Legal, Universidade de Santiago de Compostela, Spain*

WAYNE P. CHILDS • *Cellmark Diagnostics, Abingdon, UK*

JÜRGEN HENKE • *Institut für Bludgruppenforschung, Koln, Germany*

LOTTE HENKE • *Institut für Bludgruppenforschung, Koln, Germany*

MANFRED N. HOCHMEISTER • *IRM Institut für Rechtsmedizin, Universität Bern, Switzerland*

MITCHELL M. HOLLAND • *The Armed Forces DNA Identification Laboratory, Office of the Armed Forces Medical Examiner, The Armed Forces Institute of Pathology, Rockville, MD*

MAVIKY LAREU • *Instituto de Medicina Legal, Universidade de Santiago Compostela, Spain*

DENNIS A. LEE • *The Armed Forces DNA Identification Laboratory, Office of the Armed Forces Medical Examiner, The Armed Forces Institute of Pathology, Rockville, MD*

NIELS MORLING • *Department of Forensic Genetics, Institute of Forensic Medicine, University of Copenhagen, Denmark*

DAVID L. NEIL • *Department of Genetics, University of Leicester, UK*

CHRIS PHILLIPS • *Department of Haematology, St. Bartholomew's and the Royal London School of Medicine and Dentistry, London, UK*

JAMES M. ROBERTSON • *Perkin-Elmer Applied Biosystems, Foster City, CA. Present Address: Forensic Science Research and Training Center, FBI Laboratory, FBI Academy, Quantico, VA*

OSKAR RUDIN • *IFM Institut für Rechtsmedizin, Universität Bern, Switzerland*

GILLIAN RYSIECKI • *Cellmark Diagnostics, Abingdon, UK*

PETER M. SCHNEIDER • *Institut für Rechtsmedizin, Johannes Gutenberg Universität, Mainz, Germany*

ROBERT STEIGHNER • *The Armed Forces DNA Identification Laboratory, Office of the Armed Forces Medical Examiner, The Armed Forces Institute of Pathology, Rockville, MD*

ANN-CHRISTINE SYVÄNEN • *Department of Human Molecular Genetics, National Public Health Institute, Helsinki, Finland*

JIM THOMSON • *LGC, Teddington, UK*

JENNIFER WALSH-WELLER • *Perkin-Elmer Applied Biosystems, Foster City, CA*

DAVID WATTS • *Applied Biosystems Division, Perkin Elmer, Warrington, UK*

BRUCE S. WEIR • *Program in Statistical Genetics, Department of Statistics, North Carolina State University, Raleigh, NC*

JEANNE M. WILLARD • *The Armed Forces DNA Identification Laboratory, Office of the Armed Forces Medical Examiner, The Armed Forces Institute of Pathology, Rockville, MD*

1

Recovery of High-Molecular-Weight DNA from Blood and Forensic Specimens

Peter M. Schneider

1. Introduction

The isolation of genomic deoxyribonucleic acid (DNA) is a crucial step in the process of DNA profiling. The success of all subsequent genetic-typing procedures depends on the availability of sufficient amounts of highly purified DNA from biological crime stains as well as from reference blood samples. High-mol-wt DNA is usually only required for DNA-profiling protocols based on Southern blot analysis and hybridization with multi- and single-locus variable number of tandem-repeat (VNTR) probes in cases in which stain samples contain at least microgram amounts of DNA. The DNA extraction yield from biological stains is difficult to estimate in advance and depends on a number of unpredictable factors regarding the stain sample, e.g., storage conditions, exposure to heat, sunlight, moisture, or bacterial and fungal contamination, all of which influence the quality and final yield of the isolated DNA. However, the isolation of high-mol-wt DNA might also be appropriate for small stain samples that initially appear to be suitable only for polymerase chain reaction (PCR) typing. This allows a decision on the typing method after recovery of the stain DNA based on the amount and quality of the extracted DNA, thus providing more flexibility for the profiling procedure.

2. Materials

1. 50 mM KCl (for hypotonic lysis of erythrocytes).
2. Lysis buffer: 25 mM ethylenediaminetetra-acetic acid (EDTA), 75 mM NaCl, 10 mM Tris-HCl pH 7.5. Add 200 mg/mL Proteinase K immediately before use.
3. 10% (w/v) sodium dodecyl sulfate (SDS).
4. ANE buffer: 10 mM sodium acetate, 100 mM NaCl, 1 mM EDTA.

From: *Methods in Molecular Biology, Vol. 98: Forensic DNA Profiling Protocols*
Edited by: P. J. Lincoln and J. Thomson © Humana Press Inc., Totowa, NJ

5. Buffered phenol: Use only crystallized phenol, dissolve in a water bath at 65°C, and add approx 0.1–0.2 vol ANE buffer until an aqueous phase forms above the phenol; add 0.01% 8-hydroxyquinoline (w/v solid crystals).
6. Chloroform/isoamylalcohol: To 24 parts of chloroform, add one part of isoamyl-alcohol.
7. 6 *M* NaCl.
8. 3 *M* sodium acetate, pH 5.2: dissolve 40.8 g Na acetate in 80 mL distilled water, bring to pH 5.2 with concentrated acetic acid, and adjust final volume with water to 100 mL.
9. TE buffer: 10 m*M* Tris-HCl, 1 m*M* EDTA, pH 7.5.
10. 20X SSC: 175.3 g NaCl, 88.2 g Na citrate/L, adjust to pH 7.0 with NaOH.
11. 0.2 *M* sodium acetate, pH 7.0.
12. 10 mg/mL Proteinase K stock solution: dissolve 10 mg proteinase K in 1 mL of 10 m*M* Tris-HCl, pH 8.0, and store frozen in small aliquots.
13. Phosphate-buffered saline (PBS): 50 m*M* phosphate buffer, pH 7.4, 0.9% NaCl.
14. PBS/S: PBS + 2% sarcosyl.
15. PBS/PKS: PBS + 100 µg/mL proteinase K + 1% SDS (= 10 µL PK stock solution + 100 µL 10% SDS/mL PBS).
16. PBS/Lys: PBS + 100 µg/mL Proteinase K + 2% sarcosyl + 10 m*M* DTT + 25 m*M* EDTA (= 10 µL PK stock + 10 µL 1 *M* DTT stock 50 µL + 0.5 *M* EDTA stock/mL PBS/S).

3. Methods

3.1. DNA Extraction from Fresh Blood Samples

3.1.1. Purification of White Blood Cells

1. Place 10–15 mL EDTA blood into 50-mL polypropylene tube. In addition, transfer two 0.5-mL aliquots of blood into 1.5-mL microfuge tubes as reference samples and store them frozen at –20°C (*see* **Subheading 3.2.** for rapid extraction).
2. Add 50 m*M* KCl up to 50 mL, mix well, and incubate in water bath at 37°C for 10 min.
3. Spin for 10 min at 500*g* in a clinical centrifuge at room temperature.
4. Remove the supernatant with a Pasteur pipet connected to a water jet or vacuum pump; leave the pellet intact (hold the tube in front of a light source to see the white-cell pellet).
5. Repeat **steps 2–4** once or twice, until the pellet is free from red cells (*see* **Note 1**).
6. Add 15 mL lysis buffer and shake vigorously to resuspend the cell pellet.
7. Add 1.5 mL 10% SDS and mix carefully. The sample should now become very viscous because of cell lysis.
8. Incubate overnight at 37°C or at 55°C for 4–5 h.

3.1.2. Organic Extraction of DNA

1. Add 1 vol buffered phenol to the proteinase K-digested cell extract and mix the aqueous and organic phases carefully to achieve a homogeneous suspension; avoid vigorous shaking.

2. Spin for 10 min at 1500*g* in clinical centrifuge at room temperature.
3. The aqueous phase containing the DNA is on top, and the phenolic phase is below. Transfer the aqueous phase to a fresh 50-mL polypropylene tube using a wide-bore glass pipet (*see* **Note 2**). The interphase containing proteins and protein–DNA complexes may also be transferred at this step.
4. Add 1 vol of a 1:1 mixture of phenol and chloroform/isoamyl alcohol (24:1) and extract as described in **steps 1–3**.
5. Add 1 vol of chloroform/isoamylalcohol and repeat the extraction as described in **steps 1–3**. At this step, avoid transferring any residual protein debris from the interphase.

3.1.3. Inorganic Extraction of DNA

1. Add 5 mL of 6 *M* NaCl (one-third of the original volume) to the proteinase K-digested cell extract and shake vigorously for 10–15 s (*see* **Note 3**).
2. Spin for 10 min at 1500*g* in a clinical centrifuge at room temperature to separate the salt-precipitated proteins.
3. Pour the supernatant into a fresh 50-mL tube.

3.1.4. DNA Precipitation

1. After organic extraction, add 0.1 vol of 3 *M* sodium acetate (i.e., 1.5 mL for the procedure described here) and 2 vol of ice-cold absolute ethanol. After salt precipitation with 6 *M* NaCl, add only the ethanol.
2. Mix carefully without vigorous shaking. The DNA should precipitate by forming viscous strings first and finally a compact pellet, which may float on top of the solution.
3. Melt the tip of a Pasteur pipet on a Bunsen burner to form a hook. Use this tool to recover the floating DNA pellet from the solution (*see* **Note 4**).
4. Rinse the DNA pellet attached to the glass hook twice in 70% ethanol to remove excess salt.
5. Dry the pellet briefly in the air and resuspend the DNA in an appropriate volume (300–500 µL depending on the size) of 0.1X TE.
6. Incubate the sample for 1 h at 65°C or overnight at 37°C in a water bath to dissolve the DNA. If the sample is still very viscous, add more 0.1X TE and incubate again at 65°C until a homogeneous solution is obtained.

3.2. Rapid DNA Extraction
from Small Aliquots of Frozen Blood Samples

1. To a 0.5-mL blood aliquot in a 1.5-mL microcentrifuge tube add 1 mL 1X SSC, mix gently, and spin for 1 min in microfuge. The blood sample must have been frozen previously to achieve complete red-cell lysis.
2. Remove 1.2 mL of supernatant without disturbing the pellet, add 1.2 mL 1X SSC, mix gently, and spin for 1 min.
3. Remove 1.4 mL of supernatant, add 375 µL of 0.2 *M* sodium acetate, and resuspend the pellet by vortexing.

4. Add 25 µL of 10% SDS + 10 µL of 10 /mL proteinase K stock solution, vortex for 1 s, and incubate for 1 h at 56°C.
5. Spin tube for 1 s, add 120 µL of a 1:1 mixture of phenol and chloroform/ isoamylalcohol (24:1), vortex for 10 s, and spin for 2 min.
6. Transfer the aqueous upper phase (approx 400 µL) to fresh tube, add 1 mL of absolute ethanol, and mix by inverting the tube until DNA precipitate forms.
7. Spin for 15 s, remove the supernatant and add 200 µL of 0.2 *M* sodium acetate, leave the tube for 2 min at room temperature, and redissolve the pellet by vortexing for 10 s (*see* **Note 5**).
8. Add 500 µL ethanol, precipitate the DNA again, spin for 15 s, and remove the supernatant.
9. Add 1 mL of 70% EtOH, mix, and spin again for 30 s. When removing the supernatant, watch the pellet, since it may not stick to the tube wall.
10. Dry the pellet briefly by keeping the tube inverted for 15 min, and add 50 µL 0.1X TE. Incubate for 15–30 min at 65°C to dissolve the DNA.

3.3. DNA Extraction from Dried Blood Stains

1. Cut the fabric with the blood stain into small pieces. Depending on the size of the stain, transfer the fabric pieces into a 1.5-mL microcentrifuge or a 15-mL polypropylene tube and add 0.5–5 mL PBS/Lys (*see* **Note 6**). Incubate overnight at 37°C with mild agitation.
2. Add an equal volume of buffered phenol and mix the aqueous and organic phases carefully to achieve a homogeneous suspension.
3. Spin for 10 min at 1500*g* in clinical centrifuge (15-mL tubes) or at maximum speed in a microcentrifuge (1.5-mL tubes) at room temperature. The fabric pieces remain in the phenolic phase and will thus be separated from the DNA in the aqueous phase. Transfer the upper phase into a fresh tube.
4. Add 1 vol of a 1:1 mixture of phenol and chloroform/isoamylalcohol (24:1) and extract as described in **steps 2** and **3**.
5. Add 1 vol of chloroform/isoamyl alcohol and repeat the extraction as described in **steps 2** and **3**. At this step, avoid transferring any residual protein debris from the interphase.
6. Transfer the final supernatant to a Centricon-30® (Amicon, Inc., Beverly, MA) microconcentrator tube and purify DNA from salt, SDS, and contaminants with three washes with 2 mL distilled water each. Concentrate the sample to a final volume of 100–150 µL (*see* **Note 7**).

3.4. DNA Extraction from Vaginal Swabs and from Semen Stains

3.4.1. Differential Lysis Procedure for Vaginal Swabs

1. Soak swab or fabric with stain containing mixed male/female secretion in PBS/S (2–5 mL depending on size) in an appropriate tube and incubate overnight with mild agitation at 4°C (*see* **Note 8**).
2. Vortex briefly, remove the swab or fabric, and spin down cells at 2500*g* for 10 min at 4°C. Save supernatant in a separate tube (*see* **Note 9**).

3. Wash the swab or fabric in 1.5 mL PBS/S in a microcentrifuge tube. Punch a small hole in the bottom of the closed tube, insert this tube into an open second tube, and punch another hole in the lid of the tube containing the sample. Spin both tubes for 10 min at 1500*g* in a microcentrifuge. After centrifugation, the upper tube should contain the dry swab or fabric (do not discard) and the lower tube the buffer and cell pellet. Save the supernatant and combine the cell pellet with the pellet from **step 2** in 1.5 mL PBS/S.

4. Wash the pellet twice in 1.5 mL PBS/S in a microfuge tube and spin it down for 10 min at 1500*g*. Resuspend the pellet in 50 μL PBS and remove 3–5 μL for microscopic analysis.

5. Add 500 μL PBS/PKS and incubate for 2 h at 50°C with mild agitation for lysis of vaginal epithelial cells. Recover sperm heads by centrifugation at 1500*g* for 10 min at 4°C.

6. Save the supernatant (= female fraction) for further analysis of female DNA. Resuspend the pellet (containing the sperm heads) in 30 μL PBS and remove 2–3 μL for microscopic analysis.

7. Lyse the sperm heads by adding 500 μL PBS/Lys and incubate for 3 h at 50°C with mild agitation (= male fraction).

8. Separately purify DNA from female and male fractions by three subsequent extractions with an equal volume of phenol, phenol/chloroform/isoamylalcohol (25:24:1), and chloroform/isoamylalcohol (24:1), as described for blood stains (**steps 2–5** of **Subheading 3.3.**).

9. Transfer both supernatants to two separate Centricon-30® microconcentrator tubes, and purify DNA from salt, SDS, and contaminants with three washes with 2 mL distilled water each. Concentrate samples to a final volume of 100–150 μL.

3.4.2. DNA Extraction from Dried Semen Stains

1. Wash fabric with sperm stain as described in **steps 1–3** of **Subheading 3.4.1.**
2. Resuspend combined pellets in 500 μL PBS/Lys and incubate for 3 h at 50°C with mild agitation.
3. Purify sperm DNA as described in **steps 8** and **9** of **Subheading 3.4.1.**

4. Notes

1. After this step, the cell pellet can be stored frozen without buffer at –20 or –70°C. The pellets may also be shipped on dry ice for extraction by the receiving laboratory.

2. The organic-extraction protocol is designed to obtain very high-mol-wt DNA >50 kb *(1)*. Therefore, vigorous shaking of the sample has to be avoided after adding SDS. To avoid shearing of high-mol-wt DNA during pipetting, a pipetman equipped with a disposable blue tip where the narrow end has been cut off may be used to transfer the aqueous phase. Alternatively, the phenolic phase may also be removed by piercing a hole into the bottom of the 50-mL tube and holding it over a glass beaker. When the phenolic phase has been drained, the aqueous phase has to be poured immediately into a fresh tube. Wear protective gloves and goggles to avoid skin or eye contact with the caustic phenol.

3. The "salting out"-method *(2)* is routinely used in our laboratory, because it generates high-mol-wt DNA of sufficient quality both for restriction-fragment-length polymorphism (RFLP) and PCR applications without the use of hazardous organic chemicals.

4. If no DNA precipitate forms for any reason, incubate the tube at −20°C overnight and recover the DNA by centrifugation for 15 min in a clinical centrifuge at 4000g and 4°C. Remove the supernatant, rinse the pellet carefully in 70°C ethanol, and dry it briefly by keeping the tube inverted for 15 min on a paper towel. Dissolve the pellet in 100–200 μL 0.1X TE.

5. When small amounts of DNA are precipitated, the pellets may be very small or almost invisible. It is easier to locate them after centrifugation if the tubes are inserted into the microfuge rotor with the fixtures of the tube caps pointing outward.

6. If possible, remove dried-blood particles from the carrier surface (e.g., leather, wood, or plastic material) to avoid transfer of inhibitory substances interfering with restriction enzyme or *Taq* polymerase activity. The dried-blood particles can be added directly to the PBS/Lys solution for Proteinase K digestion. Alternatively, you may also wash the stain carrier first in an appropriate volume of PBS at 4°C, with occasional shaking to remove the blood cells. After this step, remove the stain carrier, spin down the cells, and add the PBS/Lys solution for Proteinase K digestion.

7. The volume reduction by spin dialysis instead of ethanol precipitation generates a higher yield of extracted DNA. Ethanol precipitation is not very efficient in solutions with a low DNA concentration and might result in poor recovery. Using spin dialysis with microconcentrator tubes, the sample should be washed very carefully to remove all residual salts and SDS, which might inhibit restriction enzyme or *Taq* polymerase activity.

8. This protocol is based on the procedures described in refs. *3* and *4*. Before beginning with the extraction procedure, it is absolutely necessary to prepare a stained smear from the swab on a microscope slide for visual analysis to check for the presence of spermatozoa and to determine the relative amount of sperm heads compared to female epithelial cells. If only a very small number of spermatozoa is present in the sample, it might be advisable to extract the sample as a whole without differential lysis and to interpret the mixed sample based on the VNTR genotype combinations. In addition, a protocol has been described for a mild differential lysis enriching the relative amount of sperm cells compared to epithelial cells *(5)*.

9. All supernatants from intermediate steps should be saved until completion of the procedure, when the presence of sufficient amounts of DNA from the swab material has been demonstrated. Thus, the supernatants can be reprocessed in case of unexpected low yields of DNA. It cannot be excluded that spontaneous lysis occurs with some cells earlier than expected because of mechanical disruption of their membranes.

References

1. Gross-Bellard, M., Oudet, P., and Chambon, P. (1973) Isolation of high molecular weight DNA from mammalian cells. *Eur. J. Biochem.* **36,** 32–38.

2. Miller, S. A., Dykes, D. D., and Polesky, H. F. (1988) A simple salting out procedure for extracting DNA from human nucleated cells. *Nucleic Acid Res.* **16,** 1215.
3. Gill, P., Jeffreys, A. J., and Werrett, D. J. (1985) Forensic application of DNA fingerprints. *Nature* **318,** 577–579.
4. Giusti, A., Baird, M., Pasquale, S., Balasz, I., and Glassberg, J. (1986) Application of deoxyribonucleic acid (DNA) polymorphisms to the analysis of DNA recovered from sperm. *J. Forensic Sci.* **31,** 409–417.
5. Wiegand, P., Schürenkamp, M., and Schütte, U. (1992) DNA extraction from mixtures of body fluid using mild preferential lysis. *Int. J. Legal Med.* **104,** 359,360.

2

Recovery of DNA for PCR Amplification from Blood and Forensic Samples Using a Chelating Resin*

Jeanne M. Willard, Dennis A. Lee, and Mitchell M. Holland

1. Introduction

A wide range of biological samples are encountered in the field of forensic science, including blood, soft tissue, semen, urine, saliva, teeth, and bone. Forensic samples are routinely found as stains on various substrates, including cotton, denim, carpet, wallboard, wood, envelopes, and cigaret butts. Prior to collection, these samples have often been exposed to severe environmental conditions, such as varying degrees of temperature and humidity, microbial and chemical contaminants, and exposure to soils and other natural substances, such as salt water.

Not only do the precollection conditions effect the quality and quantity of DNA recovered from a specimen, but the postcollection storage conditions may also have a deleterious effect on the DNA. The short- and long-term storage conditions of both evidence specimens and isolated DNA have been evaluated *(1)*. Although DNA is among the most stable biomolecules, it has been found that storage of various biological samples at different temperatures for varying amounts of time can have a marked effect on the ability to isolate high-mol-wt DNA. Consequently, this poses a problem for restriction-fragment-length polymorphism (RFLP) analysis of many forensic samples. RFLP analysis which is commonly used in forensic casework, generally requires between 20 and 100 ng of high-mol-wt DNA *(2)*. Polymerase chain reaction PCR-based typing methods, however, do not require high quantities of DNA and therefore have a

---

*The opinions and assertions expressed herein are solely those of the authors and are not to be construed as official or as the views of the United States Department of Defense or the United States Department of the Army.

From: *Methods in Molecular Biology, Vol. 98: Forensic DNA Profiling Protocols*
Edited by: P. J. Lincoln and J. Thomson © Humana Press Inc., Totowa, NJ

higher success rate with degraded samples. The selected method for isolating DNA from forensic samples, therefore, often influences the ability to successfully perform DNA analysis.

As a result, several DNA extraction methods have been developed and evaluated by the forensic community. The organic extraction method (e.g., SDS/proteinase K, phenol/chloroform, ethanol precipitation) is very successful and is routinely used for forensic samples analyzed by RFLP *(3)*. When applied to PCR-based typing methods, however, this method was found to have limitations. Although sufficient DNA was recovered for analysis, it did not always amplify *(3)*. This failure of amplification was thought to have been caused by the inability to remove inhibitors, such as heme, during the extraction process *(4)*. Greater amplification success was obtained by following organic extraction with a dialysis and concentration step using a Centricon 100® device (Amicon, Beverly, MA) rather than ethanol precipitation *(5)*. In addition to organic methods of DNA extraction, several inorganic methods have also been developed, including the use of high salt concentrations *(6)*, glass powders *(7)*, and silica-gel suspensions *(8,9)* to recover high-mol-wt DNA. Unfortunately, all of the above extraction methods are time-consuming and require several steps, washing/desalting procedures, and multiple tube transfers. An alternative DNA extraction method has been developed that involves the addition of a chelating-resin suspension directly to the sample *(10)*.

The chelating-resin-based procedure is a simple, one-tube, minimal step extraction process that requires very little time. The risk of operator-induced error, such as contamination or sample mixup, is also reduced, since the procedure requires fewer manipulations. Most importantly, the use of a chelating resin to extract DNA eliminates the use of hazardous chemicals, such as phenol–chloroform.

The simplicity of chelating-resin-extraction methods has made this extraction method very desirable in the forensic community. DNA has successfully been extracted using Chelex® 100 (Bio-Rad Laboratories, Hercules, CA), a chelating resin, from a variety of forensic samples, including whole blood, bloodstains, tissue, and bone (*see* **Note 1**). Chelex 100 scavenges metal contaminants to an extremely high degree of purity without altering the concentration of nonmetallic ions *(11)*. The resin is composed of styrene divinylbenzene copolymers containing paired iminodiacetate ions that act as chelating groups in binding polyvalent metal ions. Chelex 100 has a particularly high selectivity for divalent ions and differs from ordinary ion exchangers because of its high selectivity for metal ions and its higher bond strength *(12)*.

Using this method, the basic procedure for recovering DNA from forensic samples, such as whole blood and bloodstains, consists of an initial wash step to remove possible contaminants and inhibitors, such as heme and other proteins. For other forensic samples, such as tissue and bone, the wash step is not necessary. The samples are then boiled in a 5% suspension of Chelex 100. After a quick spin, an aliquot of the supernatant can be added directly to the amplification reaction. The alkalinity of Chelex 100 suspensions and the exposure to 100°C temperatures result in the disruption of the cell membranes and denaturation of the DNA. The exact role of Chelex 100 during the boiling process is still unclear; however, Walsh et al. *(10)* has shown that purified DNA that has been subjected to temperatures of 100°C in distilled water alone is inactive in PCR (*see* **Note 2**). Therefore, the presence of the chelating resin during the boiling step may have a protective role and prevent the degradation of DNA by chelating metal ions. These ions may act as catalysts in the breakdown of DNA in low ionic strength solutions *(13)* at high temperatures.

The effectiveness of a particular DNA extraction method may be evaluated on the DNA yield itself; however, the more important measurements of the effectiveness are the suitability of the extracted DNA for amplification and the quality of the obtained results. A variety of DNA extraction methods have been evaluated based on their ability to yield DNA from bloodstains that had been deposited on several different substrates. Included in the following study were two common extraction methods utilized in the field of forensic science: Chelex 100 and the organic method (*see* **Note 3**). Additional studies were also performed to compare PCR-typing results obtained from samples extracted by each method. The results revealed that there was no difference between the human leukocyte antigen DQα locus (HLA DQα) genotypes obtained by PCR amplification of DNA extracted by the Chelex 100-based procedure or from those extracted via the organic method (*see* **Note 4**).

Our laboratory has been able to successfully extract DNA from blood and tissue samples that have been exposed to extreme environmental conditions by using the Chelex 100 method (*see* **Notes 5** and **6**). Additionally, these samples have been successfully amplified and analyzed using several PCR-based methods. The DNA-typing systems applied include the following: HLA-DQα, PolyMarker (PM-LDLR, GYPA, HBGG, D7S8, GC), D1S80, a short-tandem-repeat (STR) quadruplex (HUMVWFA31, HUMTH01, HUMF13A01, and HUMFESFPS) and mitochondrial DNA sequencing. The AmpliType® HLA DQα *(5)* and AmpliType® PM *(14)* systems are both reverse-dot blot, colorimetric assays, based on DNA-sequence polymorphisms. The D1S80 *(15)* and STR quadruplex *(16)* systems detect length polymorphisms of repeated DNA sequences. Mitochondrial DNA sequencing detects sequence polymorphisms within two hypervariable regions *(17)*.

To conclude, although Chelex 100 is an excellent extraction method, the forensic community is continuing to investigate new automatable methods of extraction. Methods have been developed that involve the use of chemically impregnated filter papers amenable to automation. One such procedure involves the immobilization of DNA onto filter paper while cell and body-fluid contaminants are selectively removed. The filter paper containing DNA is then dehydrated by rinsing with alcohol followed by a drying step. This allows the sample to be amplified directly from the filter paper by adding amplification reagents *(18)*. Another example of a new method is one in which the chemicals impregnated on the filter paper disrupt and release DNA from nucleated cells while simultaneously discouraging the release of inhibitory substances *(19)*. The inhibitory substances are trapped on the filter paper and the DNA is captured in solution. Finally, the choice of an extraction method is an important decision and one that must be carefully considered. When processing forensic samples, selecting the appropriate method can have significant impact on the end result.

2. Materials
1. Chelex 100.
2. Sterile deionized H_2O.

3. Methods
3.1. Chelex Extraction of Whole Blood and Bloodstains
1. Pipet 1 mL of sterile deionized water into a 1.7-mL microcentrifuge tube.
2. Add approx 3 µL of whole blood or a piece of bloodstained material (approx 1/8 in.-diameter hole punch or a 3 mm² piece of material).
3. To wash the sample, mix and incubate at room temperature for 15–30 min (*see* **Note 7**).
4. Prepare a 5% Chelex solution in sterile deionized H_2O (*see* **Note 8**).
5. Centrifuge samples for 3 min at 10,000g in a microcentrifuge to pellet the red blood cells.
6. Carefully remove all but approx 20–30 µL of the supernatant and discard. Leave the substrate and pelleted material in the tube.
7. Resuspend the pellet in 5% Chelex 100 to a final volume of 200 µL (*see* **Note 9**).
8. Incubate at 56°C for 30 min (*see* **Note 10**).
9. Vortex at high speed for 5 s.
10. Incubate in a boiling water bath for 8 min.
11. Vortex at high speed for 5 s.
12. Centrifuge samples for 3 min at 10,000g to pellet the Chelex 100 resin, substrate, and remaining tissue/bone.
13. Extracts are now ready for quantitation and the PCR amplification process (*see* **Note 11**).
14. For short-term storage (up to 1 mo), store extracts at 2–8°C on the Chelex 100 resin. To use the extracts after storage, repeat **steps 11** and **12**. For long-term storage (>1 mo), the extracts should be centrifuged and the supernatant removed from the Chelex 100 resin and stored in a new tube at –20°C (*see* **Note 12**).

Table 1
Average DNA Yields (ng)/5-µL Bloodstain

	Chelex 100			Organic/ethanol		
Substrate	Average	Standard deviation	Range	Average	Standard deviation	Range
Carpet	22	14.1	0–50	32	35.6[a]	0–100
Cotton	88	72.6	25–200	30	25.0	12.5–100
Denim	26	27.6[a]	0–100	56	55.4	0–200
Nylon	34	41.1[a]	0–100	48	45.8	25–200
Wallboard	62	31.6	25–100	55	37.6	12.5–100
Wood	44	28.1	0–100	62	51.3	12.5–200

[a]The presence of both high and low yields contributed to the standard deviation being greater than the means. **Note:** Each method/substrate combination was repeated 16 times.

3.2. Chelex Extraction of Tissue and Fresh Bone

1. Prepare a 5% Chelex 100 suspension in sterile deionized H_2O.
2. Add 200 µL of 5% Chelex 100 suspension to a 1.7 mL microcentrifuge tube.
3. Cut an approx 2 mm^2 piece of tissue or, if present, scrape the marrow from inside of a fresh bone and place directly into Chelex 100 suspension (*see* **Note 13**).
4. Incubate at 56°C for 30 min.
5. Continue with protocol for whole blood and bloodstains (**steps 9–14**).

4. Notes

1. Protocols for extracting DNA using the Chelex 100 method are available for other forensic specimens, such as postcoital samples and oral swabs containing sperm, hair, semen, semen stains, and paraffin-embedded tissue *(10)*.
2. If DNA is boiled in a solution of 0.01 M Tris-HCl, pH 8.0 with 0.1 M EDTA (TE buffer), it is protected for PCR *(10)*. EDTA is a chelating agent that can attach itself to a single metal ion through six donor atoms. This theory is also in agreement with Singer-Sam et al. *(13)*, which suggests that Chelex 100 sequesters divalent heavy metals that would otherwise introduce DNA damage. An advantage of using Chelex 100 resin rather than TE buffer is that the Chelex 100 resin can be removed from the supernatant by centrifugation. This helps prevent any inhibitors or contaminants that may be bound to the resin from being introduced into the PCR. Additionally, samples that contain high concentrations of TE buffer may have a deleterious effect on the PCR by sequestering magnesium ions that are essential to Thermus aquaticus *(Taq)* polymerase. Therefore, it is advisable to use a PCR buffer with 2.0 mM $MgCl_2$ for amplifications from DNA stored in TE buffer with high concentrations of EDTA *(20)*.
3. In a study performed by the Federal Bureau of Investigation (FBI) *(21)*, several extraction methods were compared. The data shown in **Table 1** are a comparison of the two most common extraction methods in the field of forensic science:

Table 2
Results of Concordance Study in Typing Chelex
vs Organic Extraction at the HLA DQa Locus[a]

Sample type	Number	% Concordance
Bloodstains	56	100
Semen stains	9	100
Postcoital swabs	8	100
Buccal swabs	6	100
Hairs	5	100
TOTAL	84	100

[a]Number of different genotypes present: 18 of 21 possible (all except 1.2,2; 1.3,1.3, 2,2); different alleles present: 6 of 6 possible.

Chelex 100 and organic. A comparison of DNA yields was performed using the slot-blot method *(22)*. Samples extracted from either method yielded a sufficient amount of DNA for PCR. For most substrates, excluding cotton and denim, the yield of DNA was relatively consistent. A possible explanation for lower yields with the organic extraction of bloodstain on cotton may be the loss of DNA during the number of tube transfers that are required during the extraction process. The organic method resulted in a twofold difference in the concentration of DNA recovered from the bloodstain on denim, compared to the Chelex 100 method. This may be caused by the longer 56°C incubation time period (overnight for organic vs 2 h for Chelex 100), which allows more time for the cellular debris to dislodge from the fabric.

4. In a study performed by Walsh et al. *(10)*, 84 samples commonly encountered in forensics were evaluated for suitability of the DNA for amplification as well as for the quality of the results obtained. Included in these samples were bloodstains exposed to a variety of environmental conditions, blood and semen stains on several different fabric substrates, postcoital swabs, buccal swabs, and hair. Samples were originally extracted using the traditional organic method and analyzed at the HLA DQα locus. The extractions were later repeated using the Chelex 100-based procedure and reanalyzed. All samples produced typeable results with no extraneous alleles reported. As summarized in **Table 2**, there were no discrepancies in the HLA DQα genotype obtained for any sample using either extraction method.

5. DNA has been successfully extracted from a variety of tissue samples that have been exposed to harsh environmental conditions. For example, 222 extractions were performed on samples recovered from the Waco, TX incident in 1994, wherein the samples were highly incinerated and in an advanced state of decomposition *(23)*. Our strategy was to perform Chelex 100 extractions first on all of the samples. This procedure can be performed quickly, and, therefore results can be obtained in a short amount of time. In this case, 60% of the samples yielded sufficient DNA by Chelex 100 extraction to obtain results. The remaining 40%

required an organic extraction to obtain results. Overall DNA recovery was minimal. More than 50% of the extracts contained <100 ng DNA. This may be attributed to the highly degraded nature of the DNA. One of the disadvantages of the Chelex 100 method is the limited sample size. The Chelex 100 procedure does not include a purification step and, therefore, if the sample contains inhibitors and contaminants, increasing the sample size increases the concentration of inhibitors and contaminants. Larger sample sizes can be used for the organic extraction method. In this instance, increasing the sample size increased the recovery of DNA. Chelex 100 can then be added to the organic extract to bind any contaminants that may still be present in the extract. To determine if inhibition is occurring when an amplification reaction fails or partial profiles are obtained, an aliquot of the sample extract can be added to the positive control to see if the results are inhibited. For example, during the fluorescence detection of STR profiles, the peak heights of the positive control may decrease in the presence of inhibitors. Inhibition can sometimes be overcome by using 1/10 or 1/100 dilutions of the original extract. Additionally, five units of *Taq* polymerase *(24)* and/or 8 µg/mL bovine serum albumin (BSA) *(25)* can be added to the reaction to overcome inhibition.

6. Our laboratory has also been successful in extracting and analyzing DNA from samples recovered from aircraft accidents in which tissue samples were extremely charred as well as soaked with fuel. For example, results were obtained from an aircraft mishap in Alaska in which 24 individuals were killed. The aircraft crashed immediately on takeoff and was carrying approx 125,000 pounds of fuel. Our laboratory received 24 tissue samples and 20 reference bloodstain cards in order to confirm identifications using DNA analysis. All 44 samples were extracted within approx 3 h. Two samples did not produce results because of inhibition and were reextracted. The extractions were repeated using a smaller sample size and results were successfully obtained. All 44 samples were extracted and analyzed using PolyMarker analysis within 24 h. In addition, STR analysis was performed on four tissue samples and four reference specimens within an additional 24 h. Therefore, our laboratory was able to confirm the identifications of 16 individuals and provide strong evidence for identification of four other individuals within 48 h. Without the use of the Chelex 100 extraction method, this would not have been possible.

7. It has been our experience that if blood is tightly bound to the substrate or ethanol fixed to the substrate, the initial wash step should be incubated at 56°C to help facilitate the diffusion of blood off of the substrate. Porphyrin compounds derived from heme in blood have also been shown to inhibit PCR *(4)*. It has been suggested that these porphyrin rings are being washed away during the initial wash step or may bind to the Chelex 100 bead matrix itself. Additional water washes may be necessary in order to rid the sample of excess heme and inhibitors. For example, two of the 84 samples (bloodstain on a red sweatshirt and semen stain on black cotton) analyzed in the Typing Concordance Study (**Table 2**) failed to amplify initially using the Chelex 100 method. The bloodstain substrate

was a red fabric that colored the extract red. This red dye apparently inhibited the amplification. An additional 3-mm^2 piece of bloodstain was allowed to soak for 30 min at room temperature and vortexed to dislodge cellular material from the fabric. The fabric substrate was removed, and the solution was centrifuged to pellet the cellular debris. The supernatant was removed and discarded. The cellular debris was washed three times with water followed by centrifugation. All but 50 µL of the last wash was discarded and 150 µL of 5% Chelex 100 was added. The remainder of the protocol was as described previously. The sample was then successfully amplified and analyzed.

8. The quality of each new manufactured lot of Chelex 100 should be assessed. A 5% Chelex 100 suspension should be made fresh for each set of extractions. According to the manufacturer of Chelex 100, the resin has a tendency to lose its chelating capacity after more than a few hours *(12)*. The effectiveness of Chelex 100 is based on pH. The pH of the 5% Chelex 100 suspension should be between 10.0 and 11.0. Do not attempt to adjust the pH if the suspension does not fall within this range. A suggested method for assessing a new lot of Chelex 100 is to extract DNA from bloodstains or whole-blood samples (of known genotype) and analyze the extracts using the most sensitive typing system available. This will control for human contaminants. A "spiked" reagent blank (9 µL of reagent blank with 1 ng of known DNA) can also be analyzed. The "spiked" reagent blank ensures that the Chelex 100 itself is not inhibiting the PCR. All extracts and "spiked" reagent-blank controls must yield detectable PCR product and generate the appropriate DNA profile. No PCR product should be present in the reagent blanks.

9. When pipetting the Chelex 100 suspension, the resin beads must be distributed evenly in solution. This can be done by gently mixing with a stir bar in a small beaker. The pipet tip used to draw up the suspension must have a relatively large bore, p1000 (200–1000 µL tip), to ensure that a 5% suspension is maintained.

10. A study was recently performed in our laboratory to evaluate the 56°C incubation period on our typing results. Fifty bloodstains were allowed to incubate at 56°C for 30 min and an additional 50 bloodstains were incubated at 56°C for 2 h. All 100 samples were successfully analyzed by STR analysis with no significant differences.

11. Specimens extracted with Chelex 100 yield single-stranded DNA and are unsuitable for quantitation methods that use intercalating dyes, such as ethidium bromide. Therefore, a quantitation method, such as slot-blot is suggested *(22)*. In addition, the DNA is not suitable for RFLP analysis because the Chelex 100 procedure results in denatured DNA. Because of our case turnaround time, our samples are usually not quantitated prior to amplification, and therefore our laboratory routinely uses the following volumes of a bloodstain extract in the PCR: 5 µL for HLA DQα, 10 µL for PM, 10 µL for D1S80, 2 µL for STRs (British Forensic Science Service Quadruplex), and 2 µL for mtDNA sequencing. All of the aforementioned analyses can be performed multiple times on a single extract, since the final volume of a Chelex 100 extraction is 200 µL.

12. If the extracts are to be reanalyzed over a period of time, it is recommended that the supernatant be removed from the Chelex 100 resin and stored at –20°C (long-term storage). It has been suggested that the resin may break down over time, which in turn may release any previously bound inhibitors, or the resin itself may inhibit the PCR by chelating magnesium ions (12).

13. It is recommended that the sample selected for extraction be dissected from the innermost part of the tissue to avoid crosscontamination from comingled samples and inhibition from environmental factors.

References

1. Koblinsky, L. (1992) Recovery and stability of DNA in samples of forensic science significance. *Forensic Sci. Rev.* **4,** 68–87.
2. Budowle, B. and Baechtal, F. S. (1990) Modification to improve the effectiveness of restriction fragment polymorphism typing. *Appl. Theoretical Electrophoresis* 181–187.
3. Budowle, B., Waye, J. S., Shutler, G. G., and Baechtal, F. S. (1990) Hae III—A suitable restriction endonuclease for restriction fragment length polymorphism analysis of biological evidence samples. *J. Forensic Sci.* **35,** 530–536.
4. Higuchi, R. (1989) Simple and rapid preparation of samples for PCR, in *PCR Technology: Principles and Applications for DNA Amplification* (Erlich, H. E., ed.) Stockton, New York, pp. 31–38.
5. Comey, C. T. and Budowle, B. (1991) Validation studies on the HLA-DQα locus using the polymerase chain reaction. *J. Forensic Sci.* **36,** 1633–1648.
6. Dykes, D. (1988) The use of biotinylated DNA probes in parentage testing: non-isotopic labeling and non-toxic extraction. *Electrophoresis* **9,** 359–368.
7. Vogelstein, B. and Gillespie, D. (1979) Preparative and analytical purification of DNA from agarose. *Proc. Natl. Acad. Sci. USA* **76,** 615–619.
8. Boom, R., Sol, C. J. A., Salimans, M. M. M., Jansen, C. L., Wertheim-van Dillen, P. M. E., and van der Noordaa, J. (1990) Rapid and simple method for purification of nucleic acids. *J. Clin. Microbiol.* **28,** 495–503.
9. Hoss, M. and Paabo, S. (1993) DNA extraction from pleistocene bones by a silica-based purification method. *Nucleic Acids Res.* **21,** 3913,3914.
10. Walsh, P. S., Metzger, D. A., and Higuchi, R. (1991) Chelex® 100 as a medium for simple extraction of DNA for PCR-based typing from forensic material. *Biotechniques* **10,** 506–513.
11. Bio-Rad Laboratories Catalog (1996) Bio-Rad Laboratories, Hercules, CA, p. 85.
12. Bio-Rad Laboratories Chelex 100 and Chelex 20 chelating ion exchange resin instruction manual (1996), pp. 1–24.
13. Singer-Sam, J., Tangua, R. L., and Riggs, A. D. (1989) Use of Chelex to improve the PCR signals from a small number of cells. *Amplifications: A Forum for PCR Users* **3,** 11.
14. Budowle, B., Lindsey, J. A., DeCou, J. A., Koons, B. W., Giusti, A. M., and Comey, C. T. (1995) Validation and population studies of the loci LDLR, GYPA, HBGG, D7S8, and GC (PM loci), and HLA-DQα using a multiplex amplification and typing procedure. *J. Forensic Sci.* **40,** 45–54.

15. Budowle, B., Chakroborty, R., Giusti, A. M., Eisenberg, A. J., and Allen, R. C. (1991) Analysis of the VNTR locus D1S80 by the PCR followed by high-resolution PAGE. *Am. J. Human Genet.* **48,** 137–144.

16. Kimpton, C., Fisher, D., Watson, S., Adams, M., Urquhart, A., Lygo, J., and Gill, P. (1994) Evaluation of an automated DNA profiling system employing multiplex amplification of four tetrameric STR loci. *Int. J. Legal Med.* **106,** 302.

17. Holland, M. M., Fisher, D. L., Roby, R. K., Ruderman, J., Bryson, C., and Weedn, V. W. (1995) Mitochondrial DNA sequence analysis of human remains. *Crime Lab. Dig.* **22,** 109–115.

18. Belgrader, P., Del Rio, S. A., Turner, K. A., Marion, M. A., Weaver, K. R., and Williams, P. E. (1995) Automated DNA purification and amplification from blood-stained cards using a robotic workstation. *Biotechniques* **19,** 426–432.

19. Harvey, M. A., King, T. H., and Burghoff, R. (1995) Impregnated 903 blood collection paper: a tool for DNA preparation from dried blood spots for PCR amplification, in *Research and Development*, Schleicher and Schuell, Inc., Keene, NH.

20. Gyllensten, U. (1989) Direct sequencing of *in vitro* amplified DNA, in *PCR Technology: Principles and Applications for DNA Amplification* (Erlich, H. E., ed.), Stockton, New York, 45–60.

21. Comey, C. T., et al. (1994) DNA extraction strategies for amplified fragment length polymorphism analysis. *J. Forensic Sci.* **39,** 1254–1269.

22. Waye, J. S., Presley, L. A., Budowle, B., Shutler, G. G., and Fourney, R. M. (1989) A simple and sensitive method for quantifying human genomic DNA in forensic specimen extracts. *Biotechniques* **7,** 852–855.

23. Clayton, T. M., Whitaker, J. P., Fisher, D. L., Lee, D. A., Holland, M. M., Weedn, V. W., Maguire, C. N., DiZinno, J. A., Kimpton, C. P., and Gill, P. (1995) Further validation of a quadruplex STR DNA typing system: a collaborative effort to identify victims of a mass disaster. *Forensic Sci. Int.* **76,** 17–25.

24. Fisher, D. L., Holland, M. M., Mitchell, L., Sledzik, M. S., Wilcox, A. W., Wadhams, M. S. S., and Weedn, V. W. (1993) Extraction, evaluation, and amplification of DNA from decalcified and undecalcified United States Civil War bone. *J. Forensic Sci.* **38,** 60–68.

25. Pflug, W., Mai, G., Wahl, S., Aab, S., Eberspacher, B., and Keller, U. (1992) A simple method to prevent inhibition of *Taq* polymerase and Hinf*I* restriction enzyme in DNA analysis of stain material, in *Advances in Forensic Haemogenetics*, vol. 4, Springer-Verlag, Berlin, pp. 163–165.

3

PCR Analysis of DNA from Fresh and Decomposed Bodies and Skeletal Remains in Medicolegal Death Investigations

Manfred N. Hochmeister

1. Introduction

One of the greatest values of polymerase chain reaction (PCR) for the death investigator lies in the fact that even minute amounts of DNA or extensively damaged (degraded) DNA can be successfully amplified and thus become amenable for typing procedures.

In medico-legal death investigations, the types of DNA recovered from a body can be divided into two areas: DNA evidence, which adheres to the surface of the body or is present within a body cavity (e.g., blood, semen, saliva, nasal secretions, hairs, shed scalp skin, urine, feces); and biological material, which belongs to the deceased (e.g., liquid blood, soft tissues, bones, teeth, fingernails) and can be used as reference samples. In general, the success of DNA profiling depends most on the environmental conditions the body was exposed to and on the proper preservation and collecting procedures.

There are many published methods for the extraction of DNA from forensic evidence for subsequent amplification. This chapter describes methods that are robust, extensively tested, and have been used successfully in difficult and high-priority cases. Since we deal with a variety of samples, the extraction methods will first be described, followed by notes for the different samples.

2. Materials

2.1. DNA Extraction and Quantitation

1. 1X stain extraction buffer *(1)*: 10 mM Tris-HCl, 10 mM EDTA, 100 mM NaCl, 39 mM DTT, 2% sodium dodecyl sulfate (SDS), pH 8.0. Short-time storage at

From: *Methods in Molecular Biology, Vol. 98: Forensic DNA Profiling Protocols*
Edited by: P. J. Lincoln and J. Thomson © Humana Press Inc., Totowa, NJ

room temperature (SDS precipitates at 4°C). Add proteinase K fresh from frozen aliquots of 20 mg/mL.

2. Phenol/chloroform/isoamylalcohol (25:24:1); in buffer 10 mM Tris-HCl, pH 7.5–8.0, 1 mM EDTA-Na$_2$. Store at 4°C.
3. Water-saturated n-butanol. Prepare fresh before use.
4. Centricon™ 100 microconcentrator tubes (Amicon, Danvers, MA).
5. QuantiBlot™ Human DNA Quantitation Kit (Roche Molecular Systems, Inc., Branchburg, NJ) or ACES™ 2.0$^+$ Human DNA Quantitation System (Gibco-BRL, Gaithersburg, MD).

For recovery of sperm-cell and vaginal-cell DNA *(1–4)*:

6. HEPES-buffered saline: 10 mM HEPES, 150 mM NaCl, pH 7.5. Store at 4°C.
7. Phosphate-buffered saline (PBS): 137 mM NaCl, 2.7 mM KCl, 1.5 mM KH$_2$PO$_4$, 8.1 mM Na$_2$HPO$_4$, pH 7.5.
8. Sarcosyl: 20 and 10%. Store at 4°C.
9. TNE: 10 mM Tris-base, 100 mM NaCl, 1 mM Na$_2$EDTA, pH 8.0. Store at 4°C.
10. Dithiothreitol (DTT) 0.39 M. Add fresh from frozen aliquots.
11. 0.5 M EDTA, pH 8.0. Store at 4°C (for bone samples).

2.2. PCR and Detection of PCR Products

1. Standard PCR reagents (e.g., Multiplex PCR kits: AmpliType® PM PCR Amplification and Typing Kit [Roche]; GenePrint™ STR Systems PCR Amplification Kit [Promega, Madison, WI]).
2. SA 32 Electrophoresis apparatus (Gibco-BRL). Gel dimensions: 310 × 0.4 mm.
3. Gel Slick (AT Biochem, Malvern, PA) and binding solution (3 µL methacryloxypropyltrimethoxysilane in 1 mL 0.5% acetic acid in 95% ethanol) to prepare glass plates.
4. Denaturing polyacrylamide gel (4% T, 5% C, containing 7 M urea and 0.5X Tris-Borate-EDTA buffer) (210 g urea, 267 mL dH$_2$O, 25 mL 10X TBE, and 50 mL 40% acrylamide:bis [19:1]).
5. 0.5X TBE (electrophoresis buffer).
6. Silver staining reagents:
   a. Fix/stop solution (10% acetic acid);
   b. Staining solution (2 g silver nitrate and 3 mL 37% formaldehyde in 2000 mL H$_2$O);
   c. Developer solution (3 mL 37% formaldehyde, 400 µL 10 mg/mL sodium thiosulfate, 60 g sodium carbonate in 2000 mL H$_2$O).
7. Duplicating X-ray film.

3. Methods

3.1. Organic Extraction of Biological Material with Centricon Purification

1. Biological material (*see* **Notes 1–13**) is incubated overnight at 56°C in 400 µL (up to 600 µL) 1X stain-extraction buffer and 10 µL (up to 50 µL) proteinase K

(20 mg/mL) in a sterile 1.5- or 2.0-mL tube with a screw cap. If swabs or cuttings are processed, a 2.2-mL microcentrifuge tube with a spin insert (with no membrane) (Costar, Cambridge, MA) or a Spin-EASE™ Extraction tube (Gibco-BRL) is used.

2. On the following day, an additional 10 µL (up to 50 µL) proteinase K (20 mg/mL) is added and the sample is incubated for an additional 2 h at 56°C.

3. Without removing the material, 700 mL phenol/chloroform/isoamylalcohol (25:24:1) is added, and the tube is vigorously shaken by hand for 2 min to achieve a milky emulsion. The tube is subjected to centrifugation at 10,000g for 3 min. If swabs or cuttings are processed, the material is removed with sterile forceps and placed into the spin insert, which is then placed into the tube from which the swab or cutting came and centrifuged at 10,000g for 5 min. Then the spin insert with the swab or cutting is removed and discarded.

4. The aqueous phase (top layer) is transferred to a new tube with a screw cap, taking care not to transfer the interphase. The phenol-extraction step is repeated up to three times for materials with high protein content.

5. To purify the DNA, the aqueous phase (bottom layer) is transferred to a Centricon 100 microconcentrator tube containing sterile water, and the volume is brought up to 2 mL with sterile water (*see* **Note 14**). The sample reservoir is sealed with parafilm, and after punching a hole in the parafilm (using a sterile needle), the tube is subjected to centrifugation at 1000g for 30 min. Then 2 mL of sterile water is added to the sample reservoir, which is then sealed with new parafilm. Again the tube is centrifuged at 1000g for 30 min. The DNA is recovered by back centrifugation at 1000g for 5 min. The final sample volume is approx 25–40 µL.

6. Ten percent of the retentate is used to determine the quantity of human DNA by slot-blot analysis (*see* **Note 15**).

3.2. Organic Extraction with Centricon Purification for the Recovery of Sperm-Cell and Vaginal-Cell DNA

1. The swab or cuttings (*see* **Note 2**) are placed into a 2.2-mL microcentrifuge tube with a spin insert (with no membrane). After addition of 450 µL HEPES-buffered saline and 50 µL 20% Sarcosyl, the tube is vigorously shaken at 4°C for at least 2 h to overnight and then briefly centrifuged. The swab or cuttings are removed with sterile forceps and placed into the spin insert. The spin insert is placed into the tube from which the swab or cuttings came and centrifuged at 10,000g for 5 min. The swab or cuttings are removed and placed into a new 1.5-mL tube with a screw cap.

2. The supernatant fluid in the tube is discarded, leaving behind approx 20 µL; 1 µL of the cell pellet is spotted on a microscopic slide and stained for the presence of sperm cells and epithelia cells (*2*).

3. The swab or cuttings are placed back into the original tube, 200 µL Tris-EDTA-NaCl, 50 µL 10% SDS, 245 µL sterile water, and 5 µL proteinase K (20 mg/mL) is added, and the tube incubated at 37°C for 2 h to isolate the DNA from epithelia cells. Again, the swab or cuttings are removed with sterile forceps and placed

into a spin insert. The spin insert is placed into the tube from which the swab or cutting came and centrifuged at 10,000*g* for 5 min.

4. The swab or cuttings are removed and discarded.
5. The supernatant fluid is saved as "epithelial-cell fraction."
6. The sperm-cell pellet is washed by adding 500 µL PBS, vortexing, and centrifuging at 10,000*g* for 5 min. This washing step is repeated three times.
7. To the washed sperm-cell pellet, 200 µL TNE, 125 µL 10% sarcosyl, 50 µL 0.39 *M* dithiothreitol (DTT), 115 µL sterile water, and 10 µL proteinase K (20 mg/mL) are added and the tube incubated at 37°C for 2 h to isolate the DNA from the sperm cells. This fluid is called "sperm-cell fraction." Both fractions are extracted with an equal volume of phenol/chloroform/isoamylalcohol and the procedure continued as described in **Subheading 3.1.**

3.3. Organic DNA Extraction from Bone

1. The DNA is extracted from 5-g fragment of the femur bone (*see* **Note 12**). The bone is cleaned with sandpaper to remove the outer layer, broken into small pieces, and pulverized into a fine powder using liquid nitrogen. The bone powder is transferred into a sterile 50-mL polypropylene tube and decalcified in 40 mL of 0.5 *M* EDTA, pH 8.0, on a rotator at 4°C for 24 h. After centrifugation at 2000*g* for 15 min, the supernatant is discarded. The powder is washed with 40 mL of extraction buffer (0.5 *M* EDTA, 0.5% sarcosyl, pH 8.0) to remove excess calcium. Then the sample is centrifuged at 2000*g* for 15 min and the supernatant discarded.
2. The DNA is extracted by adding prewarmed (37°C) extraction buffer (0.5 *M* EDTA, 0.5% sarcosyl, pH 8.0) to a final volume of approx 7 mL. Then 100 µL of proteinase K (20 mg/mL) is added and the tube incubated at 37°C for 12 h. Subsequently, 100 µL proteinase K (20 mg/mL) is added and incubation continued at 37°C for an additional 12 h.
3. The solution is extracted two to three times with phenol/chloroform/isoamylalcohol (25:24:1).
4. One extraction with 20 mL water-saturated *n*-butanol is carried out to remove traces of phenol.
5. The aqueous phase is concentrated using a Centricon 100 microconcentrator tube that is subjected to several 30-min centrifugation steps at 1000*g*. Finally, the retentate is washed three times with 2 mL of sterile water. The final sample volume is approx 25–40 µL. Bone extraction is carried out independently in duplicate, and a sample containing no bone serves as a reagent negative control sample.
6. Ten percent of the retentate is used to determine the quantity of human DNA by slot-blot analysis (*see* **Note 15**).

4. Notes

1. Blood: Blood on the skin is removed with a cotton swab moistened with sterile water. The area is swabbed carefully in a circular motion for approx 30 s. The swab is air-dried for 6 h or immediately frozen, and DNA is extracted from the swab using the method described in **Subheading 3.1.**

2. Semen: Evidence is removed as described above. DNA is extracted from the swab in one tube using the differential lysis method described in **Subheading 3.2.** *(1,3,4)*.

3. Saliva: Bite marks or areas where a perpetrator might have kissed or sucked a victim (e.g., nipples) are swabbed as described above *(5)*. DNA is extracted from the swabs using the method described in Chapter 4.

4. Nasal secretions: The material is extracted using the method described in **Subheading 3.1.** or in the same manner as described for saliva stains.

5. Hair: DNA is isolated from 1 cm of the root portion from a single hair with attached sheath material using the method described in **Subheading 3.1.**

6. Fingernails: The fingernail is chopped into very fine pieces and DNA extracted using the method described in **Subheading 3.1.** Nails can be boiled for 5 min in sterile water and then chopped into fine pieces to help the nails digest faster *(6)*. It is also possible to extract DNA without this step.

7. Fingernail debris: In a case where a victim has scratched the perpetrator, fingernail debris might contain nucleated cells or even small amounts of blood. Fingernail debris can be recovered by swabbing all nails of a hand with a single cotton swab moistened with sterile water. The material is extracted using the method in **Subheading 3.1.**

8. Urine: Extraction and amplification of DNA from urine stains is seldom successful. Methods for isolating DNA are provided by several authors *(7–10)*.

9. Feces: There is no reference that indicates that an individual was identified from fecal material. A method for preparation of fecal DNA suitable for PCR is provided by Deuter et al. *(11)*.

10. Liquid blood from a cadaver is not a reference material of choice, even if the cadaver is fresh, since after death blood clots form that trap nucleated blood cells. During the autopsy, often hemolyzed serum containing only a few nucleated blood cells is collected, which makes DNA isolation cumbersome. In our experience, DNA extraction from 300–500 µL of cadaver blood using standard protocols often fails to yield enough DNA. If cadaver blood is the only reference material, best results are obtained by using approx 5–10 mL of blood and the organic extraction method provided by Maniatis *(12)*.

11. Soft tissues are an excellent reference material, if the cadaver is fresh *(13)*. The material should be stored frozen. Approximately 100 mg of soft tissue (preferable a fat-free lymph node from the neck or a piece of muscle tissue) are cut with a sterile scalpel blade on a microscope slide and processed using the method in **Subheading 3.1.** Fifty microliters of proteinase K (20 mg/mL) are added to the sample and the solution extracted two to three times with an equal amount of phenol/chloroform/isoamylalcohol.

12. Bone: From fresh cadavers or decomposed bodies approx 5 cm of the femur bone should be routinely collected as reference material and stored frozen *(14)*. From recovered skeleton remains any available bone may be used, but compact bone is preferred. The material should be removed by cutting a long bone in a wedge shape in order to keep the length of the bone intact for future measurements. DNA can be extracted using various methods *(14–21)*; we use routinely the method in **Subheading 3.3.** In our experience, it is necessary to start with approx

5 g of bone powder in order to extract enough human DNA (sometimes only a few nanograms). Before the bone is pulverized into a fine powder, the metal blender must be thoroughly cleaned with ethanol. During pulverization, the blender should be covered with a piece of cardboard. Special care must be taken not to contaminate the powder during the process. In all bone cases, the DNA extraction is carried out independently in duplicate (preferably from different bones of the body), and a sample containing no bone serves as a reagent-negative control sample. If no amplification is achieved, the Centricon-100 extract shoud be subjected to a silica purification as described by Boom et al. *(26)*. This procedure takes approx 1 h and has led to amplification of samples that did not yield any product without this step.

13. Teeth: Dental DNA yield can be maximized by crushing the entire specimen *(22–24)*. A broken tooth is placed in 1.5 mL of buffer (0.5 M EDTA, pH 9.0; 0.01 g/mL SDS; 1 mg/mL proteinase K) and incubated for 1–3 d at 55°C. The solution is extracted two to three times with an equal amount of phenol/chloroform/isoamylalcohol (25:24:1). Then, one extraction with water-saturated n-butanol is carried out to remove traces of phenol. The aqueous phase is subsequently purified by centrifugation through a Centricon 100 microconcentrator tube as described in **Subheading 3.3.**

14. Centricon 100 purification: Each Centricon 100 microconcentrator tube is checked by adding 1.0 mL sterile water, sealing the tube with parafilm, punching a hole, and centrifuging the tube at 1000g for 3 min. In case of leaking (defect in the membrane), the water would pass through the membrane faster than in a normal tube. During the Centricon 100 purification step, it is important to remove the old parafilm and seal the tube with new parafilm every time sterile water is added. If water is added through the hole in the parafilm, contamination might occur. The Centricon 100 purification step should not be carried out in a centrifuge regularly used for centrifugation of blood samples. We could amplify DNA by swabbing the rotor of such centrifuges. The centrifuge should always be cleaned thoroughly with ethanol.

15. For storage, we transfer the retentate to 1.5 mL Saarstedt tubes with screw caps. Determination of the final sample volume is done by by weighing.

16. Compared to the Chelex method *(25)*, we obtained better results from evidentiary material using the organic extraction methods described above.

17. All forensic samples are routinely amplified in the presence of BSA (albumin bovine fraction V; Sigma A 4503) in the amplification mix (16 µg/100 µL) *(14,16,17)* in order to overcome inhibition.

References

1. Budowle, B. and Baechtel, F. S. (1990) Modifications to improve the effectiveness of restriction fragment length polymorphism typing. *Appl. Theoretical Electrophoresis* **1**, 181–187.
2. Oppitz, E. (1969) Eine neue Färbemethode zum Nachweis der Spermien bei Sittlichkeitsdelikten. *Archiv Kriminol.* **144**, 145–148.

3. Gill, P., Jeffreys, A. J., and Werrett, D. J. (1985) Forensic application of DNA fingerprints. *Nature* **318,** 577–579.

4. Giusti, A., Baird, M., Pasquale, S., Balazs, I., and Glassberg, J. (1986) Application of deoxyribonucleic acid (DNA) polymorphisms to the analysis of DNA recovered from sperm. *J. Forensic Sci.* **31,** 409–417.

5. Sweet, D. J., Lorente, J. A., Valenzuela, A., and Villanueva, E. (1995) Forensic identification using DNA recovered from saliva on human skin. *Abstracts 16th Int. Cong. Int. Soc. Forensic Haemogenetics*, EP 30.

6. Tahir, M. A. and Watson, N. (1995) Typing of DNA HLA-DQ alpha alleles extracted from human nail material using polymerase chain reaction. *J. Forensic Sci.* **40,** 634–636.

7. Prinz, M., Grellner, W., and Schmitt, C. (1993) DNA typing of urine samples following several years of storage. *Int. J. Legal Med.* **106,** 75–79.

8. Gasparini, P., Savoia, A., Pignatti, P. F., Dallapiccola, B., and Novelli, G. (1989) Amplification of DNA from epithelia cells in urine. *New Engl. J. Med.* **320,** 809.

9. Medintz, I., Chiriboga, L., McCurdy, L., and Kobilinsky, L. (1994) Restriction fragment length polymorphism and polymerase chain reaction—HLA DQA1 analysis of casework urine specimens. *J. Forensic Sci.* **39,** 1372–1380.

10. Brinkmann, B., Rand, S., and Bajanowski, T. (1992) Forensic identification of urine samples. *Int. J. Legal Med.* **105,** 59–61.

11. Deuter, R., Pietsch, S., Hertel, S., and Müller, O. (1995) A method for preparation of fecal DNA suitable for PCR. *Nucleic Acids Res.* **23(18),** 3800,3801.

12. Maniatis, T., Fritsch, E. F., and Sambrook, J. (1989) *Molecular Cloning: A Laboratory Manual.* Cold Spring Harbor Laboratory, Cold Spring Harbor, N.Y.

13. Bär, W., Kratzer, A., Mächler, M., and Schmid, W. (1989) Postmortem stability of DNA. *Forensic Sci. Int.* **39,** 59–70.

14. Hochmeister, M., Budowle, B., Borer, U. V., Eggmann, U., Comey, C. T., and Dirnhofer, R. (1991) Typing of deoxyribonucleic acid (DNA) extracted from compact bone from human remains. *J. Forensic Sci.* **36,** 1649–1661.

15. Fisher, D. L., Holland, M. M., Mitchell, L., Sledzik, P. S., Webb Wilcox, A., Wadhams, M., and Weedn, V. W. (1993) Extraction, evaluation and amplification of DNA from decalcified and undecalcified United States civil war bone. *J. Forensic Sci.* **38,** 60–68.

16. Hagelberg, E., Sykes, B., and Hedge, R. (1989) Ancient bone DNA amplified. *Nature* **342,** 485,486.

17. Hagelberg, E., Gray, I. C., and Jeffreys, A. J. (1991) Identification of the skeletal remains of a murder victim by DNA analysis. *Nature* **352,** 427–429.

18. Holland, M. M., Fischer, D. L., Mitchel, L. G., Rodriquez, W. C., Canik, J. J., Merril, C. R., and Weedn, V. W. (1993) Mitochondrial DNA sequence analysis of human skeletal remains: identification of remains from the Vietnam war. *J. Forensic Sci.* **38,** 542–553.

19. Höss, M. and Pääbo, S. (1993) DNA extraction from pleistocene bones by a silica-based purification method. *Nucleic Acids Res.* **21,** 3913,3914.

20. Jeffreys, A. J., Allen, M. J., Hagelberg, E., and Sonnberg, A. (1992) Identification of the skeletal remains of Josef Mengele by DNA analysis. *Forensic Sci. Int.* **56,** 65–76.

21. Kurosaki, K., Matsushita, T., and Ueda, S. (1993) Individual DNA identification from ancient human remains. *Am. J. Hum. Genet.* **53,** 638–643.
22. Gunther, C., Issel-Tarver, L., and King, M. C. (1992) Identifying individuals by sequencing mitochondrial DNA from teeth. *Nature Genet.* **2,** 135–138.
23. Pötsch, L., Meyer, U., Rothschild, S., Schneider, P. M., and Rittner, C. (1992) Application of DNA techniques for identification using human dental pulp as a source of DNA. *Int. J. Legal Med.* **105,** 139–143.
24. Schwartz, T., Schwartz, E. A., Mieszerski, L., McNally, L., and Kobilinsky, L. (1991) Characterization of deoxyribonucleic acid (DNA) obtained from teeth subjected to various environmental conditions. *J. Forensic Sci.* **36,** 979–990.
25. Singer-Sam, J., Tanguay, R. L., and Riggs, A. D. (1989) Use of Chelex to 0improve the PCR signal from a small number of cells. *Amplifications* **3,** 11.
26. Boom, R., Sol, C. J. A., Salimans, M. M. M., Jansen, C. L., Wertheim-van Dillen, P. M. H., and Noorda-van der, J. (1990) Rapid and sample method for purification of nucleic acids. *J. Clin. Microbiol.* **28,** 495–503.

4

PCR Analysis from Cigaret Butts, Postage Stamps, Envelope Sealing Flaps, and Other Saliva-Stained Material

Manfred N. Hochmeister, Oskar Rudin, and Edda Ambach

1. Introduction

The polymerase chain reaction (PCR) has offered the forensic scientist a new range of sensitivity in the examination of forensic samples. PCR has been successfully used to amplify specific DNA fragments from extremely small amounts of DNA present on cigaret butts *(1,2)*, postage stamps *(3)*, envelope sealing flaps, and other saliva-stained materials *(4)*. In addition to DNA typing results, it is at times desirable to confirm the presence of saliva by the simultaneous employment of an amylase assay. This chapter describes methods that are routinely and successfully used on forensic casework samples.

2. Materials

2.1. Amylase Assay

1. Amylase buffer: 50 mM KCl, 50 mM phosphate buffer, pH 6.8. Dissolve 3.728 g KCl in 1.0 L dH$_2$O (50 mM KCl). Solution A: Dissolve 3.402 g KH$_2$PO$_4$ in 500 mL 50 mM KCl. Solution B: Dissolve 3.549 g Na$_2$HPO$_4$ in 500 mL 50 mM KCl. To 500 mL of solution A add approx 400–450 mL of solution B to bring the pH to 6.8. Store at 4°C.
2. α-Amylase Uni-Kit I Roche (Hoffmann-La Roche AG, Basel, Switzerland) or α-amylase Granutest® 3 Merck (Merck, Darmstadt, Germany). Store at 4°C.
3. Filter photometer or spectrophotometer (405 nm).

2.2. DNA Extraction and Quantitation

1. 3X stain extraction buffer modified from *(5)*: 30 mM Tris-HCl, 30 mM EDTA, 300 mM NaCl, 6% sodium dodecyl sulfate (SDS), pH 10.2. Short-time storage at

From: *Methods in Molecular Biology, Vol. 98: Forensic DNA Profiling Protocols*
Edited by: P. J. Lincoln and J. Thomson © Humana Press Inc., Totowa, NJ

room temperature (SDS precipitates at 4°C). Add proteinase K fresh from frozen aliquots of 20 mg/mL.

2. 1X stain extraction buffer (5): 10 mM Tris-HCl, 10 mM EDTA, 100 mM NaCl, 2% SDS, pH 8.0. Short time storage at room temperature (SDS precipitates at 4°C). Add proteinase K fresh from frozen aliquots of 20 mg/mL.
3. Phenol/chloroform/isoamylalcohol (25:24:1), in buffer, 10 mM Tris-HCl, pH 7.5–8.0, 1 mM EDTA-Na$_2$. Store at 4°C.
4. Water-saturated n-butanol. Prepare fresh before use.
5. Centricon™ 100 microconcentrator tubes (Amicon Inc., Danvers, MA).
6. QuantiBlot™ Human DNA Quantitation Kit (Roche Molecular Systems, Inc., Branchburg, NJ) or ACES™ 2.0$^+$ Human DNA Quantitation System (Gibco-BRL, Gaithersburg, MD).

2.3. PCR and Detection of PCR Products

1. Standard PCR reagents (e.g., Multiplex PCR kits: AmpliType® PM PCR Amplification and Typing Kit [Roche]; GenePrint™ STR Systems PCR Amplification Kit [Promega Corporation, Madison, WI]).
2. SA 32 Electrophoresis Apparatus (Gibco-BRL). Gel dimensions: 310 × 0.4 mm.
3. Gel Slick (AT Biochem, Malvern, PA) and binding solution (3 μL of methacryloxypropyltrimethoxysilane in 1 mL 0.5% acetic acid in 95% ethanol) to prepare glass plates.
4. Denaturing polyacrylamide gel (4% T, 5% C, containing 7 M urea and 0.5X Trisborate-EDTA buffer) (210 g urea, 267 mL dH$_2$O, 25 mL 10X TBE, and 50 mL 40% acrylamide:bis [19:1]).
5. 0.5X TBE (electrophoresis buffer).
6. Silver staining reagents:
   a. Fix/stop solution: 10% acetic acid;
   b. Staining solution: 2 g silver nitrate and 3 mL 37% formaldehyde in 2000 mL H$_2$O; and
   c. Developer solution: 3 mL 37% formaldehyde, 400 μL 10 mg/mL sodium thiosulfate, 60 g sodium carbonate in 2000 mL H$_2$O.
7. Duplicating X-ray film.

3. Methods

3.1. Amylase Assay

3.1.1. Postage Stamps and Envelope Sealing Flaps

1. The evidence is handled with forceps at all times (*see* **Note 1**). One-half of a postage stamp with the attached part of the envelope or approx 1 cm^2 of an envelope sealing flap is cut into very small pieces and placed in a sterile 1.5-mL Saarstedt tube with a screw cap (*see* **Note 2**). Alternatively, a 2.2-mL microcentrifuge tube with spin insert (with no membrane) (Costar, Cambridge, MA) or a Spin-EASE™ extraction tube (Gibco-BRL) can be used.
2. After addition of 400 μL amylase buffer, the tube is vigorously shaken at 4°C for at least 2 h to overnight and then subjected to centrifugation at 10,000g for 5 min.

3. For the determination of the enzymatic activity of α-amylase, the α-amylase Uni-Kit I Roche or the α-amylase Granutest 3 Merck Kit is used. The kits contain vials of reagents (2-chloro-4-nitrophenyl-β-D-maltoheptaoside, α-glucosidase, and β-glucosidase) and a solvent (50 mM KCl, 50 mM phosphate buffer, pH 6.8). The contents of one vial are dissolved using, respectively, 2.5 mL (Roche) and 3 mL (Merck) of the solvent.

4. To 1 mL of prewarmed (37°C) α-amylase reagent, 10 μL of the supernatant from the sample is added in a 1-cm light path, thermostated cuvet. The contents are mixed, and after 5 min the absorbance is measured against the solvent and again read after exactly 1, 2, and 3 min.

5. The ΔA/min is calculated. If ΔA/min exceeds 0.16, the sample is diluted 1:10 and the assay repeated. To determine the enzyme activity (U/L) in the sample, ΔA/min is multiplied with a factor provided in the kits (9099).

6. A reaction is considered positive if the enzyme activity in the sample exceeds three times the enzyme activity of the blank reagent control.

7. A negative-control sample (containing no saliva-stained material) and a positive control sample (one half of a stamp licked by a known person) are processed in the same manner.

3.1.2. Cigaret Butts, Chewing Gum, and Other Saliva-Stained Materials

In contrast to postage stamps or envelope sealing flaps, the actual determination of the presence of saliva on these evidentiary items may be deemed unnecessary, since it is almost certain to be present. However, the amylase test can be applied exactly as described above, since it neither consumes parts of the sample nor adversely affects the yield of DNA.

1. Cigaret butts are handled with forceps at all times. From the end of the cigaret butt that would have been in contact with the mouth, three cross-sectional slices, each 3–5 mm wide, are made using a sterile scalpel blade. The outer paper covering from the three sections is removed using sterile forceps, cut into small pieces and placed in a single 1.5-mL Saarstedt tube with a screw cap. Alternatively, a 2.2-mL microcentrifuge tube with a spin insert (with no membrane) can be used. If an amylase assay is desired, follow **steps 2–7** of the procedure above (**Subheading 3.1.**).

2. A portion of the chewing gum is cut into very small pieces and placed in a single 1.5-mL Saarstedt tube with a screw cap or the 2.2-mL tube described above. If an amylase assay is desired, follow **steps 2–7** of the procedure above (**Subheading 3.1.**).

3. Other saliva-stained materials are treated in the same manner.

3.2. DNA Extraction and Quantitation

1. If an amylase assay was required, 200 μL of 3X stain-extraction buffer and 15 μL proteinase K are added to the 390 μL amylase buffer (total vol 605 μL) and the tube is incubated overnight at 56°C (*see* **Notes 3** and **4**).

2. If no amylase assay is required, 600 μL of 1X stain-extraction buffer, pH 8.0, and 15 μL proteinase K, 20 mg/mL are added to the tube and the tube is incubated overnight at 56°C.

3. On the following day, an additional 15 µL proteinase K (20 mg/mL) is added and the sample incubated for an additional 2 h at 56°C.

4. Without removing the saliva-stained material (*see* **Note 7**), 700 mL phenol/chloroform/isoamylalcohol (25:24:1) is added and the tube vigorously shaken by hand for 2 min to achieve a milky emulsion in the tube. The tube is subjected to centrifugation at 10,000g for 3 min. If the 2.2-mL microcentrifuge tube with a spin insert is used, the material is removed with sterile forceps and placed into the spin insert. The spin insert is placed into the tube from which the material came and centrifuged at 10,000g for 5 min. Then the spin insert with the material is removed and discarded.

5. The aqueous phase (top layer) is transferred to a new 1.5-mL Saarstedt tube with a screw cap, taking care not to transfer the interphase.

6. To the aqueous phase, 700 µL of water-saturated *n*-butanol is added in order to remove traces of phenol. The tube is vigorously shaken by hand for 2 min and then subjected to centrifugation at 10,000g for 3 min.

7. To purify the DNA, the aqueous phase (bottom layer) is transferred to a Centricon 100 microconcentrator tube (see **Note 6**). After the transfer the volume is brought up with sterile water to 2.0 mL. The sample reservoir is sealed with parafilm, and after punching a hole in the parafilm (using a sterile needle) the tube is subjected to centrifugation at 1000g for 30 min. Then 2 mL of sterile water is added to the sample reservoir and the reservoir sealed with new parafilm. Again the tube is centrifuged at 1000g for 30 min. The DNA is recovered by back centrifugation at 1000g for 5 min. The final sample volume is approx 25–40 µL.

8. To determine the quantity of human DNA, 10% of the retentate is used in slotblot analysis (*see* **Note 7**).

4. Notes

1. For postage stamps, postcards, and envelopes, methods for the detection of fingerprints *(7)* should be applied before any DNA testing is carried out.

2. The saliva-stained material should be cut in small pieces ($\approx 3 \times 5$ mm), to be completely submerged in the phenol phase during the phenol-extraction step. If the pieces are too big, removing the aqueous phase becomes difficult. When no spin insert is used, the stained material is better submerged in 1.5 mL tubes than in 2.0 mL tubes.

3. The amylase buffer is the same as the solvent provided in the amylase test kits.

4. Adding 200 µL of 3X stain-extraction buffer, pH 10.2, to 390 µL amylase buffer, pH 6.8, and 15 µL proteinase K, changes the pH to ≈ 8.0. This is equal to the pH of 1X stain-extraction buffer. The 3X stain extraction buffer *(5)* is modified to meet these requirements; also, it does not contain dithiothreitol (DTT). It must be prewarmed before use.

5. In contrast to a previously published method *(1)*, a step is omitted in this new protocol. This step involved cleaning of the outside of the tube with ethanol and punching a hole with a sterile needle in the bottom of the tube. The solution was then transferred to a new tube by the piggyback method via centrifugation at

1000*g* for 5 min. After removing the upper tube, leaving behind the paper, the solution was extracted. In the new protocol, phenol extraction is carried out without removing the stained material. Alternatively, a microcentrifuge tube with a spin insert can be used.

6. Each Centricon 100 microconcentrator tube is checked by adding 1.0 mL sterile water, sealing the tube with parafilm, punching a hole, and centrifuging the tube at 1000*g* for 3 min. In case of leaking (defect in the membrane), the water would pass through the membrane faster than in a normal tube. During the Centricon 100 purification step, it is important to remove the old parafilm and seal the tube with new parafilm every time sterile water is added. If water is added through the hole in the parafilm, crosscontamination might occur. The Centricon 100 purification step should not be carried out in a centrifuge regularly used for centrifugation of blood samples. We could amplify DNA by swabbing the rotor of such centrifuges with cotton swabs. The centrifuge should be cleaned regularly and thoroughly with ethanol.

7. For storage, we transfer the retentate to 1.5-mL Saarstedt tubes with screw caps. Determination of the final sample volume is by weighing.

8. The success rate of the procedure described above on cigaret butts up to 10 yr old is > 95%; also, we detected amylase activity and extracted and typed DNA from postage stamps up to 33 yr old.

9. Compared to the previously published Chelex method *(1,8)*, we obtained better results from old evidentiary material using the organic extraction method described above.

10. All forensic samples are routinely amplified in the presence of bovine serum albumin (BSA) (albumin bovine fraction V; Sigma A 4503) in the amplification mix (16 µg/100 µL) *(5,9–11)* in order to overcome inhibition.

References

1. Hochmeister, M., Budowle, B., Borer, U. V., Comey, C. T., and Dirnhofer, R. (1991) PCR-based typing of DNA extracted from cigarette butts. *Int. J. Legal Med.* **104,** 229–233.
2. Schmitter, H. and Sonntag, M. L. (1995) STR-analysis on cigarette butts. Experiences with casework material. *Klin. Lab.* **41,** 177–180.
3. Hopkins, B., Williams, N. J., Webb, M. B. T., Debenham, P. G., and Jeffreys, A. J. (1994) The use of minisatellite variant repeat-polymerase chain reaction (MVR-PCR) to determine the source of saliva on a used postage stamp. *J. Forensic Sci.* **39,** 526–532.
4. Walsh, D. J., Corey, A. C., Cotton, R. W., Forman, L., Herrin, G. L., Word, C. J., and Garner, D. D. (1992) Isolation of deoxyribonucleic acid (DNA) from saliva and forensic science samples containing saliva. *J. Forensic Sci.* **37,** 387–395.
5. Budowle, B. and Baechtel, F. S. (1990) Modifications to improve the effectiveness of restriction fragment length polymorphism typing. *Appl. Theoret. Electrophoresis* **1,** 181–187.

6. Budowle, B., Chakraborty, R., Giusti, A. W., Eisenberg, A. J., and Allen, R. C. (1991) Analysis of the VNTR locus D1S80 by the PCR followed by high-resolution PAGE. *Am. J. Hum. Genet.* **48,** 137–144.
7. Lee, H. C. and Gaensslen, R. E. (eds.) (1991) *Advances in Fingerprint Technology.* Elsevier, New York.
8. Singer-Sam, J., Tanguay, R. L., and Riggs, A. D. (1989) Use of Chelex to improve the PCR signal from a small number of cells. *Amplifications* **3,** 11.
9. Hagelberg, E., Sykes, B., and Hedge, R. (1989) Ancient bone DNA amplified. *Nature* **342,** 485,486.
10. Hagelberg, E., Gray, I. C., and Jeffreys, A. J. (1991) Identification of the skeletal remains of a murder victim by DNA analysis. *Nature* **352,** 427–429.
11. Hochmeister, M., Budowle, B., Borer, U. V., Eggmann, U., Comey, C. T., and Dirnhofer, R. (1991) Typing of deoxyribonucleic acid (DNA) extracted from compact bone from human remains. *J. Forensic Sci.* **36,** 1649–1661.

5

Quantification of DNA By Slot-Blot Analysis

Julia Andersen

1. Introduction

Quantification of template DNA is an essential step in the analysis of samples using polymerase chain reaction (PCR). Once a PCR reaction has been optimized, the amplification of too little genomic DNA may yield only partial results, and the addition of too much template may increase the tendency for amplification of artifact products.

Extracts obtained from items submitted for forensic analysis often only contain a low concentration of DNA, which may be degraded. Extraction with Chelex® resin renders DNA partially single-stranded *(1)*. Any method used for quantifying the amount of DNA present in such extracts must be designed to detect subnanogram quantities of denatured and possibly degraded DNA. In addition, preparations from samples other than fresh blood will probably contain components from bacteria and yeasts as well as human genomic DNA. A reliable assay of template DNA for PCR therefore needs to detect human genomic DNA specifically, rather than the total DNA present.

These requirements are met by using a slot-blot (or dot-blot) hybridization procedure, as described by Walsh et al. *(2)*. This method involves the hybridization of a biotinylated oligonucleotide probe to DNA samples immobilized on a nylon membrane and subsequent binding of streptavidin-horseradish peroxidase conjugate to the captured biotin molecules. With the addition of chemiluminescent detection reagents, hydrogen peroxide is reduced by the peroxidase bound indirectly to the DNA samples. This reaction is coupled to the oxidation of luminol, and the photons emitted are detected using autoradiography film *(3)*. The size and density of the dots or slots produced on the film are related to the amount of DNA immobilized on the membrane in each position. Therefore, it is possible to estimate the quantity of DNA present in the sample extract

From: *Methods in Molecular Biology, Vol. 98: Forensic DNA Profiling Protocols*
Edited by: P. J. Lincoln and J. Thomson © Humana Press Inc., Totowa, NJ

slots or dots by comparison with the dots produced from a dilution series of a standard DNA sample.

This methodology can be used with Chelex-extracted DNA, because the denaturation of samples is central to the procedure. The sequence of the oligonucleotide probe is complementary to an α satellite repeat region, D17Z1, found only in higher primates *(4)*, so the assay is "human" DNA-specific. Degraded DNA can be detected using this method because the probe is a 40-mer and will therefore hybridize to small fragments of DNA. In addition, this method of quantification is sensitive to 0.1 ng of DNA. It is simple and relatively quick to perform and only requires a minimum of laboratory equipment.

2. Materials

1. Biodyne B membrane (Gibco-BRL, Gaithersburg, MD).
2. 5 N NaOH (store at room temperature for up to 3 mo). **Caution:** 5 N NaOH is highly caustic.
3. 0.5 M EDTA, pH 8.0 solution (Ultrapure reagent, available from BDH Pharmaceuticals, London, UK).
4. 1 M Tris-HCl, pH 8.0 (AnalaR grade Tris, available from BDH; autoclave and store at room temperature for up to 3 mo).
5. TE buffer: 10 mM Tris-HCl, 0.1 mM EDTA, pH 8.0. Mix 1 mL 1 M Tris-HCl, pH 8.0, 20 µL 0.5 M EDTA, pH 8.0; make up to 100 mL with distilled water (autoclave in 10-mL aliquots and store at room temperature for up to 3 mo; discard remnants of an aliquot once opened).
6. Prewetting solution: 0.4 N NaOH, 25 mM EDTA. Mix 80 mL 5 N NaOH, 50 mL 0.5 M EDTA, pH 8.0/L (autoclave and store at room temperature for up to 4 wk).
7. Hybridization tray and lid (ABD, Perkin-Elmer, Norwalk, CT).
8. 0.04% (w/v) bromothymol blue (prepare using sterile distilled water, aliquot, and store at 4°C for up to 3 mo).
9. Spotting buffer: 0.4 N NaOH, 25 mM EDTA, 0.00008% (w/v) bromothymol blue. Mix 6 mL 5 N NaOH and 3.75 mL 0.5 M EDTA, made up to 75 mL with sterile distilled water and autoclaved before adding 150 µL 0.04% (w/v) bromothymol blue (store at 4°C for up to 1 wk).
10. Standard DNA solution: 0.2 µg/µL K562 DNA solution (Gibco-BRL), diluted to 100 pg/µL (*see* **Note 1**) (store at 4°C for up to 3 mo).
11. Slot-blot/dot-blot apparatus: the Convertible™ Filtration system (Gibco-BRL) (*see* **Note 2**).
12. Vacuum source with a pressure of at least 8–10 in. Hg (*see* **Note 3**).
13. 20X SSPE solution (Ultrapure reagent, available from Gibco-BRL): 3 M NaCl, 200 mM NaH$_2$PO$_4$, 20 mM EDTA, pH 7.4.
14. 10% (w/v) SDS solution (Ultrapure reagent, available from BDH).
15. Hybridization solution: 5X SSPE, 0.5% (w/v) SDS. Mix 250 mL 20X SSPE, 50 mL 10% (w/v) sodium dodecyl sulfate (SDS)/L (autoclave and store at room temperature for up to 4 wk).

16. 5' Biotinylated oligonucleotide probe: Dilute to 15 pmol/μL and store at –20°C in aliquots. Once an aliquot is opened, store at 4°C and discard after 2 wk. The oligonucleotide can be ordered from Oswel, sequence (5'-3'): TAG AAG CAT TCT CAG AAA CTA CTT TGT GAT GAT TGC ATT C.
17. 30% hydrogen peroxide (store at 4°C).
18. Shaking water bath, adjustable to 50°C.
19. Wash solution: 1.5X SSPE, 0.5% (w/v) SDS. Mix 75 mL 20X SSPE, 50 mL 10% (w/v) SDS/L (autoclave and store at room temperature for up to 4 wk).
20. Streptavidin-horseradish peroxidase conjugate (ABD, Perkin-Elmer) (store at 4°C).
21. Orbital shaker.
22. Citrate buffer: 0.1 M sodium citrate, pH 5.0. Adjust pH to 5.0 with 20% HCl (autoclave and store at room temperature for up to 4 wk).
23. ECL™ Detection solutions (Amersham, Arlington Heights, IL) (store at 4°C).
24. Antistatic acetate sheets (18 × 24 cm) (Cadillac Plastics).
25. 95% ethanol.
26. Autoradiography cassette.
27. Hyperfilm™ autoradiography film (Amersham).

3. Method

1. Wearing clean gloves, cut a piece of Biodyne B membrane to fit the slot-blot apparatus and place in 50 mL of prewetting solution in a hybridization tray. Allow to equilibrate for up to 30 min while preparing the samples.
2. Prepare samples for assay by adding an aliquot of each to 150 μL of spotting buffer (*see* **Note 4**).
3. Prepare standard DNA samples by adding the following volumes of standard DNA solution to 150 μL of spotting buffer: 1 μL (0.1 ng), 2.5 μL (0.25 ng), 5 μL (0.5 ng), 10 μL (1 ng), 20 μL (2 ng), 30 μL (3 ng), 40 μL (4 ng), 50 μL (5 ng), 70 μL (7 ng), 100 μL (10 ng) (*see* **Note 5**). Also prepare a negative control sample by adding 100 μL of TE, used to prepare the standard DNA solution, to 150 μL of spotting buffer.
4. Vortex all tubes briefly and then spin for 5–10 s (*see* **Note 6**).
5. Using clean forceps, place the wetted membrane on the gasket of the slot-blot apparatus and turn on the vacuum source. On the manifold, turn the clamp vacuum on and the sample vacuum off, then press down on the top plate to ensure that a seal is formed.
6. Pipet each sample into a different well of the slot-blot apparatus, taking care not to introduce air bubbles or to touch the surface of the membrane with the pipet tip (*see* **Note 7**).
7. When all samples and standards have been pipeted, turn the manifold sample vacuum on slowly and leave on until all samples have been drawn through onto the membrane (30 s–1 min), giving a uniform blue slot or dot.
8. On the manifold, turn the clamp vacuum off, while leaving the sample vacuum on, and remove the top plate (*see* **Note 8**).
9. Without allowing the membrane to dry out, make an orientation mark on the DNA side of the blot in pencil, then remove from the apparatus (*see* **Note 9**).

10. Immediately transfer the membrane to 100 mL of prewarmed hybridization solution (*see* **Note 10**) in a hybridization tray. Add 5 mL of 30% H_2O_2, place the lid on the tray, and incubate at 50°C (*see* **Note 11**) in a shaking water bath (*see* **Note 12**) for 15 min.

11. Pour off the solution and add 30 mL of prewarmed hybridization solution to the tray. Tilt the liquid in the tray to one side and add 1 μL of probe solution to it. Replace the lid and incubate at 50°C in a shaking water bath for 20 min.

12. Pour off the liquid and rinse the blot briefly in 100 mL of prewarmed wash solution by rocking the tray for 5–10 s. Pour off the solution.

13. Add 30 mL of prewarmed wash solution to the tray. Tilt the liquid in the tray to one side and add 90 μL of streptavidin-horseradish peroxidase conjugate to it. Replace the lid and incubate at 50°C in a shaking water bath for 10 min.

14. Pour off the solution and rinse the membrane in 100 mL of prewarmed wash solution by rotating the tray on an orbital shaker (*see* **Note 13**) for 1 min at room temperature. Repeat this rinse step.

15. Pour off the solution and add a further 100 mL of prewarmed wash solution to the tray. Replace the lid and rotate the tray on an orbital shaker for 15 min at room temperature.

16. Pour off the solution and rinse the membrane briefly in 100 mL of citrate buffer by rocking the tray.

17. Pour off the solution and prepare the ECL-detection solution by adding 5 mL of ECL reagent 1 to 5 mL of ECL reagent 2 (*see* **Note 14**).

18. Add the 10 mL of ECL-detection solution to the tray and shake for exactly 1 min at room temperature.

19. Using forceps, remove the membrane from the tray and place between two sheets of acetate. Smooth over the surface of the acetate with a ruler to remove any air bubbles and excess ECL-detection reagent from the blot and then clean the outside of the acetate sheets with 95% ethanol (*see* **Note 15**).

20. Secure the blot in an autoradiograph cassette, DNA-side uppermost.

21. Under safe red-light illumination, place a piece of Hyperfilm in the cassette, on top of the blot, taking care not to move the film once in contact with the blot. Close the cassette.

22. Expose the film to the blot for approx 45 min at room temperature.

23. Develop the film either manually or by using an automatic film processor (*see* **Note 16**).

24. Check that the negative control sample did not give a detectable slot or dot (*see* **Note 17**).

25. Estimate the quantity of DNA present in the sample aliquots by comparing the intensity and size of the slots or dots produced with those produced from the DNA standards (*see* **Note 18**).

4. Notes

1. We prepare the DNA standard solution by diluting the stock solution purchased to 1 ng/μL in TE buffer. This solution is then diluted further to the working

concentration of 100 pg/μL. Mix the solutions well at each stage, but avoid excessive use of the vortex mixer. Refer to the specification sheet sent out with the stock solution for the accurate starting concentration of the product. Having prepared a new batch of DNA standard solution, we check its performance against the batch in use by loading both onto one blot and assessing the compatibility of results obtained.

2. This apparatus allows for a number of alternative blot formats, depending on the choice of top plate. We use the top plate that gives 96X 3-mm dots, which, in our experience, are easier to "read" than slots.

3. We use a diaphragm vacuum pump (Vacuubrand).

4. We assay samples in duplicate—1 and 2 μL aliquots of extracts from fresh blood or hair roots, 4 and 7 μL aliquots of extracts from all other types of stain.

5. We load two sets of standards onto each membrane and check for the compatibility of results. Alternative concentrations of standard DNA samples may be prepared, depending on the test samples to be assayed.

6. If using a 12 × 8 array of sample wells with the manifold, it is useful to place the tubes in a 12 × 8 rack at this stage in the same positions as the samples are to be pipeted onto the membrane.

7. If the sample is not pulled onto the membrane, it may be gently pipeted back into a clean tip and repipeted onto the membrane.

8. Soak the dot-blot apparatus in 0.1% (w/v) SDS solution after each use. Pay special attention to cleaning the gasket and plate that come into contact with the membrane. Rinse well and allow to air-dry.

9. Do not allow the membrane to dry out at any stage in the procedure.

10. Solutions may be prewarmed by placing in a 50°C water bath or incubator and should be at 37–50°C before use.

11. The temperature of the water in the water bath should be 50°C ± 1°C.

12. The water-bath platform should be set to shake at 190–200 strokes/min.

13. The orbital shaker should be set at 50–60 rpm.

14. Do not prepare the ECL-detection solution more than 5 min before use.

15. Immediately after each use, wash the hybridization tray and lid in water and rinse with distilled water, then air-dry.

16. We use an X-OGRAPH (Malmesbury, UK) Compact X2.

17. If a detectable dot is produced from the negative control sample, this would suggest that either a solution used was contaminated or that the manifold was not clean before use.

18. It may be necessary to reassay a sample using a larger or more dilute aliquot so that the result obtained is within the range of the DNA standards. To allow easy visualization of very weak slots or dots, the autoradiography film may be exposed overnight to the blot.

19. The (Quantiblot™ Human DNA Quantitation kit) (ABD, Perkin–Elmer) contains enough streptavidin-horseradish peroxidase and Quantiblot™ D17Z1 probe for 10 blots. The protocol provided with the kit is similar to that described here, except that the autoradiography film requires only a 15 min exposure, because of the design of the Quantiblot™ probe.

19. It is possible to automate the "reading" of quantification blots using a flatbed scanner and image-processing software *(2,5)*. However, we have not found a flatbed scanner capable of resolving the OD range (typically 0–2 OD units) developed on the autoradiography film when using the DNA standards described here.

References

1. Walsh, P. S., Metzger, D. A., and Higuchi, R. (1991) Chelex® 100 as a medium for simple extraction of DNA for PCR-based typing from forensic material. *Biotechniques* **4,** 506–513.
2. Walsh, P. S., Varlaro, J., and Reynolds, R. (1992) A rapid chemiluminescent method for quantitation of human DNA. *Nucleic Acids Res.* **20,** 5061–5065.
3. Waye, J. S. and Willard, H. F. (1986) Structure, organisation and sequence of alpha satellite DNA from human chromosome 17: evidence for evolution by unequal crossing-over and an ancestral pentamer repeat shared with the human X chromosome. *Molec. Cell. Biol.* **6,** 3156–3165.
4. Whitehead, T. P., Thorpe, G. H. G., Carter, T. J. N., Groucutt, C., and Kricka, L. J. (1983) Enhanced luminescence procedure for sensitive determination of peroxidase-labelled conjugates in immunoassay. *Nature* **305,** 158,159.
5. Velleman, S. G. (1995) Quantifying immunoblots with a digital scanner. *Biotechniques* **18,** 1056–1058.

6

Rapid Assessment of PCR Product Quality and Quantity by Capillary Electrophoresis

John M. Butler

1. Introduction

With the growing applications for the polymerase chain reaction (PCR) in human identity testing, a need exists for more rapid and automated forms of assessing amplification success. Methods that allow the characterization of a sample without consuming much material are helpful, especially in forensic cases. In addition, information regarding the quality and quantity of PCR product can be valuable in some applications. This chapter will focus on quantitating PCR products using capillary electrophoresis (CE). This CE method has proven effective prior to sequence analysis of polymorphisms in the D-loop region of mitochondrial DNA *(1)*.

Traditional means of analyzing PCR products are often limited in the amount of information that can easily be obtained. For example, the most common method of PCR-product detection involves gel electrophoresis and subsequent staining for detection purposes *(2)*. Although the quality of the same may be assessed (i.e., Do extra bands appear besides the target sequence?), quantitation of the PCR product requires an extra step (e.g., densiometric scanning). Likewise, many procedures that provide quantitative information, such as hybridization to immobilized probes *(2)* or fluorescence spectrophotometry *(3)*, fail to provide qualitative information. Performing multiple characterizations of the sample in order to obtain both qualitative and quantitative information is time-consuming, labor-intensive, and more importantly, may consume significant portions of the sample. Thus, a method that consumes only a small portion of the sample and provides both qualitative and quantitative information would be valuable to characterizing PCR products.

From: *Methods in Molecular Biology, Vol. 98: Forensic DNA Profiling Protocols*
Edited by: P. J. Lincoln and J. Thomson © Humana Press Inc., Totowa, NJ

CE can serve as an effective tool for PCR-product analysis because it is a rapid and quantitative technique in addition to being automated *(4)*. Direct, on-column detection may be performed with laser-induced fluorescence (LIF) of an appropriate intercalating dye placed in the run buffer *(5)*. No precolumn derivatization is necessary because the DNA fragments bind the dye when traveling through the column. High sensitivity is possible because the dye alone gives rise to very little signal. On interacting with the DNA fragments, a significant enhancement of fluorescent signal is observed *(5,6)*. The use of an intercalating dye allows the PCR products to be tested with high sensitivity without having fluorescent tags already attached, which may interfere with future characterization of the sample (e.g., sequencing). Using LIF and intercalating dyes, CE has recently been applied to quantitating PCR products from HIV-1 DNA *(7)*, the polio virus *(8)*, the hepatitis C virus *(9)*, and mitochondrial DNA *(1,4)*. The method described in this chapter uses an internal DNA standard of known concentration to correct for variation between injections. By comparing the peak area of the PCR product to the peak area of the internal-standard DNA fragment, the relative amount of the PCR product may be calculated *(4,10)*.

When using a PCR product as the DNA template in a sequencing reaction, the DNA fragment's concentration and purity can impact on the quality of the sequencing itself *(1,11)*. Knowledge of a poor DNA template, prior to sequencing, can thus save time, effort, and expense. For example, when polymorphisms are examined in the D-loop of mitochondrial DNA, contaminating PCR products can generate a mixture of signals at positions throughout the chromatogram where the sequence of the contaminant differs from the template of interest. In addition, failure to remove unreacted PCR primers following the initial PCR reaction may result in mispriming of the DNA template during cycle sequencing *(1)*. Both of these scenarios increase the noise in a sequence chromatogram, which makes it more difficult to unambiguously determine the sequence. Thus, it is desirable to have a presequencing method that can detect any contaminating DNA fragments (e.g., nonspecific PCR products) along with showing the presence of excess PCR primers and the concentration of the PCR product of interest. Wilson et al. *(1)* noted that an optimum template concentration of 20–35 ng exists for PCR-product templates used in sequencing the D-loop region of mitochondrial DNA. The CE method described here has been used to rapidly assess the quality and quantity of the PCR-amplified mitochondrial DNA prior to cycle sequencing.

2. Materials

1. The work described in this chapter was performed on a Beckman P/ACE™ 2050 with a Laser Module 488 argon ion laser (Beckman, Fullerton, CA). System Gold software (Beckman) was used to collect and integrate the peak information.

2. Capillary column, 50 μm id × 27 cm, DB-17 coated (J&W Scientific, Folsom, CA); *see* **Note 1**. The distance from injection to detection is 20 cm because the detection window is located 7 cm from the outlet end of the capillary.
3. Sieving buffer: 100 m*M* Tris-borate, 2 m*M* EDTA, pH 8.2, with 1% hydroxyethyl cellulose (HEC), HEC viscosity: 86–113 cP for a 2% solution at 25°C (Aldrich, Milwaukee, WI). The HEC (5 g) was typically stirred overnight at room temperature in 500 mL of the Tris-borate-EDTA solution.
4. Intercalating dye: 50 ng/mL YO-PRO-1 (Molecular Probes, Eugene, OR), 3.2 μL of the 1-m*M* stock solution into 40 mL of buffer solution. Fresh dye-containing buffer solutions were typically prepared every 2–3 d to avoid problems with degradation. The dye is light-sensitive and a possible carcinogen and should be treated appropriately.
5. Internal DNA quantitation standard: 200 bp DNA fragment (QS-200) at a concentration of 100 ng/μL (GenSura, Del Mar, CA).
6. Methanol, HPLC grade.
7. The PCR products, which were evaluated using the method in **Subheading 3.**, came from the D-loop region of mitochondrial DNA (**Fig. 1**). The primer sequences for the hypervariable regions HV1 and HV2 are shown below *(1,12)*:
   HV1
   L15997   5'-CAC CAT TAG CAC CCA AAG CT-3'   **(A1)**
   H16395   5'-CAC GGA GGA TGG TGG TCA AG-3'   **(B1)**
   HV2
   L047   5'-CTC ACG GGA GCT CTC CAT GC-3'   **(C1)**
   H408   5'-CTG TTA AAA GTG CAT ACC GCC A-3'   **(D1)**
   The combination of primers A1 and B1 generates a PCR product over the HV1 region that is 437 bp in length. Likewise, the combination of C1 and D1 produces a 402 bp DNA fragment for HV2. The entire D-loop region can be amplified by combining A1 and D1 for a 1021-bp amplicon.
8. Vials: both amber and clear, 4 mL wide-mouth with threads.
9. Sample vials: 0.2-mL MicroAmp™ Reaction tubes (Perkin-Elmer, Norwalk, CT).

3. Methods

1. Prepare the capillary by cutting it to the desired length and removing approx 5 mm of the polyimide coating for the detection window (*see* **Note 1**). Place the capillary in an LIF capillary cartridge (Beckman). The cartridge will allow liquid to flow around the capillary and maintain a constant-temperature environment.
2. Using a transfer pipet, fill three 4-mL amber vials with the HEC buffer containing the intercalating dye. Be sure to remove all bubbles from the solution surface because they may interfere with the flow of electrical current. Two buffer vials will be used as the inlet and outlet vials during the separation. The third vial will be used to fill the capillary with fresh separation media between each run.
3. Dilute the 200-bp quantitation standard. Typically, the 100 ng/μL 200-bp fragment was diluted to 0.400 ng/μL by placing 2 μL of the 200-bp standard into 498 μL of deionized water. The diluted quantitation standard was prepared in 500-μL

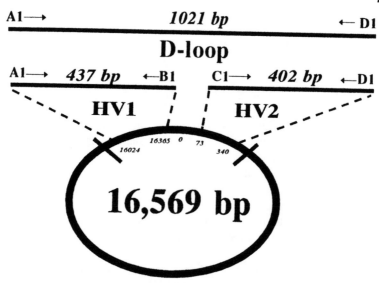

Fig. 1. PCR-products generated from human mitochondrial DNA (control region). The sizes of PCR products generated from amplifying the D-loop of mitochondrial DNA (16,569 bp in total length) using the primer pairs A1/B1 and C1/D1; A1 and B1 define the first hypervariable region (HV1), whereas C1 and D1 define the other (HV2). The primer sequences and positions are described in the text.

volumes so that it could be used in multiple analyses (e.g., up to 20 CE samples) and thus facilitate better reproducibility between runs.

4. Prepare the CE sample by adding 1 µL of the DNA sample generated by PCR into 24 µL of the 0.400 ng/µL diluted 200-bp standard (**Note 2**). Mix the sample well by drawing it into and out of the pipet tip several times.
5. Place the sample vial on a spring inside a 4-mL wide-mouth vial. Screw a silicon rubber cap on the 4-mL vial (to prevent evaporation) and load the samples into the autosampler.
6. Program the method for CE analysis:
   a. Set the detector to collect data at a rate of 10 points/s.
   b. Set the column temperature at 25°C.
   c. Rinse the capillary for 1 min with methanol.
   d. Fill the capillary with run buffer containing the entangled HEC polymer and YO-PRO-1 intercalating dye for 2 min.
   e. Dip the tip of the capillary inlet in a vial of deionized water for 5 s.
   f. Inject the sample under 0.5 psi (low pressure on Beckman P/ACE unit) for 45 s; (*see* **Note 3**).
   g. Apply a separation voltage of 15 kV (556 V/cm) (*see* **Note 4**). The current should rise to ~20 µA.
   h. Repeat **steps c–g** for each sample.

7. Enter the sample names into the computer to relate the sample with the appropriate autosampler position.
8. Start the sample sequence. During application of voltage to the first sample (**step g** above) watch to see that the current rises to ~20 μA and remains stable. If, within the first minute, the current does not rise, the capillary is plugged and needs to be further rinsed with methanol and run buffer.
9. Following the completion of the CE separations (or while the next sample is being processed), data analysis may be performed:
   a. The data-collection software will integrate the area of each peak. Use this peak information in the equation shown below. The primer peaks should pass the detector around 2 min, the 200-bp internal standard at ~2.5 min, and the HV1 or HV2 DNA fragments at ~3 min.
   b. Calculate the concentration of the PCR product (in this case, HV1 or HV2 regions of mtDNA):

$$[\text{PCR Product}] = (25\ \mu\text{L}/1\ \mu\text{L}) \times (\text{Area PCR Product})/$$
$$(\text{Area 200 bp Stnd}) \times (0.384\ \text{ng}/\mu\text{L}) \qquad (1)$$

   where (0.384 ng/μL) is the amount of 200 bp DNA in the CE sample.
10. **Figure 2** illustrates the typical result for a negative control (top frame) and a successful amplification of HV1 from a hair shaft (bottom frame). No primers were observed because they were effectively removed by Microcon 100 centrifugation following PCR. Further cleanup may be necessary if large amounts of primers are observed by CE. Excessive PCR primer present in the sequencing reactions has been shown to result in noisy, uninterpretable sequencing chromatograms *(1)*.

4. Notes

1. The capillaries may be purchased in 10-m rolls and cut to size by the user. In this case, the polyimide outer coating of the fused silica capillary must be removed to allow on-column detection. This detection window may be made by etching a short section of the polyimide coating with hot fuming sulfuric acid and cleaning with ethanol as described by Boček and Chrambach *(13)*. This method protects the integrity of the inner-wall coating. Alternatively, some researchers use the flame from a match to remove the polyimide outer coating. Precut capillaries are also available and may be purchased with detection windows already prepared, albeit at a higher price.
2. Glass sample vials are commercially available. However, modified PCR tubes work well and can be disposed of after use. To make the CE sample tubes, carefully remove the top portion of a 0.2-mL MicroAmp Reaction Tube (Perkin–Elmer) with a scalpel.
3. A long injection time benefits reproducibility and sensitivity. A 10-s pressure injection of water, prior to injecting the sample, may be used to improve resolution *(4,10)*. Hydrodynamic (pressure) injections are more reproducible than electrokinetic injections and are thus preferred in quantitative work *(4)*.

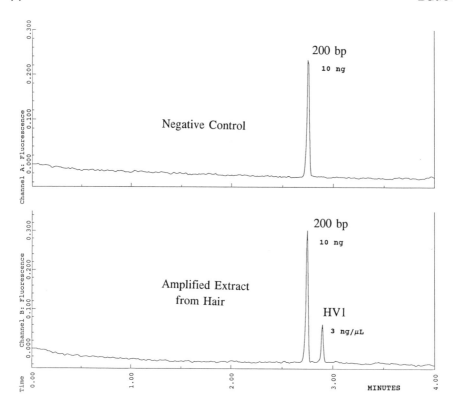

Fig. 2. Quantitation of PCR-amplified mtDNA. The 200-bp fragment was added prior to analysis to serve as an internal standard. The negative control contained no template DNA. The quantity of the HV1 PCR product (437 bp) was determined using the equation described in the text through comparison of its peak area to the peak area of the 200-bp internal standard. Conditions: *Capillary*: 50 μm id × 27 cm DB-17; *Buffer*: 1% HEC (Aldrich), 100 m*M* TBE, pH 8.2, 50 ng/mL YO-PRO-1; *Temp.*: 25°C; *Injection*: 45 s @ 0.5 psi; *Separation*: 15 kV (~20 μA); *Sample*: 1 μL PCR product + 24 μL 200 bp (0.4 ng/μL) in water.

4. Separation speed in CE is directly related to the applied voltage and the capillary length. A shorter capillary would permit faster separations, but the Beckman instrument design is limited to 20 cm (injection to detection distance; 27 cm total length).

5. The CE method described in this chapter is very similar to that detailed in Chapter 10. However, two differences exist: The applied voltage here is 15 kV (556 V/cm) rather than 5 kV (185 V/cm) and the concentration of intercalating dye is 50 ng/mL of YO-PRO-1 rather than 500 ng/mL. Both of these differences benefit separation speed, at the sacrifice of resolution, for the rapid quantitation method described here.

6. A regular check of column resolution is recommended to maintain good-quality results. Running a daily standard restriction digest works well. The 4-min method described here should split the 271 and 281 bp fragments of the ϕX174 *Hae*III restriction digest. Occasionally a capillary may not be adequately coated on the inside, which leads to electro-osmotic flow and affects both reproducibility and resolution. In addition, after extended use (typically over 1000 runs if the capillary is maintained well), the inner coating of the capillary may degrade, which leads to a rapid decrease in column performance. In either case, simply replace the capillary with a new one.

7. The methanol wash and buffer rinse between each run help to maintain column integrity. In addition, storing the capillary in deionized water overnight or during periods of instrument inactivity benefits column lifetime. Following the final run of the day, water may be pushed through the capillary for several minutes. Both capillary ends should be left in water vials to prevent the tips from drying out.

8. The piercing levers, which come in contact with the HEC buffer, may become sticky with extended use. They should be cleaned once a week, or as needed, to remove any HEC residue. Likewise, the silicon vial caps need to be regularly cleaned. Failure to keep the instrument clean will result in buffer vials sticking, which may prematurely halt a run sequence.

9. It is important to keep in mind that when using intercalating dyes for detection, the fluorescent-signal intensity is related to the number of fluorophores bound to the DNA rather than the actual quantity of DNA present. Peak heights and areas increase incrementally as DNA size increases because more intercalating sites exist for larger DNA fragments. Thus, the length of the DNA fragments should be considered when noting the limit-of-detection or when absolute measurements are being made *(10)*. A factor may be included in the quantitation equation shown above to adjust for differences between internal standard and the PCR-product fragment lengths. However, this factor was circumvented in this work since the PCR-product concentration was always determined relative to the internal standard concentration.

10. Buffer vials used for quantitative work should be replaced approximately every 20 runs to avoid problems from buffer depletion. A gradual loss of intercalating dye and other buffer ions from the outlet buffer vial can lead to a loss in signal intensity over a series of multiple injections *(4)*. With lower levels of fluorescent intercalating dye available in the buffer, less dye will bind to the DNA and result in a lower fluorescent signal. Frequent changes of the buffer vials minimize this problem *(10)*.

11. In determining the number of samples that can be processed in a certain period of time, the rinse steps between separations must be considered. Thus, the time to completely process a sample in a routine fashion is 8 min (1 min methanol wash + 2 min buffer fill + 45 s injection + 4 min run).

12. Other PCR products can also be rapidly analyzed with this same method to verify amplification or to evaluate/optimize a multiplex PCR reaction (**Fig. 3**). With the quantitative capability of CE, primer concentrations and PCR conditions can be easily modified to balance amplification of multiple systems.

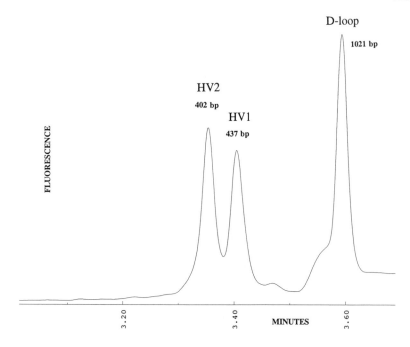

Fig. 3. Evaluation of multiplex PCR conditions. All three PCR products were generated from simultaneously amplifying the HV1 and HV2 regions. The D-loop 1021 bp
amplicon resulted from the extension of primers A1 and D1. Conditions as in **Fig. 2**.

Acknowledgments

This method was developed to benefit sample processing and sequencing of
mitochondrial DNA while the author was working at the FBI Laboratory's
Forensic Science Research Unit (Quantico, VA). The input of Bruce McCord,
Mark Wilson, Deborah Polanskey, and Martin McDermott to the development
of this method is much appreciated. Names of commercial manufacturers are
provided for identification purposes only and inclusion does not imply endorsement by the National Institute of Standards and Technology.

References

1. Wilson, M. R., Polanskey, D., Butler, J. M., DiZinno, J. A., Replogle, J., Budowle,
   B. (1995) Extraction, PCR amplification, and sequencing of mitochondrial DNA
   from human hair shafts. *Biotechniques* **18,** 662–669.
2. Jenkins, F. J. (1994) Basic methods for the detection of PCR products. *PCR Meth.
   Appl.* **4,** S77–S82.
3. Rye, H. S., Dabora, J. M., Quesada, M. A., Mathies, R. A., and Glazer, A. N.
   (1993) Fluorometric assay using dimeric dyes for double- and single-stranded
   DNA and RNA with picogram sensitivity. *Anal. Biochem.* **208,** 144–150.

4. Butler, J. M., McCord, B. R., Jung, J. M., Wilson, M. R., Budowle, B., and Allen, R. O. (1994) Quantitation of polymerase chain reaction products by capillary electrophoresis using laser fluorescence. *J. Chromatogr. B* **658,** 271–280.

5. Schwartz, H. E. and Ulfelder, K. J. (1992) Capillary electrophoresis with laser-induced fluorescence detection of PCR fragments using thiazole orange. *Anal. Chem.* **64,** 1737–1740.

6. Glazer, A. N. and Rye, H. S. (1992) Stable dye-DNA intercalation complexes as reagents for high-sensitivity fluorescence detection. *Nature* **359,** 859–861.

7. Lu, W., Han, D. S., Yuan, J., and Andrieu, J. M. (1994) Multi-target PCR analysis by capillary electrophoresis and laser-induced fluorescence. *Nature* **368,** 269–271.

8. Rossomando, E. F., White, L., and Ulfelder, K. J. (1994) Capillary electrophoresis: separation and quantitation of reverse transcriptase polymerase chain reaction products from polio virus. *J. Chromatogr. B* **656,** 159–168.

9. Felmlee, T. A., Mitchell, P. S., Ulfelder, K. J., Persing, D. H., and Landers, J. P. (1995) Capillary electrophoresis for the post-amplification detection of a hepatitis C virus-specific DNA product in human serum. *J. Cap. Elec.* **2(3),** 125–130.

10. Butler, J. M. (1995) Sizing and quantitation of polymerase chain reaction products by capillary electrophoresis for use in DNA typing. Ph. D. dissertation. University of Virginia, Charlottesville.

11. Hyder, S. M., Hu, C., Needleman, D. S., Sonoda, Y., Wang, X.-Y., and Baker, V. V. (1994) Improved accuracy in direct automated DNA sequencing of small PCR products by optimizing the template concentration. *Biotechniques* **17,** 478–482.

12. Wilson, M. R., Stoneking, M., Holland, M. M., DiZinno, J. A., and Budowle, B. (1993) Guidelines for the use of mitochondrial DNA sequencing in forensic science. *Crime Lab. Dig.* **20,** 68–77.

13. Boček, P. and Chrambach, A. (1991) Capillary electrophoresis of DNA in agarose solutions at 40°C. *Electrophoresis* **12,** 1059–1061.

7

Southern Blotting of Genomic DNA for DNA Profiling

Jim Thomson

1. Introduction

The investigation of DNA polymorphisms has revolutionized forensic-identity analysis over the past decade. Since the first papers by Jeffreys on "DNA fingerprinting" appeared in 1985, DNA-based profiling techniques have been adopted by the worldwide forensic community. Until very recently, the primary method for DNA analysis was the detection of restriction fragment-length polymorphisms (RFLPs) generated at a number of highly polymorphic loci. These loci are characterized by the presence of variable numbers of tandemly repeated conserved DNA sequences (and are hence termed VNTR loci). The RFLPs are detected by hybridization of Southern blots of total genomic DNA with a probe homologous to the conserved repeated sequence.

Methods for the recovery of genomic DNA from source materials, such as blood, semen, hair, and other potential forensic samples, are discussed elsewhere in this volume. Details of probes for VNTR regions, used under both stringent (single-locus) and nonstringent (multilocus) conditions, are also covered in other chapters. This chapter serves specifically to provide a method for the production of the Southern-blot membrane. Restriction of the DNA, assessment of restriction, electrophoresis, and finally Southern blotting are all covered.

Although these techniques are familiar in every molecular-biology laboratory, two points must be borne in mind. First, many forensic scientists have little or no background in basic molecular-biology techniques, although this is changing as DNA technologies continue to develop in the field. Second, the requirements of forensic science, in terms of reproducibility, reliability, and overall method validation, far outweigh those of the average research laboratory using these techniques. The apparently straightforward process of South-

From: *Methods in Molecular Biology, Vol. 98: Forensic DNA Profiling Protocols*
Edited by: P. J. Lincoln and J. Thomson © Humana Press Inc., Totowa, NJ

ern blotting must be seen as a vital link in the chain that could ultimately convict or exonerate a suspect.

2. Materials

1. Genomic DNA (0.5–2 μg high-mol-wt human DNA isolated from reference sample or evidential material).
2. Restriction endonuclease, e.g., *Hin*fI, *Hae*III (Promega [Madison, WI] or numerous other suppliers).
3. 10X restriction buffer (supplied with enzyme).
4. Agarose (Seakem LE, FMC Bioproducts, Rockland, ME).
5. 1X TBE buffer: 0.089 M Tris-HCl, 0.089 M boric acid, 0.02 M EDTA.
6. Ethidium bromide: 10 mg/mL in H_2O.
7. Minigel (10–12 cm) horizontal gel-electrophoresis apparatus (e.g., Gibco-BRL, Gaithersburg, MD, Horizon 58).
8. Load buffer #1: 125 mg/mL Ficoll 400, 0.4 mg/mL bromophenol blue.
9. Load buffer #2: 125 mg/mL Ficoll 400, 2 mg/mL bromophenol blue, 2 mg/mL xylene cyanol.
10. DNA molecular weight reference ladder (e.g., Gibco-BRL DNA analysis ladder).
11. Control DNA samples (prepared from reference samples or commercially available cell line DNA; e.g., K562).
12. Analytical gel (20 cm) horizontal gel-electrophoresis apparatus (e.g., Gibco-BRL Horizon 20/25).
13. Gel denaturation buffer: 1.5 M NaCl, 0.4 M NaOH.
14. Gel neutralization buffer: 1.5 M NaCl, 1 M Tris-HCl, pH7.4.
15. (20X standard saline citrate) (20X SSC): 3 M NaCl, 0.3 M Na citrate, pH 7.0.
16. Nylon membrane; uncharged (e.g., GenePrint Light DNA typing grade membrane, Promega).
17. 3MM Chr paper—46 × 57 cm size (Whatman, Clifton, NJ).
18. 3MM Chr paper—20 × 20 cm size (Whatman).
19. Absorbent blotting pads (e.g., Quickdraw 20 × 20 cm, Sigma, St. Louis, MO).
20. Vacuum blotting apparatus (e.g., Vacuum Blotter Appligene, Illkirch, France).
21. UV crosslinker apparatus.

3. Method

3.1. Restriction

1. The genomic DNA to be analyzed should ideally be at a concentration of 50–200 ng/μL. The methods described here will detect fragments of as little as 50 ng of high-mol-wt DNA, but 0.5–2 μg represents the optimal quantity for a strong signal without any danger of overloading the gel. It is sensible to balance DNA quantities across the gel. For example, if only 100 ng of DNA is obtained from an evidential crime stain, it is advisable to load a similar quantity of the reference-sample DNA.
2. Calculate the volume required of each DNA sample to be digested, based on the guidelines given in **step 1**.

3. Transfer this volume to a sterile, labeled 1.5 mL reaction tube. Adjust the volume to 34 μL with sterile water (*see* **Note 1**).
4. Add 4 μL of 10X restriction buffer (supplied with enzyme). Add 25 U of the appropriate restriction enzyme (*see* **Note 2**).
5. Incubate the reaction tubes at 37°C for 6–20 h. Stop reaction by incubation at 65°C for 10 min.

3.2. Assessment of Restriction

1. Prepare a 1% agarose/TBE minigel as follows. Weigh out 0.3 g agarose (Seakem LE). Place in a 100 mL conical flask with 30 mL 1X TBE, 0.5 μg/mL ethidium bromide (*see* **Note 3**). Weigh flask and contents. Microwave for 1 min or until agarose is fully dissolved. Return flask to balance and add water to restore original weight. Allow to cool to about 60°C and pour into prepared minigel tray (**Note 4**). Allow the gel to set for at least 30 min before use.
2. Briefly centrifuge digested DNA samples to collect condensation. Vortex to mix. Take 2 μL aliquots of each sample into sterile, labeled 0.4 mL tubes. Add 3 μL of load buffer #1. Vortex to mix.
3. Add 1X TBE, 0.5 μg/mL ethidium bromide to the gel tank until the surface of the gel is submerged beneath 1–2 mm of buffer. Remove combs and load samples into wells using a Gilson P10 pipet. Run the gel at 70 V constant for 30 min.
4. Visualize the DNA in the gel by laying the gel directly on a UV transilluminator (*see* **Note 5**). Photograph the gel using a Polaroid camera and appropriate hood at 1/15 s at f4 (Polaroid type 667 monochrome film).
5. Examine all samples on the gel photograph to assess completeness of restriction and uniformity of fluorescence intensity (*see* **Note 6**). All digested samples should appear as smooth smears. A small amount of high-mol-wt DNA, seen as a tight band near to the origin, is acceptable, but if smearing is seen from the origin, or if there is a heavy band of high-mol-wt DNA visible, then the sample should be redigested before continuing (*see* **Note 7**).

3.3. Analytical Gel

1. Prepare a 0.7% agarose/TAE gel as follows: Place 2.45 g agarose (Seakem LE) in a 1 L conical flask with 350 mL 1X TAE. Note that no ethidium bromide is added to the analytical gel or running buffer. Bring to a boil in a microwave oven. Add a stirring bar, transfer to a stirrer, and allow to cool with constant stirring to about 60°C. Meanwhile, prepare a 20 × 25 cm gel tray. Position comb, pour gel, and allow to set for at least 1 h.
2. Prepare the molecular-weight ladder (Gibco-BRL DNA analysis ladder) as follows: prepare three or four (*see* **Note 8**) 1.5 mL tubes, each containing 5 μL ladder DNA, 35 μL TE buffer, and 10 μL load buffer #2.
3. To each digested DNA sample, add 10 μL load buffer #2. Arrange the samples and ladders in a rack in the desired order (*see* **Note 9**).
4. Position the gel in the electrophoresis rig and add 1X TAE buffer to submerge the gel to a depth of 2–3 mm. Remove the comb carefully. Load the prepared 50 μL

samples using a Gilson pipet. Return each tube to the same position in the rack after loading. Run the gel at 40 V constant for 22 h.
5. Record the identity and position of each sample by reference to the empty tube rack (*see* **Note 10**).

3.4. Southern Blotting

Southern blotting can be carried out using a capillary-transfer method or by a vacuum-transfer method. Both methods are described here. **Steps 1** and **2** are common to both methods.

1. Gel denaturation: The DNA within the gel must first be denatured to render it single-stranded before blotting. Carefully slide the gel from the gel tray into a plastic tray (*see* **Note 11**). Add enough gel denaturation buffer to completely cover the gel. Agitate gently on a rocking platform for 15 min at room temperature. Carefully remove the used denaturation buffer (*see* **Note 12**).
2. Proceed with two further washes of 15 min each with gel neutralizing buffer.

3.4.1. Capillary Transfer

1. Cut two 20 × 50 cm pieces of 3MM paper to serve as wicks. Cut one 20 × 20 cm piece of transfer membrane (GenePrint Light Membrane, Promega) (*see* **Note 13**). Label the membrane by writing the gel identification number firmly and clearly in biro in the top righthand corner (*see* **Note 14**).
2. Lay a 25 × 25 cm glass plate over the top of an appropriately sized plastic tray. Lay both wicks together over the glass plate. Fold the wicks under the glass at each end so that they rest against the bottom of the tray. Wet the wicks thoroughly with 10X SSC. Smooth the wicks onto the glass plate to remove bubbles or wrinkles by rolling with a 10 mL plastic pipet. Pour about 500 mL 10X SSC into the tray.
3. Very carefully slide the gel from the plastic tray onto the wicks. The gel is very liable to slip so great care must be taken. Lay the membrane onto the gel, starting flush with the bottom end (furthest from the origin). Smooth gently with a gloved hand to ensure that no air bubbles are trapped. Cut off the few centimeters of excess gel at the origin end and discard. Wet one 20 × 20 cm sheet of 3MM paper with 10X SSC. Carefully lay this sheet on top of the membrane, smoothing again to ensure no bubbles are trapped. Lay three more sheets of dry 3MM paper on top, followed by 10 20 × 20 cm sheets of thick blotting paper (Sigma Quickdraw). Lay a glass sheet on top of these and a 500 g (approx) weight on top of the plate. Allow the transfer to proceed for 4–20 h (*see* **Note 15**). The capillary transfer assembler is illustrated in **Fig. 1**.
4. Dismantle the blot down to the level of the membrane. Discard all blotting papers and wicks. Rinse the membrane in 2X SSC for 5 min. Allow to dry at room temperature on a sheet of 3MM paper.

3.4.2. Vacuum Blotting

The following protocol is for the Appligene (Illkritch, France) Vacuum Blotter. Apparatus from alternative manufacturers may require slightly different protocols.

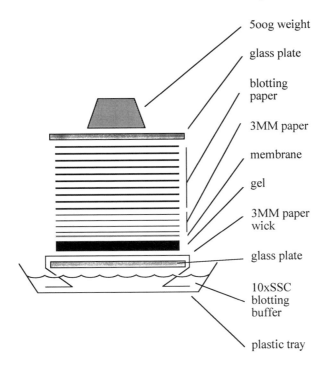

5oog weight

glass plate

blotting paper

3MM paper

membrane

gel

3MM paper wick

glass plate

10xSSC blotting buffer

plastic tray

Fig. 1. Diagram showing the construction of a capillary blot.

1. Cut a 25 × 25 cm piece of 3MM paper; wet with water and place on top of the porous plate on the vacuum blot apparatus. Place the plastic window gasket over this. Turn on the vacuum for a few seconds to remove bubbles and wrinkles. Cut and label the membrane as described above (**Subheading 3.4.1.**). Wet the membrane and place label side down on the 3MM paper. Again, turn on the vacuum briefly to remove bubbles and wrinkles.
2. Slide the gel carefully from the plastic tray so it overlays the membrane and overlaps the edges of the sealing window (*see* **Note 16**). Fasten the sealing frame in place. Overlay the gel with a few milliliters of 20X SSC. Turn on the vacuum and adjust to 50 bar. As the vacuum takes effect, the gel will become concave at the edges, allowing further 20X SSC to be added (*see* **Note 17**). Allow the transfer to proceed for 1 h.
3. Turn off the vacuum. Remove gel and membrane. Rinse membrane in 2X SSC for 5 min and allow to dry at room temperature on a sheet of 3MM paper.

3.5. Fixing

The DNA must be fixed on the membrane. The methods for this vary according to the membrane type. Generally they involve baking at 80°C, exposure to UV light or, in some cases, the use of an alkaline transfer buffer. For an uncharged membrane, such as Promega GenePrint Light Membrane, the preferred method is UV fixing in a dedicated UV crosslinker apparatus (Appligene).

1. Place the dried membrane into the UV crosslinker, ensuring that the DNA side of the membrane is face up.
2. Turn on the crosslinker, set the timer to the desired exposure (*see* **Note 18**) and start the program.

4. Notes

1. The total reaction volume can be adjusted to suit the comb to be used in the analytical electrophoresis gel. If the DNA samples are of sufficiently high concentration, then a smaller volume can be digested. This, in turn, allows the use of a thinner comb, which may produce tighter bands on the final lumigraph or autoradiograph. If the reaction volume is reduced in this manner, the amount of added restriction enzyme should be kept constant.
2. The most commonly used restriction enzymes are *Hae*III (in the United States) and *Hin*fI (in Europe). Other enzymes in occasional use are *Alu*I and *Pst*I.
3. Ethidium bromide is a powerful mutagen; a direct result of its ability to intercalate with DNA that is exploited in the UV fluorescence visualization. Gloves and protective clothing should be worn when handling gels or solutions containing ethidium bromide. All such materials should be disposed of in accordance with local safety procedures.
4. Gel trays may be sealed with tape, or gels may be poured *in situ* depending on the design of the apparatus.
5. Shortwave UV irradiation can damage eyes and skin. Suitable shielding, either on the transilluminator or in the form of a face mask, should be used.
6. Although the DNA has been quantified prior to restriction, it may be that this post-restriction minigel will show some variation in intensity of fluorescence between samples. Further adjustments can be made at this stage if any samples appear excessively bright in comparison to the others. The samples in question can either be requantified, or an adjustment to the volume loaded may be made using the judgment and experience of the analyst.
7. This method for assessing the completeness of restriction, based on the appearance of the sample on a minigel, is largely reliant on the experience and judgment of the analyst. However, it has proved to be a reliable and informative method in experienced hands. Other methods have been developed in the past, including hybridization of the Southern blot with a probe homologous to a locus known to be resistant to complete restriction. If bands additional to those expected are seen with this probe, then the sample has not been completely restricted. If no additional bands are seen, then, given the resistant nature of this locus, the analyst can be assured that the VNTR loci of interest will also have been fully restricted. The disadvantage of this method, apart from the time required for an additional hybridization, is that any partially restricted samples will already have been run on the analytical gel. In many forensic situations, the size of the available sample

is limiting: Such a procedure that uses a large proportion of the available sample is clearly unsatisfactory.

8. At least three ladder lanes should be run. Two of these should be at the extreme edges of the gel, flanking all of the samples. The third should be positioned approximately in the center of the gel. The use of three ladders gives an indication of internal distortions within the gel and so improves the accuracy of fragment-size estimations. These distortions may be a result of slight variations in the electric field during electrophoresis or actual physical distortion of the gel during the postelectrophoresis processing and blotting.

9. The order in which the samples are loaded on the gel depends largely on the nature of the case under investigation. In paternity testing, it is advisable to load putative family trios in adjacent lanes with the child's sample between the named parents (i.e., putative father:child:mother). This allows easy visual identification of matching parental/child fragments. However, in identity cases attempting to link an evidential sample with a reference sample (for example, a semen stain and a suspected rapist), the samples must not be loaded in adjacent lanes because of the danger of overspill from one well to another. This is particularly relevant since in these cases the amount of DNA in the reference sample may greatly exceed that in the stain sample so a very small amount of overspill from the reference lane could give a weak result, consistent with that expected from the stain, in an adjacent lane.

10. A reliable record of the order in which samples are loaded onto the analytical gel is clearly paramount, particularly because the gel itself cannot be clearly labeled. Laboratories must devise suitable procedures of witnessed loading and subsequent tube-order checks to satisfy the demands of evidential continuity.

11. The best trays to use are those with angled sides and rounded corners to allow the gel to be slid in and out easily.

12. Buffer can be removed in two ways. The safest is to use a water-line aspirator to suck off the buffer. Alternatively, the gel can be held in place with the palm of the hand and the buffer carefully tipped away. Remember that the gel is very slippery, so great care must be taken, or it will end up in the sink.

13. Other blotting membranes may be suitable. For alkaline-phosphatase-based detection methods, uncharged nylon membranes are advised, because charged membranes may result in higher background. For radio-isotopic detection, charged membranes may well provide the most sensitive results.

14. Labeling of the membrane is important both for identification and for orientation purposes. Firmly written biro will fade but remain legible through a number of probe/strip cycles. It is recommended that the membrane always be labeled in the same corner with respect to the gel. This allows the analyst to identify both the orientation of the membrane and the side of the membrane onto which the DNA is blotted.

15. Great care is required to prevent the gel sliding rapidly from the tray. In our laboratory, a customized gel tray has been made, cut off at one end to give a large open-ended scoop. The gel can be transferred from the standard tray to this scoop and then slid easily onto the platform of the vacuum blotter.

16. The vacuum blotter requires only a small amount of transfer buffer for successful blotting. Once the concave edges have been formed a minute or so after the vacuum is applied, then a single application of approx 50 mL of 20X SSC will suffice for a 1-h blot.

17. DNA fixation using UV irradiation can be carried out using dedicated apparatus or on a standard UV transilluminator. In both cases, it is necessary to calibrate the apparatus to give optimum exposure times. A calibration gel, consisting of paired lanes of ladder DNA and a standard control DNA, should be run. The blotted membrane can be cut into strips and the strips exposed to the UV source for varying exposures (e.g., 15 s, 30 s, 1 min, 2 min, 5 min, 10 min). The strips can then be probed, stripped, and reprobed to give an indication of the optimum exposure time. The apparatus should be recalibrated on a regular basis, because the exposure characteristics of the UV tubes may change with age.

8

Preparation and Use of Alkaline-Phosphatase-Conjugated Oligonucleotide Probes for Single-Locus and Multilocus VNTR Analysis

Wayne P. Childs, Michael B. T. Webb, and Gillian Rysiecki

1. Introduction

Nonisotopic probes have been widely adopted for DNA fingerprinting and DNA profiling because of their ease and speed of use and obvious safety and environmental advantages *(1)*.

Nonisotopic DNA probes designed to detect variable number of tandem-repeat (VNTR) sequences are typically single-stranded oligomers of 20–30 nucleotides, with sequence complementary to the target tandem-repeat sequence. There are two types of probes used for the analysis of VNTR sequences: multilocus probes (MLP) and single-locus probes (SLP).

MLPs consist of tandem repeats containing a minisatellite "core" sequence that can simultaneously detect a number of highly polymorphic loci to generate individual-specific DNA "fingerprints" *(2,3)*. These probes have found a number of applications in the field of identity analysis, including forensic and paternity testing *(4,5)* and cell-line verification *(6)*. When hybridized to Southern blots under conditions of low stringency, each multilocus probe will detect a family of minisatellites that all share the same "core" sequence. This produces the multiband DNA "fingerprint" pattern. Several different MLPs have been isolated, and most can be used to detect minisatellites in a wide range of species.

Single-locus probes (SLPs), on the other hand, each detect only one minisatellite locus. SLPs are, however, extremely sensitive and tolerant of degraded-target DNA sequence. They are widely used for forensic analysis as well as for paternity testing *(7,8)*. Many useful SLPs have been identified, and by using several in series a "DNA Profile" can be constructed.

From: *Methods in Molecular Biology, Vol. 98: Forensic DNA Profiling Protocols*
Edited by: P. J. Lincoln and J. Thomson © Humana Press Inc., Totowa, NJ

The process of probe design and preparation is not always simple or straight-forward if high-quality results are to be guaranteed. A probe used for VNTR analysis should exhibit high sensitivity, high specificity, low backgrounds, reproducibility, and robustness. The cost of setting up probe manufacturing systems that incorporate the process and quality-control steps required to guar-antee high performance can be significant. High-quality VNTR probes can be purchased from specialist manufacturers prelabeled and ready to use. Hybrid-ization is more rapid than with traditional radioactive systems; typically, mem-branes can be hybridized and lumigraphs produced within a normal working day. Hybridized probes can be detected on the membrane by a range of chemi-luminescent substrates *(9,10)* which are activated by the alkaline phosphatase label on the probe. This enzymatic reaction results in the emission of light that can be recorded on X-ray-sensitive film.

2. Materials

Note: Autoclaved, purified water must be used in all solutions. The follow-ing procedures are optimized for use with NICE™ probes, but can be success-fully applied to alkaline phosphatase-labeled oligonucleotide VNTR probes prepared by alternative methods.

1. 0.5 M Disodium hydrogen phosphate buffer, pH 7.2: 71 g/L Na_2HPO_4. Adjust pH with concentrated orthophosphoric acid and autoclave.
2. Standard saline citrate (20X SSC): 88.2 g/L trisodium citrate $Na_3C_6H_5O_7 \cdot 2H_2O$. 175.3 g/L NaCl.
3. 0.1 M magnesium chloride: 20.3 g/L $MgCl_2 \cdot 6H_2O$. Autoclave.
4. Wash solution 2: 11.6 g/L maleic acid 8.7 g/L NaCl. Adjust pH to 7.5 with concentrated NaOH and autoclave.
5. Membrane blocking reagent. Dissolve 100 g/L casein (Hammarsten Grade) in wash solution 2 by heating at 50–70°C for 1 h. Autoclave and store at –20°C.
6. 10% Sodium dodecyl sulfate (SDS): 100 g/L Sodium lauryl sulfate.

Prepare the following solutions on the day of use from sterile stock solutions.

7. Prehybridization buffer (1 L): 990 mL 0.5 M Na_2HPO_4 pH 7.2, 10 mL 10% SDS. Heat to 50°C before use.
8. Hybridization buffer (1 L): 900 mL prehybridization buffer, 100 mL membrane blocking reagent. Heat to 50°C before use.
9. Wash solution 1: For MLP: 160 mL/L 0.5 M Na_2HPO_4 pH 7.2, 10 mL/L 10% SDS. For SLP: 20 mL/L 0.5 M Na_2HPO_4, pH 7.2, 10 mL/L 10% SDS. Dilute Na_2HPO_4 before adding SDS. Heat to 50°C before use.
10. Membrane-stripping solution: 10 mL/L 10% (SDS). Heat to 80°C before use.
11. CDP-*Star*™ assay buffer: 10 mL/L 0.1 M $MgCl_2$, 10.5 mL/L diethanolamine (cor-rosive). Adjust pH to 9.5 with concentrated hydrochloric acid. Store at 4°C.

12. Alkaline phosphatase labeled oligonucleotide VNTR probes (e.g., NICE probes, Cellmark Diagnostics). Store at –20°C.
13. CDP-*Star* concentrate (Tropix Inc., Bedford, MA); store at 4°C. Protect from light and heat.
14. Lumi-Phos® 530 (Lumigen Inc., Detroit, MI) store at 4°C. Protect from light and heat.
15. Polyester sheets (e.g., Melinex 539, 100 µm thickness, ICI, Cadillac Plastics, Swinden, UK).
16. Spray gun (e.g., Camlab, Cambridge, UK, cat no. CHR0057).
17. X-ray film (e.g., Hyperfilm, Amersham Int., Bucks, UK).
18. X-ray film cassette (e.g., Hypercassette, Amersham).

3. Method

1. Prehybridize up to 10 membranes (prepared by Southern blotting) by wetting in 1X SSC and placing, DNA side down, in 500 mL of prehybridization buffer at 50°C. Gently agitate for 20 min at 50°C.
2. Add 160 mL of hybridization buffer at 50°C to a hybridization chamber or sandwich box, and add the volume of probe(s) as recommended by the manufacturer.
3. Using forceps, individually transfer the membranes, DNA side down, to the hybridization buffer, ensuring there are no air bubbles. Gently agitate for 20 min at 50°C.
4. Wash the membranes by individually transferring them, DNA side down, to 500 mL of prewarmed wash solution 1. Gently agitate for 10 min at 50°C.
5. Repeat **step 4** with fresh wash solution 1.
6. Rinse the membranes by individually transferring them, DNA side down, to 500 mL of Wash solution 2 at room temperature. Gently agitate for 10 min at room temperature.
7. Repeat **step 6** with fresh wash solution 2.
8. Using CDP-*Star*:
   a. Rinse the membranes by individually transferring them, DNA side down, to 500 mL of CDP-*Star* assay buffer at room temperature. Rinse for <10 s.
   b. Dilute an appropriate amount of the CDP-*Star* concentrate with CDP-*Star* assay buffer (1:100 dilution). Add the membranes individually, DNA side down, and incubate at room temperature for 5 min.
   c. Remove the membranes individually from the CDP-*Star* and drain carefully. It is important that treated membranes do not come into direct contact with each other.

   or
   Using Lumi-Phos 530:

   a. Place each membrane DNA side up on a glass or Perspex plate, transfer to a vented fume cabinet, and support at an angle of approx 80° from horizontal.
   b. Using a spray gun, spray each membrane evenly with 3–4 mL of Lumi-Phos 530. Spray over the whole membrane but do not oversaturate.
9. Sandwich each membrane between two polyester sheets. Using the straight edge of a ruler (or equivalent), squeeze out any excess substrate. Avoid getting any substrate on the outside of the polyester sheets.

10. Secure the membrane/polyester "sandwich" with a piece of tape along each edge, and place in a light-proof cassette against X-ray-sensitive film. Intensifying screens are not required. Incubate at 30°C.

11. Develop the film after 0.5–4 h (CDP-*Star*) or 2–18 h (Lumi-Phos 530), depending on the quantity of DNA on the membranes. Chemiluminescence continues for several days; therefore, re-exposure is possible, although longer exposure times may then be required.

12. Stripping and Reprobing: Probes can be removed from hybridized membranes by agitating for 15 min in 0.1% SDS at 80°C, thus allowing multiple sequential hybridizations of a panel of probes to the same membrane. Rinse in 1X SSC prior to rehybridization or storage.

4. Notes

1. The quality of the prelabeled probe can affect overall performance characteristics. A poorly designed probe sequence will never produce sensitive or specific results. Ineffective purification of probe conjugate will result in poor detection sensitivity because of competing "unlabeled" probe. Background problems and overall quality of the autoradiograph can also be affected if probe conjugates used in hybridizations are contaminated with "free" unconjugated alkaline phosphatase and oligonucleotide.

2. Neutral membranes (e.g., MagnaGraph, Micron Separations Inc., Westborough, MA) are recommended for use with this protocol. Removal of probes from charged membranes can be difficult, and backgrounds may be high.

3. The short duration of the hybridization and washing steps in this procedure may not allow full temperature equilibration, which can result in an incorrect hybridization stringency. Strict temperature control during the hybridization/washing reactions is essential for high-quality results. Particular problems may be encountered when using dry-air incubators, because heat exchange is poor. The actual temperature attained within the hybridization reaction should be monitored carefully (**Fig. 1**).

4. Contamination with microbial alkaline phosphatase is a common cause of heavy uniform background masking the specific probe signal (**Fig. 2A**). Low levels of microbial contamination may produce spots of background. All solutions should be sterilized or prepared immediately before use from sterile stock solutions. Sterile pure water should be used to prepare working solutions and to clean all laboratory apparatus.

5. Insufficient agitation during hybridization may cause membranes to stick together, preventing hybridization solution and/or wash solutions from reaching the entire membrane surface. The resulting autoradiographs are characterized by the presence of uneven areas of background (**Fig. 2B**). This problem is particularly common when hybridizing multiple membranes in rotary ovens, where membranes overlap.

6. The formation of bubbles between membranes during the addition of membranes to hybridization solutions may result in unhybridized patches.

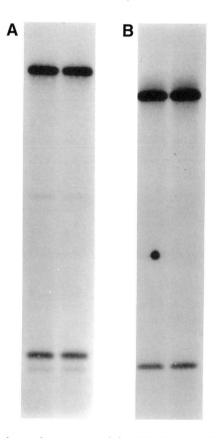

Fig. 1. Two identical membranes containing K562 genomic DNA digested with *Hin*fI and probed with NICE™ G3 (Cellmark Diagnostics). Hybridization and washing were carried out at **(A)** 40°C and **(B)** 50°C (the recommended temperature). Nonspecific secondary bands are visible when membranes are hybridized at suboptimal temperatures, because of a lowering of hybridization stringency. Similar results may also be produced if the phosphate concentration in wash 1 is too high (>10 mM). **Note:** If the temperature of hybridization is too high (>55°C) or the wash 1 phosphate concentration is too low (<10 mM), a loss of sensitivity will be observed on the autoradiograph because of the increase in hybridization stringency.

7. Uneven application of Lumi-Phos 530 may produce patchy autoradiographs with dark and light areas (**Fig. 2C**).
8. DNA fixation by ultraviolet radiation should be standardized to ensure consistent results. It is essential that the ultraviolet output of the irradiation equipment be regularly monitored. Output of 80,000–160,000 μJ/cm² is usually sufficient for MagnaGraph membranes (MSI). Each batch of membranes should be optimized for maximum sensitivity.
9. Excess Lumi-Phos 530 remaining on hybridized membranes will cause elevated backgrounds. It is essential to firmly squeeze Lumi-Phos 530-treated membranes

C

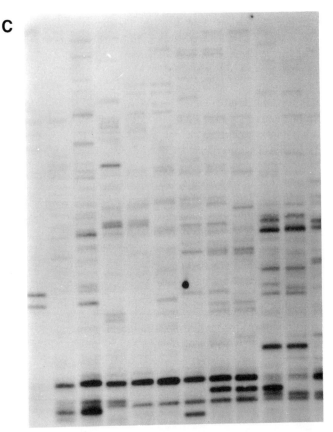

Fig. 1. (*continued from previous page*) Typical appearance of multilocus "DNA fingerprints" hybridized at incorrect stringency (**C**). Membrane containing human genomic DNA (1–4 μg) was hybridized to multilocus probe 33.6 at 57°C rather than 50°C. Similar patterns may be caused by low phosphate concentration (<80 mm).

to remove any excess. The presence of excess CDP-*Star* substrate is not as problematic. However, excessive removal of either substrate will produce a generalized loss of sensitivity.

10. Photobleaching of chemiluminescent substrates and the exposure of treated membranes to temperatures >37°C will reduce sensitivity. It is essential that membranes are protected from light following treatment with chemiluminescent substrate.

11. Membranes are prone to surface damage, especially during hybridization (**Fig. 2D**), and should be treated with care at all stages (avoid sharp metal forceps).

12. Fungal growth during storage can ruin a membrane (**Fig. 2E**). Membranes should be rinsed in 1X SSC and completely dried prior to storage at room temperature.

13. Exposure times depend on the quantity and quality of DNA on the membranes, the type of chemiluminescent substrate used, and the length of time since hybridization. Chemiluminescence slowly decreases over several days.

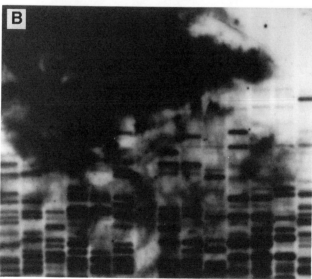

Fig. 2. Autoradiograph illustrating the effects of commonly encountered problems: **(A)** Use of nonsterile solutions, resulting in high background chemiluminescence from microbial alkaline phosphatase activity; **(B)** membranes stuck together during washing resulting in poor washing (*continued*).

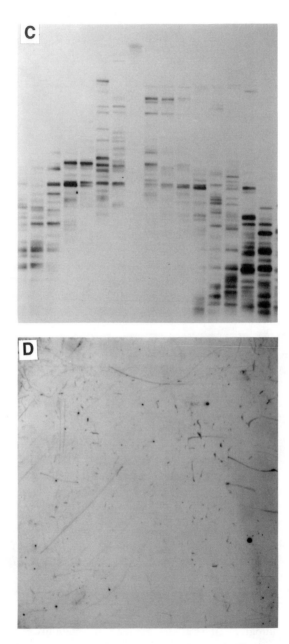

Fig. 2. *(continued from previous page)* Autoradiograph illustrating the effects of commonly encountered problems: **(C)** Uneven application of Lumi-Phos 530; **(D)** membrane scratched with forceps during hybridization.

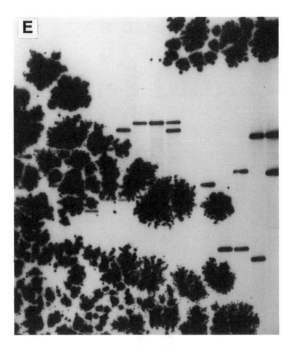

Fig. 2. *(continued)* Autoradiograph illustrating the effects of commonly encountered problems: **(E)** Fungal growth on a membrane.

14. Complete digestion of sample DNA with a restriction endonuclease (usually *Hin*fI) is essential for successful multilocus "DNA fingerprinting" (**Fig. 3A**). Incomplete or partial digestion can produce anomalous results (**Fig. 3B**) or prevent the formation of the DNA fingerprint at all (**Fig. 3C, D**).
15. Sample DNA should be highly purified and of high molecular weight (e.g., phenol chloroform extraction). The DNA should be completely resolvated and ideally quantitated prior to digestion with restriction enzyme.
16. Because of the large number of bands, standardization of DNA migration is absolutely necessary for reliable interpretation of multilocus DNA fingerprints. A DNA migration marker such as λ *Hin*dIII digest (Life Technologies, Bethesda, MD), is recommended. Gels should be run until the 2.3-kb λ fragment has migrated to a fixed distance from the origin of sample loading (usually approx 20 cm). The portion of the DNA fingerprint pattern below 3.5 kb is not normally analyzed because of poor resolution of the bands.

Acknowledgments

The single-locus probes discovered by Prof. Sir Alec Jeffreys are claimed in U.K. Patent No. 2188323 and corresponding worldwide patents. The multilocus probes discovered by Prof. Sir Alec Jeffreys are claimed in U.K. patent No.

A B C D

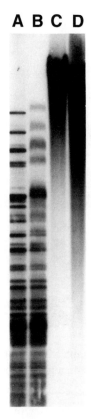

Fig. 3. MLP autoradiograph illustrating complete and partial digestion. Human genomic DNA (1–4 µg) was **(A)** completely digested and **(B–D)** partially digested with *Hin*fI restriction endonuclease, prior to hybridization with the MLP probe 33.6.

2166445 and corresponding worldwide patents. Lumi-Phos 530 is a proprietary product of Lumigen Inc. and is the subject of the following patents Europe: 254051B1, 352712B1; Australia: 603736; Korea: 69259; Taiwan: 46563; Japan: 5-45590; and corresponding worldwide patent property. CDP-*Star* is a proprietary product developed and produced exclusively by Tropix Inc. and is the subject of US Patent No. 5,326,882 and corresponding worldwide patent property.

References

1. Giles, A. F., Booth, K. J., Parker, J. R., Garman, A. J., Carrick, D. T., Akhaven, H., and Schaap, A. P. (1990) Rapid, simple non-isotopic probing of Southern blots for DNA fingerprinting. *Adv. Forensic Haemogenetics*, vol. 3 (Polesky, H. F. and Mayr, W. R., eds.), Springer-Verlag, 40–42.

2. Jeffreys, A. J., Wilson, V., and Thein, S. L. (1985) Hypervariable "minisatellite" regions in human DNA. *Nature* **314**, 67–73.
3. Jeffreys, A. J., Wilson, V., and Thein, S. L. (1985) Individual specific "fingerprints" of human DNA. *Nature* **316**, 76–79.
4. Jeffreys, A. J., Brookfield, J. F. Y., and Semenoff, R. (1985) Positive identification of an immigration test-case using human DNA "fingerprints." *Nature* **317**, 818,819.
5. Gill, P., Jeffreys, A. J., and Werrett, D. J. (1985) Forensic application of DNA "fingerprints." *Nature* **318**, 577–579.
6. Stacey, G. N., Bolton, B. J., and Doyle, A. (1992) DNA fingerprinting transforms the art of cell line authentication. *Nature* **357**, 261,262.
7. Wong, Z., Wilson, V., Patel, I., Povey, S., and Jeffreys, A. J. (1987) Characterization of a panel of highly variable minisatellites cloned from human DNA. *Ann. Hum. Genet.* **51**, 269–288.
8. Smith, J. C., Anwar, R., Riley, J., Jenner, D., Markham, A. F., and Jeffreys, A. J. (1990) Highly polymorphic minisatellite sequences: allele frequencies and mutation rates for five locus-specific probes in a Caucasian population. *J. Forensic Sci. Soc.* **30**, 19–32.
9. Schaap, A. P., Akhaven, H., and Romano, L. (1989) Chemiluminescent substrates for alkaline phosphatase. *Clin. Chem.* **35**, 1863,1864.
10. Bronstein, I., Olesen, C. E. M., Martin, C. S., Schneider, G., Edwards, B., Sparks, A., and Voyta, J. C. (1994) Chemiluminescent detection of DNA and protein with CDP and CDP-*Star*™ 1,2-dioxetane enzyme substrates, in *Bioluminescence and Chemiluminescence: Fundamentals and Applied Aspects* (Campbell, A. K., Kricka, L. J., and Stanley, P. E., eds.), Wiley, Chichester, UK, pp. 269–272.

9

Preparation and Use of ^{32}P-Labeled Single-Locus VNTR Probes in Identity Testing

Jürgen Henke and Lotte Henke

1. Introduction

Nucleic acids labeled with the radioisotope phosphorus-32 [^{32}P] are used as probes for various purposes; coupled with autoradiographic detection, they provide a high degree of sensitivity.

All forms of nucleic acid can be labeled with isotopes. The majority of labeling techniques are based on enzymatic incorporation of a nucleotide labeled with, e.g., [^{32}P] into the DNA. Labeling techniques can be classified into those that lead to uniformly labeled probes and into those that result in end-labeled ones. Uniformly labeled DNA probes incorporate more dNTP than end-labeled probes. The first category is commonly employed in hybridization analysis, whereas end-labeled DNA is often used in DNA sequencing.

1.1. Precautions

Purchase, possession, and use of radioactive isotopes are generally strictly regulated by national law. Many countries have imposed laws that regulate use, storage, and disposal of radioactive materials. Managers and other laboratory staff must possess appropriate training and competence, because radiochemicals require special handling. Even in the most sophisticated laboratory it is inevitable that radioactive aerosols or dusts are generated. Where radiochemicals are not regulated by law, it is strongly recommended to monitor the exposure to isotopes by means of personal dosimetry badges and by health checkups twice a year, which should include the investigation of staff's urine. Shielding material (e.g., Lucite) must be placed between the analysts and the radioactive materials.

From: *Methods in Molecular Biology, Vol. 98: Forensic DNA Profiling Protocols*
Edited by: P. J. Lincoln and J. Thomson © Humana Press Inc., Totowa, NJ

Table 1
Choice of Isotopes and Some of Their Characteristics

Isotope	Half-life	Particle emitted	Applications	Energy
^{32}P	14.3 d	β	Filter hybridization	High
^{125}J	60.0 d	c	*In situ* hybridization	Medium
^{3}H	12.35 yr	β	*In situ* hybridization	Low
^{35}S	87.4 d	β	Filter/*in situ* hybridization	Medium

1.2. Labeling Nucleic Acids for Use as Probes

The choice of the radioactive isotope depends on the desired application. The isotopes that are those most commonly used in nucleic acid hybridizations are shown in **Table 1**.

For nucleic acid hybridizations in solution and on filters, ^{32}P is the isotope of choice because of its high energy, which results in short scintillation counting times and short autoradiographic exposures. Each of the four nucleotides is available in an [α^{32}P]-labeled form suited for incorporation into DNA using one of the polymerase reactions. In addition, [γ^{32}P] ATP is also available for 5'-end-labeling DNA using polynucleotide kinase. For 3'-end-labeling [α^{32}P], cordycepin triphosphate can be used. Probes labeled with ^{32}P should be thawed rapidly; they also should be used as soon as possible after preparation because of the isotope's short half-life and also because of its high energy, which can deteriorate the probe's structure.

1.3. Theory of Nick-Translation

The method called nick-translation was described by Rigby et al. *(1)*. It is well suited to produce uniformly labeled DNA of high specific activity *(2)*. It utilizes DNase I to create single-strand nicks in double-stranded DNA. (Although double-stranded DNA in any form can be used for nick-translation when recombinant plasmid probes are used, the probe insert is typically cut out, purified, and nick-translated). The 5' ⇒ 3' exonuclease and 5' ⇒ 3' polymerase actions of *Escherichia coli* DNA polymerase I are then used to remove stretches of single-stranded DNA starting at the nicks and replace them with new strands made by the incorporation of labeled deoxyribonucleotides *(1)*. As a result, each nick moves along the DNA strand being repaired in a 5' ⇒ 3' direction. The method is illustrated in **Fig. 1** (after **ref. 3**).

Nick-translation can utilize any deoxyribonucleotide labeled with ^{32}P in the α-position ([^{125}J],[^{3}H], and biotinylated nucleotides can also be incorporated). With [α^{32}P]-labeled nucleotides, final (specific) activities of 5×10^8 dpm/μg DNA can be achieved.

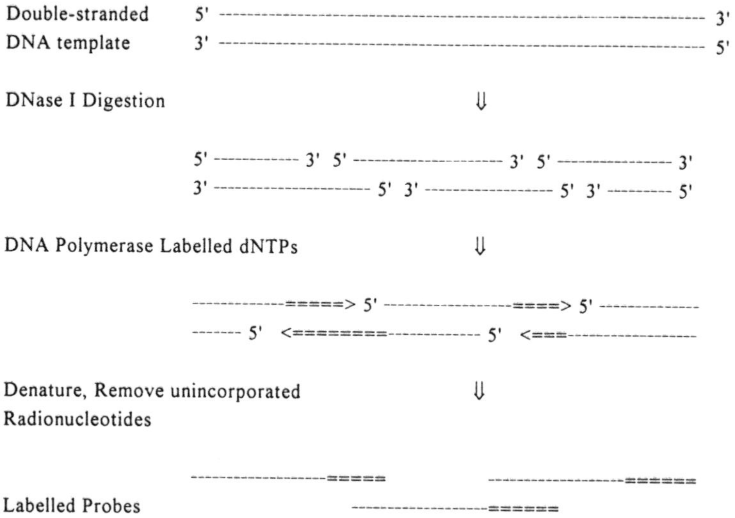

```
Double-stranded   5' ---------------------------------------------------------- 3'
DNA template      3' ---------------------------------------------------------- 5'


DNase I Digestion                                    ⇓


          5' ----------- 3' 5' -------------------- 3' 5' ---------------- 3'
          3' -------------------- 5' 3' ----------------- 5' 3' --------- 5'


DNA Polymerase Labelled dNTPs                        ⇓


          --------------=====> 5' --------------------====> 5' --------------
          ------- 5' <=========--------------- 5' <===-------------------


Denature, Remove unincorporated                      ⇓
Radionucleotides


          --------------------=====       --------------------=====
Labelled Probes           --------------------======
```

Fig. 1. Schematic illustration of nick-translation of DNA (*see* **ref. 3**). Double-stranded DNA is subjected to limited DNase I digestion and incubated with DNA polymerase I in the presence of an $[\alpha^{32}P]dNTP$ precursor. The polymerase primes DNA synthesis at the nick and incorporates the isotope into a newly synthesized chain in the 5' to 3' direction, whereas the 3'–5' exonuclease digests the pre-existing strand. Following denaturation, the salts, enzyme, and unincorporated radionucleotides are removed from the ^{32}P-labeled probe.

The advantages of "nick-translation" as a labeling method are:

1. The simplicity of the reaction,
2. The uniform labeling of the probe, and
3. The high specific activity.

The disadvantage, however, is the nicking itself, because it results in short single-stranded probe molecules in the hybridization reaction. A variation of the protocol was described by Pardue (*4*). The standard as well as the modified protocols are given below.

1.4. Theory of Radioactive Labeling of Oligonucleotide Probes by Means of T4 DNA Polymerase (see ref. 5)

Oligonucleotide probes are employed in various disciplines. T4 DNA polymerase reaction (*6*) is among the strategies available to label oligonucleotide probes.

The T4 DNA polynucleotidekinase labeling method has some advantages over the nick-translation procedure:

1. T4 DNA technique yields intact double-stranded molecules with no nicks; it can be cut with restriction enzyme,

2. Defined regions of DNA can be labeled by controlling the reaction conditions,
3. DNA can be labeled to extremely high specific activity (10^9 dpm/µg) if high specific-activity nucleotides (2000 Ci/mM) are used, and
4. Labeling of one strand of a double-stranded DNA fragment can be achieved.

T4 DNA polymerase has two activities:

1. A 3' ⇒ 5' exonuclease activity and
2. A 5' ⇒ 3' polymerase *(7,8)*.

The exonuclease is active if the exogeneous deoxyribonucleoside triphosphates are absent. Its activity is ≈ 200 × higher than the exonuclease of DNA polymerase. Oligonucleotide probes can be labeled by means of ^{32}P-phosphate transfer from the [γ^{32}P]-ATP to the 5'-OH-end in the presence of T4 polynucleotide kinase (T4 DNA polymerase).

1.5. Theory of Labeling DNA by Random Priming (Oligopriming)

This procedure for labeling DNA fragments using oligonucleotide primers and the Klenow fragment of *E. coli* DNA polymerase I was developed by Feinberg and Vogelstein *(9,10)* as an alternative to nick-translation to produce uniformly labeled probes. Random priming offers a number of advantages over nick-translation. In this procedure, random-sequence hexanucleotides are hybridized to the heat-denatured double-stranded or single-stranded template at multiple sites along the DNA as is illustrated in **Fig. 2** (after ref. *3*). The 3'-OH-end of the hexanucleotides serve as primers for the 5' ⇒ 3' polymerase activity of the Klenow fragment of DNA polymerase I *(11)*. If radiolabeled deoxynucleotides are present in the reaction mixture, the hexanucleotide primers are extended to generate double-stranded DNA that is uniformly radiolabeled on both strands. Because the Klenow fragment lacks the 5' ⇒ 3' exonuclease activity, its use in primer extension avoids loss of incorporated label. The 3' ⇒ 5' exonuclease activity is greatly diminished by the use of a pH 6.6 buffer, permitting synthesis of highly labeled probes. For random priming, [α^{32}P]-deoxynucleoside triphosphates can be used.

1.6. Theory of Filter Hybridization

Filter hybridization is of considerable importance in molecular biology. The technique is derived from the classical experiments conducted by Gillespie and Speigelman *(12)*. Denatured DNA, for example, is immobilized on a nitrocellulose filter in such a way that self-annealing is prevented. The filters are first prehybridized with hybridization buffer without the probe. Nonspecific DNA binding sites on the filter are thus saturated with carrier DNA, synthetic polymers, or proteins. The bound base sequences are available for specific hybridization with a (single-stranded) DNA probe. The latter is often labeled

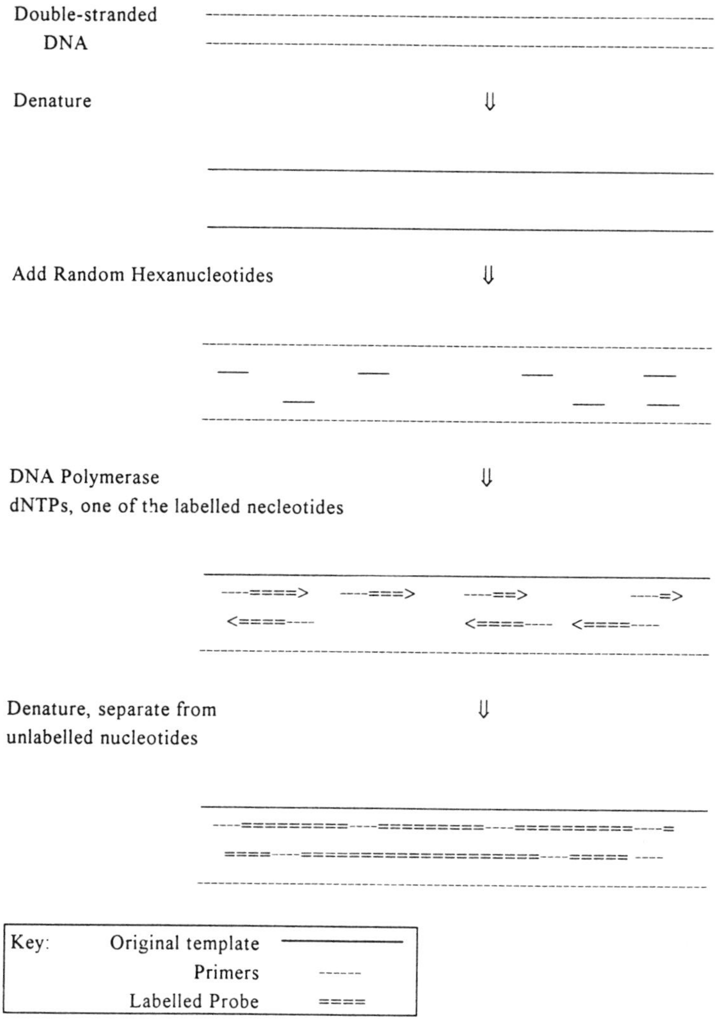

Fig. 2. Random primer labeling of DNA (*see* **ref. 3**). Double-stranded DNA is denatured and annealed with random oligonucleotide primers (6-mers). The oligonucleotides serve as primers for the 5'–3' polymerase, which synthesizes labeled probes in the presence of an [α-^{32}P]dNTP precursor.

with ^{32}P. Hybridization is followed by extensive washing of the filter, which is required to remove superfluous probe. Detection of hybridization reactions is usually achieved by means of autoradiography after a preliminary scintillation counting. The procedure is widely applied for various methods (plaque and colony hybridization, Northern- and Southern-blot hybridization *[13]*, dot blot, and so forth).

2. Materials

2.1. Reagents Required
for the "Standard" Nick-Translation (see ref. 1)

1. 10X nick-translation buffer: 500 mM Tris-HCl, pH 7.8, 50 mM MgCl$_2$, 100 mM 2-mercaptoethanol, 1 mg/mL nuclease-free bovine serum albumin (BSA).
2. 10X dNTP without dATP (depending on the target sequence): 1 mM dCTP, 1 mM dGTP, 1 mM dTTP.
3. [α-^{32}P] dCTP (3000 Ci/mM).
4. STOP buffer: 10 mM Tris-HCl, pH 7.5, 10 mM EDTA, 0.1% (w/v) sodium dodecylsulfate (SDS).
5. DNA polymerase I: follow manufacturer's instructions.
6. DNase I: The enzyme is commercially available at different concentrations. A stock solution can be prepared at a concentration of 1 mg/mL. Aliquots of 10 µL can be stored deep-frozen. For storage at –20°C dissolve DNase I in a solution of 20 mM Tris-HCl, pH 7.5, 1 mM MgCl$_2$, and 50% glycerol. Enzyme should be dissolved without vortexing and thawed on ice. It cannot be used a second time.

2.2. Reagents Required for Modified Nick-Translation

1. 10X nick-translation buffer: 500 mM Tris-HCl, pH 7.8, 50 mM MgCl$_2$, 0.5 mg/mL BSA.
2. 1 µL 0.5 mM dGTP.
3. 1 µL 0.5 mM dCTP.
4. 1% 2-mercaptoethanol.
5. 1 µL DNase I (commercially available DNase [1 mg/mL stock] is stored in aliquots at –20°C. Before use, the enzyme is diluted by a factor of 10^5 in distilled water).
6. 1 µL DNA polymerase I (10 U/µL).
7. Distilled water.

2.3. Reagents Required for Labeling with T4 DNA Polymerase

1. 1 mL solution of oligonucleotide probe (10 pmol/µL).
2. 2.8 µL distilled water.
3. 1 µL 10X kinase buffer: = 500 mM Tris-HCl, pH 7.4, 50 mM MgCl$_2$, 20 mM dithiothreitol (DTT), 1.0 mM spermidine.
4. 5 µL [c^{32}P] ATP (370 MBq/mL, 5 mM).
5. 0.2 µL T4 polynucleotide kinase (2 U).

2.4. Required Reagents for Oligopriming (see ref. 3)

1. 10X DNA polymerase I (Klenow fragment) buffer: 500 mM Tris-HCl, pH 6.6, 100 mM MgCl$_2$, 10 mM DTT, 0.5 mg/mL nuclease-free BSA, fraction V.
2. 10X dNTP without dCTP: 0.5 mM dATP, 0.5 mM dGTP, 0.5 mM dTTP.
3. Random hexanucleotides, which are commercially available.
4. Large fragment of *E. coli*. DNA polymerase I. Follow supplier's instructions.
5. 0.5 M EDTA.

6. 10 mg/mL yeast tRNA.
7. TE buffer: 10 mM Tris-HCl, pH 7.6, 1 mM EDTA.

2.5. Reagents Required
for the Prehybridization and Hybridization of Filters

1. Prehybridization solution 1: 10X Denhardt's solution.
2. 100X Denhardt's solution: 10 g Polyvinylpyrrolidone (PVP), 10 g BSA, 10 g Ficoll 400, brought to a total volume of 500 mL with distilled water. The solution can be aliquoted and stored at –20°C.
3. Prehybridization solution 2: 1X SSC, 10X Denhardt's solution, 6% polyethylene-glycol 6000 (PEG 6000), 0.1% SDS, 50 µg/mL sheared herring sperm DNA (Sigma D-6898, Sigma Chemical Co., St. Louis, MO), 5 µg/mL sheared human-placenta DNA (Sigma D-7011).
4. Sheared herring-sperm DNA; Prepare as follows: Cut 300 mg DNA with scissors and suspend in 40 mL distilled water. After a mechanical shearing in a syringe, dissolve DNA for approx 2 d at 37°C. After two phenol- and one chloroform-extractions, precipitate DNA by adding sodium acetate (final concentration 0.2 M) and ethanol. Dissolve DNA in 20 mL TE buffer. Confirm degradation of herring-sperm DNA on a separate gel. Before use, boil sperm DNA for 5 min and chill on ice.
5. Sheared human-placenta DNA; The placenta DNA is prepared as follows: Dilute 25 mg placenta DNA in 2 mL 0.3 M NaOH, 20 mM EDTA and boil for 5 min. Follow with neutralization in 2 mL 0.3 M HCl, precipitation with ethanol, and centrifugation (3000g). Remaining salts can be removed by washing in 70% etha-nol. Dried DNA can be rediluted in 25 mL distilled water and stored aliquoted at –20°C. Prior to adding the placenta DNA to the hybridization solution, boil the DNA for 5 min and cool on ice.

3. Methods
3.1. "Standard" Nick-Translation

1. Combine 200–300 ng DNA probe, 0.4 µL 1 M dCTP, dGTP, and dTTP, 1.6 µL 10X nick translation buffer, 2 µL [α^{32}P]-dATP (5 mM, 370 MBq/mL), 0.3 µL polymerase I (5 × 103 U/mL), and sufficient distilled water to bring the volume to 16 µL.
2. Mix components and incubate at 15°C for 1.5 h.
3. Stop the reaction by adding 34 µL STOP buffer.
4. Separate labeled probe from unlabeled DNA by using a "Nick-translation col-umn" (Pharmacia, Uppsala, Sweden).

3.2. "Modified" Nick-Translation

1. Dry down 150 pmol of [^{3}H] dTTP and 150 pmol of [^{3}H] dATP in a plastic micro-centrifuge tube.
2. Add 1 µL 0.5 mM dGTP, 1 µL 0.5 mM dCTP, 1 µL 10X 0.5 M Tris-HCl, pH 7.8, 50 mM MgCl$_2$, 0.5 mg/mL BSA, 1 µL 1% 2-mercaptoethanol, 0.1 µg cloned DNA, 1 µg DNase I (commercial DNase I [1 mg/mL stock] is stored in aliquots

at –20°C. DNase is to be diluted by a factor of 10^5 in distilled water prior to use), and 1 μg *E. coli* DNA polymerase I (10 U/μL) water to 10 μL final volume.
3. Incubate the mixture at 14°C for 1–2 h.
4. Add 90 g of carrier DNA (*E. coli* DNA sheared by sonication and denatured by boiling for 10 min) and adjust the total volume to 100 μL with distilled water.
5. Add 3 μL 0.1 *M* spermine. Mix and leave on ice for 15 min.
6. Centrifuge at 10,000*g* for 10 min. Discard the supernatant.
7. Resuspend the pellet in 75% ethanol containing 0.3 *M* sodium acetate and 10 m*M* magnesium acetate. Vortex. Leave on ice for 1 h, mixing frequently.
8. Centrifuge for 10 min (10,000*g*) to recover the DNA.
9. Resuspend the labeled double-stranded DNA probe in water.

3.3. Labeling with T4 DNA Polymerase

1. Mix reactants 1–5 (*see* **Subheading 2.**) are be mixed and incubated at 37°C for 60 min.
2. The reaction is stopped by adding 1 μL 0.5 *M* EDTA, pH 8.0, and 89 μL TE buffer, pH 8.0.
3. The solution is then stored at 0°C.

3.4. Oligopriming

1. Keep all components on ice.
2. DNA probes (e.g., single-locus probes YNH24, MS1) are denatured (to single-stranded DNAs) at 95°C and mixed with random hexanucleotides after cooling to 0°C.
3. The reaction tube is centrifuged (microfuge) for a few seconds to collect the DNA solution as a single drop in the bottom of the tube. Place the tube on ice.
4. Combine: 2.5 μL 10X dNTP minus dCTP, 2.5 μL 10X Klenow buffer, 5 μL α^{32}P-dCTP (3000 Ci/m*M*) = 50 μCi, and 1 μL Klenow fragment (3–8 U).
5. Add the reaction mix to the denatured DNA; adjust the reaction volume to 25 μL with distilled water and mix briefly.
6. Incubate the mixture at room temperature for 1–2 h.
7. Stop the reaction by adding 1 μL of 0.5 *M* EDTA and dilute the reaction mixture to 100 μL with TE buffer.
8. Separate the labeled DNA from unincorporated radioactive precursors by column chromatography.
9. Ethanol-precipitate the labeled DNA-probe fragments if desired, and estimate the activity as described above.

Encompassing descriptions of the procedure are also given by the manufacturers (e.g., Oligolabeling Kit, Pharmacia).

3.5. Separation by Means of Column Chromatography

1. Unincorporated [γ^{32}P]-ATP has to be separated by means of column chromatography with DE-52 cellulose (Whatman). Cellulose must be washed in TE buffer, pH 8.0, several times until the fresh TE buffer remains at this pH value.

2. Preparation of columns: Polypropylene Econo Columns (Bio-Rad Laboratories, Hercules, CA) are filled to the 0.3 mL mark with washed DE-52 cellulose.
3. The labeling solution is poured onto the surface of the cellulose and allowed to entirely soak. Thereafter, unincorporated [^{32}P]-ATP is eluted by means of 4 mL TE buffer and 4 mL 0.2 M NaCl in TE buffer. The labeled probe is then eluted by adding 2X 500 µL 0.5 M NaCl in TE buffer. The first two drops of the eluate are discarded.

3.6. Control of Incorporation

By means of ^{32}P-solutions of known activities, a reference curve can be established. The success of labeling reactions can be estimated using a scintillation counter at a constant distance *(6,14)*.

3.7. Hybridization Procedures
Using ^{32}P-Labeled Single-Locus Probes

Prehybridization is carried out in as follows:

1. Incubate nylon filters for 1 h (at 53°C) in prehybridization solution 1 and thereafter for 2 h in solution 2 (at 62°C) by adding 50 µL/mL herring-sperm DNA plus 5 µg/mL placenta DNA.
2. The radiolabeled DNA probe is denatured by boiling for 5 min and cooled on ice for 10 min.
3. At least 10^6 counts per minute (cpm) of ^{32}P-labeled probe per milliliter of prehybridization solution 2 is introduced to produce the hybridization solution. The probe should have a specific activity of at least 10^8 cpm/µg (*see* **Subheading 3.6.**). An incorporation rate of $4–5 \times 10^9$ cpm/µg DNA is recommended.
4. Hybridization is carried out overnight at 62°C. All hybridizations are preferably carried out in tubes with 4–8 cm diameter and a length of 28 cm that are rotated in a hybridization oven. Up to 12 filters can be hybridized simultaneously *(6)*. **Caution:** Air pressure within the heated tubes is raised during hybridization. Great care should be taken when opening.
5. After hybridization, carry out the following washes:
    a. 2 × 10 min in 2X SSC, 1.5% SDS at room temperature.
    b. 1 × 15 min in 1X SSC, 1.0% SDS at 62°C.
    c. 2 × 10 min in 0.1X SSC, 0.1% SDS at 62°C.
6. After the last washing, the solution is drained carefully, and excess moisture is blotted from the filter. The filters are then wrapped in clear plastic wrap. Wrinkles are smoothed out.

3.8. Autoradiography Procedure

^{32}P emits X- and β-rays, which are able to blacken films. Intensity of blackening depends on the rate of isotope decomposition.

1. In a dark room, the filter is placed against a sensitive X-ray film in a film cassette *(6,15)*. Employment of two selected intensifying screens is recommended,

Fig. 3. Exposure arrangement (*see* **ref. *15***).

because this allows working at –20 instead of –70°C without any loss of information. Filter and film are sandwiched between the intensifying screens (*see* **Fig. 3**).
2. Expose the film either at –20 or –70°C for the chosen amount of time.
3. Remove the cassette from the freezer and allow equilibration to room temperature. Remove the film in the dark.
4. Process the film by using standard X-ray development techniques.

3.9. Reprobing of Filters

After hybridization and autoradiography, the filters can be stripped of the first probe by means of a high-stringency wash and hybridized again with a second probe. How often filters can be stripped and reprobed depends on the membrane material. Nitrocellulose membranes are quite fragile. It is essential that filters do not dry out completely after hybridization, because this may bind the initially used probe to the filter. Stripped filters must be prehybridized again in order to block nonspecific DNA-binding sites. Short-term storage of filters is possible if they are stored moist in sealed bags at 4°C until reprobed. Long-term storage is recommended at –20°C.

1. After autoradiography the filter is immersed in 500 mL high-stringency-wash buffer and incubated at 95°C for 20 min under gentle shaking.
2. The buffer is poured off and replaced with a second 500-mL preheated buffer; incubation at 95°C for 20 min is continued.
3. The filter is placed in a hybridization tube that contains 20 mL prehybridization solution 2 and incubated at 62°C overnight.
4. The filter is removed from the prehybridization solution. Excess fluid is blotted off. The filter is ready for second hybridization.

4. Notes

4.1. Factors Affecting the Rate of Filter Hybridization

Anderson and Young *(16)* have compiled variously important factors that affect the rate of filter hybridization. For gaining detailed information, it is recommended to read their article. The following list is a rough compilation of factors:

1. Concentration of the probe: When single-stranded probes are used, no reassociation of single-stranded nucleic acids in solution is to be expected unless regions

of extensive self-complementarity exist. Probe concentration must not be increased without limit, because if more than approx 100 mg ^{32}P labeled probe/mL is used, nonspecific and irreversible binding to the filters will be observed.

2. Molecular weight of the probe: Two situations exist: When the concentration of filter-bound nucleic acid sequence [$=C_f$] is low compared to the concentration in solution [$=C_s$], the rate of hybridization is independent of the molecular weight. However, when [C_f] is high compared with [C_s], the rate of hybridization is roughly inversely proportional to the molecular weight of the probe.

3. Base composition: Increased percentage of G + C increases the rate of hybridization.

4. Temperature affects the rate of any hybridization reaction. Typically, a Gaussian curve is observed: At 0°C, hybridization proceeds extremely slowly. The rate increases to a maximum, which is 20–25°C below T_m. At higher temperatures, the duplex molecules tend to dissociate, and at T_m –5°C, the rate is extremely low.

5. Formamide has several practical advantages, because it decreases the T_m of duplex molecules. By introducing 30–50% formamide to the hybridization solution, the incubation temperature can be reduced to 30–42°C. Usually, a probe is more stable at lower temperatures. However, we always try to avoid formamide, because it is considered to be hazardous. Hybridization can be carried out in trays, plastic bags, or tubes in hybridization ovens *(6)*. Working with tubes in an oven is highly advantageous, because of easier handling and reduction of contamination.

4.2. Some Background Information
Concerning DNA Reassociation (see ref. 17)

The rate at which complementary strands of nucleic acid form stable base-paired duplexes depends on a variety of factors.

6. Length of nucleic acids: Wetmur and Davidson *(18)* theoretically predicted that the rate of reannealing would be proportional to the length *(L)* of the fragments involved. Appropriate experiments, however, revealed that the reannealing rate is proportional to the square root of *L (17)*.

7. Composition of bases: It has been demonstrated by Hutton and Wetmur *(19)* that the base composition of nucleic acids has a negligible effect on the rate of either DNA–DNA annealing or RNA–DNA hybridization. This was also demonstrated by Bishop *(20)* by means of hybridization experiments.

8. Ionic strength: It has been shown by Wetmur and Davidson *(18)* that the DNA reannealing rate depends strongly on the salt concentration. At concentrations up to 0.2 *M*, the rate has been shown to be proportional to the cube of ionic strength *(21)*. Britten et al. *(22)* have compiled experimental data on the variation of DNA reassociation rates with ionic strength.

9. Viscosity: If considering the effect of viscosity, one has to distinguish between "microscopic" and "macroscopic" viscosity. The first refers to the microenvironment around the DNA bases, which can be altered by the addition of, e.g., sucrose and glycerol. The "macroscopic" viscosity depends on the presence of polymers, which will have no effect on the microenvironment. A detailed study by Chang et

al. *(23)* revealed, e.g., that in a 5.7% Ficoll solution, the rate of phage T4 DNA renaturation was increased by 50%. Dextran sulfate is now widely used to accelerate reactions.

10. Denaturing agents: The optimal temperature for nucleic-acid reassociations (in aqueous salt solutions) lies between 60 and 75°C. However, the extended incubation at optimal temperatures can cause thermal strand scissions. Therefore, it is desirable to reduce the temperature while maintaining the stringency of nucleic-acid reorganization. This can be achieved by introducing a reagent (e.g., formamide), which destabilizes double-stranded nucleic acid. In other words, a 1% increase in formamide concentration lowers the melting temperature (T_m) of double-stranded DNA by 0.72°C. Reagents other than formamide that can reduce the T_m are sodium perchlorate, tetramethylammonium chloride, tetraethylammonium chloride, and urea.

11. Mismatching: It is common knowledge that the presence of mismatched base pairs reduces the thermal stability of DNA duplex molecules. Thus, it was expected that imperfect base complementarity will influence the rate of reassociation. From several studies (quoted in **ref. *17***) it is clear that mismatching of DNA that results in a decrease in the T_m of 15°C will reduce the reannealing rate by a factor of 2.

12. Temperature: Marmur and Doty *(24)* observed that, as the temperature was reduced from the T_m, the rate of reassociation increased until a maximum is reached, which lies about 25°C below the T_m.

Acknowledgments

The authors are indebted to Sabine Cleef, Malgorzata Tahar, and Inge Kops for skillful technical assistance.

References

1. Rigby, P. W. S., Dieckmann, M., Rhodes, C., and Berg, P. (1977) Labelling deoxyribonucleic acid to high specific activity in vitro by nick-translation with DNA polymerase I. *J. Mol. Biol.* **113,** 237–251.

2. Sambrook, J., Fritsch, E. F., and Maniatis, T. (1989) *Molecular Cloning: A Laboratory Manual*, 2nd ed. Cold Spring Harbor Laboratory, Cold Spring Harbor, NY.

3. Keller, G. H. and Manak, M. M. (1989) *DNA Probes.* Stockton Press, New York.

4. Pardue, M. L. (1985) In-situ-hybridisation, in *Nucleic Acid Hybridisation: A Practical Approach* (Hames, B. D. and Higgins, S. J., eds.), IRL, Oxford, UK, pp. 179–202.

5. Arrand, J. E. (1985) Preparation of nucleic acid probes, in *Nucleic Acid Hybridisation: A Practical Approach* (Hames, B. D. and Higgins, S. J., eds.), IRL, Oxford, UK, pp. 17–45.

6. Henke, L. (1994) Validierung einiger molekularbiologischer Methoden zur Individualisierung und zur Abstammungsbegutachtung des Menschen. Ph.D. dissertation, Univ. Frankfurt/M.

7. Morris, C. F., Hama-Inaba, H., Mace, D., Sinha, N. K., and Alberts, B. (1979) Purification of the gene 43, 44, 45 and 62 proteins of the bacteriophage T4 DNA replication apparatus. *J. Biol. Chem.* **254,** 6787.

8. O'Farrell, P. (1981) Replacement synthesis method of labelling DNA fragments. *BRL Focus* **3**, 1.

9. Feinberg, A. P. and Vogelstein, B. (1983) A technique for radiolabelling DNA restriction endonucleose fragments to high specific activity. *Anal. Biochem.* **132**, 6–13.

10. Feinberg, A. P and Vogelstein, B. (1984) A technique for radiolabelling DNA restriction endonucleose fragments to high specific activity. *Add. Anal. Biochem.* **137**, 266,267.

11. Klenow, H., Overgard-Hanson, K., and Patkar, S. (1971) Proteolytic cleavage of native DNA polymerase into two different catalytic fragments. *Eur. J. Biochem.* **22**, 371–381.

12. Gillespie, D. and Speigelman, S. (1965) A quantitative assay for DNA–RNA hybrids with DNA immobilized on a membrane. *J. Mol. Biol.* **12**, 829.

13. Southern, E. (1975) Detection of specific sequences among DNA fragments separated by gel electrophoresis. *J. Mol. Biol.* **98**, 503–517.

14. Schmitter, H., Berschick, P., Henke, L., and Henke, J. (1995) Molekularbiologische Untersuchungsverfahren in der forensischen Spurenanalyse (DNA-Analysen, DNA-Profiling), in *Humanbiologische Spuren* (Schleyer, F., Oepen, I., and Henke, J., eds.), Kriminalistik Verlag, Heidelberg, pp. 181–229.

15. Henke, J., Henke, L., and Cleef, S. (1988) Comparison of different x-ray films for ^{32}P autoradiography using various intensifying screens at –20°C and –70°C. *J. Clin. Chem. Clin. Biochem.* **26**, 467,468.

16. Anderson, M. L. M. and Young, B. D. (1985) "Quantitative filter hybridisation," in *Nucleic Acid Hybridisation: A Practical Approach* (Hames, B. D. and Higgins, S. J., eds.), IRL, Oxford, UK, pp. 73–111.

17. Young, B. D. and Anderson, M. L. M. (1985) "Quantitative analysis of solution hybridisation in *Nucleic Acid Hybridisation: A Practical Approach* (Hames, B. D. and Higgins, S. J., eds.), IRL, Oxford, UK, pp. 47–71.

18. Wetmur, J. G. and Davidson, N. (1968) Kinetics of renaturation of DNA. *J. Mol. Biol.* **31**, 349.

19. Hutton, J. R. and Wetmur, J. G. (1973) Renaturation of bacteriophage Ø x 174 DNA–RNA hybrid: RNA length effect and nucleation rate constant. *J. Mol. Biol.* **77**, 495.

20. Bishop, J. O. (1972) Molecular hybridisation of ribonucleic acid with a large excess of deoxyribonucleic acid. *Biochem. J.* **126**, 171.

21. Studier, F. W. (1969) Effects of the conformation of single stranded DNA on renaturation and aggregation. *J. Mol. Biol.* **41**, 199.

22. Brittan, R. J., Graham, D. H., and Neufeld, B. R. (1974) in *Methods in Enzymology*, vol. 29 (Grossmann, L. and Moldave, K., eds.), Academic, London, UK, pp. 363ff.

23. Chang, C. T., Hain, T. C., Hutton, J. R., and Wetmur, J. G. (1974) Effects of microscopic and macroscopic viscosity on the rate of renaturation of DNA. *Biopolymers* **13**, 1847.

24. Marmur, J. and Doty, P. (1961) Thermal renaturation of deoxyribonucleic acids. *J. Mol. Biol.* **3**, 585.

10

Statistical Methods Employed
in Evaluation of Single-Locus Probe Results
in Criminal Identity Cases

Bruce S. Weir

1. Introduction

Evaluation of single-locus probe results in criminal-identity cases is best presented within the standard forensic framework for handling trace evidence. A comprehensive account was given by Aitken *(1)*, and the language of that text will be used here. For simplicity, suppose a stain from the scene of the crime has DNA profile type A that is known to be from the perpetrator P of the crime. Person S, who is a suspect in the crime, has profile B. In the conventional treatment, S is removed from suspicion if $B \neq A$, although this match-based approach can be avoided with a continuous treatment *(2)*. If there is a match between the stain and suspect profiles, $A = B$, the evidence of the match is denoted by E. There are two explanations for E:

1. $\underline{C}$: Person S is the perpetrator P (usually the prosecution explanation).
2. $\bar{C}$: Person S is not the perpetrator P (usually the defense explanation).

If S denies being the perpetrator, then the issue needing to be determined by the trier(s) of fact is whether C or $\bar{C}$ is true. Before the evidence of matching profiles is found, there may be prior probabilities, $\Pr(C)$ and $\Pr(\bar{C})$, for the two explanations. The ratio of these two is called the "prior odds." It is unlikely that a forensic scientist can, or should, assign a value to this ratio. It may even be that each trier of fact has his or her own prior odds. Under carefully stated assumptions, however, the scientist can present probabilities of the evidence if either of the explanations is true. These quantities are written as $\Pr(E|C)$ and $\Pr(E|\bar{C})$. It needs to be stressed that the forensic scientist is not claiming that

From: *Methods in Molecular Biology, Vol. 98: Forensic DNA Profiling Protocols*
Edited by: P. J. Lincoln and J. Thomson © Humana Press Inc., Totowa, NJ

either explanation *is* true, but is merely pointing to the consequences of either being true. The ratio of probabilities

$$L = (\Pr[E|C])/(3\Pr[E|\overline{C}]) \tag{1}$$

is called a likelihood ratio. In paternity testing it is called the "paternity index."

It is a consequence of Bayes' theorem that the posterior odds is the product of the likelihood ratio and the prior odds:

$$(\Pr[C|E])/(\Pr[\overline{C}|E]) = L \times (\Pr[C])/(\Pr[\overline{C}]) \tag{2}$$

The calculation of L, however, is not a "Bayesian analysis," because this term usually implies the assignment of prior probabilities.

The likelihood ratio leads to statements such as "The evidence is L times more likely if the prosecution explanation is correct than if the defense explanation is correct," or "The evidence is L times more likely if S left the stain than if someone unrelated to S left the stain." It is both dangerous and easy to transpose the conditional to produce statements like "After considering the evidence of a match, it is L times more likely that S left the stain than it is that someone else did." The forensic scientist needs to guard against the tendency of both prosecutors and defense attorneys to make such statements, and simple analogies are useful. The probability that a man weighs over 250 lb. if he is professional football player is high, but the probability that a man is a professional football player if he weighs over 250 lb. is quite low and depends on the (low) proportion of the population who play football for a living.

If the DNA profiling technique is error-free, then a match is guaranteed when the suspect is actually the perpetrator. The evidence of a match is certain: $\Pr(E|C) = 1$. Furthermore, if the suspect is not the perpetrator and these two people have independent chances of being of DNA profile type A, then the denominator of L reduces to the probability of a random member of the population having profile A. In other words, L is just the reciprocal of the profile frequency and it makes little difference whether A is said to have a frequency of 1 in a million, or is said to give a likelihood ratio of one million. Although this is expedient, it glosses over some issues and is of no help in accommodating population structure, relatives, or mixed stains.

The rest of this chapter is concerned with evaluating the probability of the evidence when explanation $\overline{C}$ is true. The determination of matching profiles will be accepted as being reliable for these analyses. Issues of laboratory error, evidence mishandling, or fraud that may lead to false declarations of a match are ignored.

1.1. Conditional Frequencies

The evidence E is actually the compound event that both S and P have profile A; this will be written in an informal way as S_A, P_A. The two explanations refer to S and P either being the same or different people, written here as $S = P$ or $S \neq P$. Then the likelihood ratio is

$$L = (\Pr[S_A, P_A | S = P])/(\Pr[S_A, P_A | S \neq P])$$
$$= \{[\Pr(P_A|S_A, S = P)]/[\Pr(P_A|S_A, S \neq P)]\} \times \{[\Pr(S_A|S = P)]/[\Pr(S_A|S \neq P)]\} \quad (3)$$
$$= 1/[\Pr(P_A|S_A, S \neq P)]$$

The last line follows because P is certain to have profile A if S does when they are the same person, and the probability of S having profile A is the same regardless of whether S and P are the same person. This algebra emphasizes that the interpretation of single-contributor stains requires the conditional frequency with which one person (P) has a particular profile, given that another person (S) has been seen to have that profile. Only under the special circumstance of independent profile frequencies can this conditional frequency be replaced by an unconditional frequency:

$$\Pr(P_A|S_A, S \neq P) = \Pr(P_A) \quad (4)$$

and the likelihood ratio can also be expressed in terms of this frequency

$$L = 1/[\Pr(P_A)] \quad (5)$$

1.2. Typing Errors

The symbols S_A and P_A in the **Subheading 1.1.** mean that S or P had profile A. Strictly, they mean that these people are declared to have profile A, and it is possible that one or both of these declarations is in error. Although there has been discussion regarding how such errors could be quantified *(3)*, it is difficult to envisage how this could be done. A DNA-typing laboratory can participate in various proficiency tests and could report the numbers of successes and failures from such tests. The fact that a failure is likely to cause changes in protocols or personnel makes it difficult to speak about error "rates." A rate indicates a proportion of some outcome under repeated trials, and a change in conditions following an error means that trials are not being repeated exactly. The fact that the laboratory either did or did not make errors in these tests is, of course, relevant information to the trier of fact.

Trying to incorporate proficiency-test results into likelihood ratios faces the additional complication that each case is unique. Circumstances surrounding the analysis of a case may be quite different from those surrounding a proficiency test. Known (person S) and query (person P) profiles may be determined some months apart, for example. There is no repetition of the analysis that could give rise to a rate at which errors arose—an error was either made or not made. Having said that, it must be acknowledged that the forensic scientist can discount the possibility of errors with greater confidence when multiple items of evidence provide the same matching profiles or when different laboratories either get duplicate results or each type different members of matching profile pairs *(4)*.

2. Matching Criteria

For loci with discrete alleles, there is unlikely to be any doubt regarding whether two profiles match. Matching requires that every allele in one profile has an equal allele in the other. For loci in which alleles are interpreted by measurement, such as migration distance on an electrophoretic gel, however, there is a need for a matching criterion.

Experiments are needed in which two DNA samples from the same person are typed; these samples could be blood and semen from the same man or blood and vaginal epithelial cells from the same woman, for example. Suppose a series of pairs of electrophoretic bands, representing the same allele but from different samples, have estimated lengths x, y that never differ by more than a multiple 2α of their mean:

$$|x - y| \le 2\alpha(x + y)/2 \tag{6}$$

Typically, α has values in the range of 2 ~ 3. Any future pair of bands can be declared to match if they have estimated lengths that satisfy the same relation. The relation can be manipulated to give:

$$x \le [(1 + \alpha)/(1 - \alpha)]\, y, \text{ if } x \ge y$$
$$y \le [(1 + \alpha)/(1 - \alpha)]\, x, \text{ if } y \ge x \tag{7}$$

or

$$[(1 - \alpha)/(1 + \alpha)]\, y \le x \le [(1 + \alpha)/(1 - \alpha)]\, y$$
$$[(1 - \alpha)/(1 + \alpha)]\, x \le y \le [(1 + \alpha)/(1 - \alpha)]\, x \tag{8}$$

Providing α is small, these relations may be approximated by

$$(1 - 2\alpha)y \le x \le (1 + 2\alpha)y \tag{9}$$
$$(1 - 2\alpha)x \le y \le (1 + 2\alpha)x \tag{10}$$

In other words, each band of a matching pair must lie within a window of total width 4α centered on the other band. Each band is considered to point to an interval of total length 2α, centered on its estimated length, that contains the true length of the DNA fragment producing that band.

Other ways of determining α are possible. A laboratory may find that α differs for different lengths, for example. The criteria need to be based on observations made under conditions employed by the laboratory and need to be established prior to examination of bands in a particular case.

3. Binning Strategies

Assigning frequencies to individual alleles in matching profiles for loci with discrete alleles is straightforward. All those n_i alleles in a sample of size n alleles that are of type i provide a maximum likelihood estimate $\tilde{p}_i$ of the frequency:

$$\tilde{p}_i = n_i/n \tag{11}$$

For loci in which alleles are inferred from measurements, it is unlikely that any sample will contain alleles with any specific estimated length. The *floating bin* approach uses the number n_i of sample alleles that satisfy the matching criterion to provide an estimate of n_i/n. This procedure has the virtue of simplicity, but the unusual feature that the frequencies assigned to each band in a sample by this method will have a sum >1.

The *fixed bin* method developed by the Federal Bureau of Investigation *(5)* first divides a sample of bands into a number of discrete categories, whose boundaries are given by a sizing ladder. Each sample band is therefore assigned to one of these categories or bins. If a future band, along with its associated interval of width 2α, lies within a bin, the sample frequency of that bin is assigned to the band. If the interval overlaps a bin boundary, the practice is to assign the higher of the two adjacent bin frequencies. This procedure will be satisfactory unless the sample has bands clumped near the boundary in question, when the assigned frequency would underestimate the frequency of bands that would match the band in question. The sample can be checked for clumping, but it should be noted that fixed bins are designed not to give matching frequencies but to give conservative indicators of band frequencies, with the degree of conservatism increasing with the bin widths. The fixed bin should have a width of at least 4α.

4. Profile Frequencies

Assigning a numerical value to the likelihood ratio has the dual problem of needing to be based on sound science and to convey meaning to the trier of fact. The most straightforward demonstration of the meaning of the index can be provided by experiments conducted on databases.

Simulating the case of the prosecution, explanation C being true can be shown by evaluating the likelihood ratio for every individual represented in a database and regarding that person as both suspect and perpetrator. For the case in which the defense explanation is true, each pair of people in a database can be regarded as representing suspect and perpetrator. A good DNA typing system should give large likelihood ratios when suspect and perpetrator are the same person and low values when they are unrelated people. Both the system and an appreciation of the range of likelihood ratios can be explained by reference to these experiments. The value for the case at hand can be put into context with a minimum of assumptions or background theory.

Table 1
One-Locus Frequency Estimates for a Matching Profile

	Database				
Locus	African American	Caucasian	SE Hispanic	SW Hispanic	Total
D1S7	2/359	6/595	3/305	2/288	13/1547
D2S44	2/475	3/792	2/300	0/284	7/1851
D4S139	2/448	12/594	2/311	3/265	19/1618
D5S110	4/353	2/511	4/286	1/165	11/1315
D10S28	3/288	4/429	1/230	6/283	14/1230
D14S13	0/524	0/751	1/306	0/187	1/1768

This simple experiment will work only if matching profiles occur in a database, because otherwise the index is not defined in the second experiment. Evett et al. *(2)* got around this problem for VNTR loci by using a continuous analysis that does not require a declaration of matching. For the less-discriminating STR loci, four-locus matches were found in some databases in the United Kingdom *(6)*.

The point is that profiles at several loci tend to be so rare that duplicates are not found in samples of a few hundred, or even a few thousand, people. Likewise, the profile(s) of interest in a particular case are not found in these samples. Such procedures as adding the profile to a database of n profiles will give an estimated frequency of $1/(n + 1)$ under C or $2/(n + 2)$ under $\bar{C}$, but this has the unsatisfactory aspect of giving the same answer whether the profile not seen in a database was based on a single locus or several loci. Any numerical statements must recognize that a DNA profile is not a single entity, but is actually a composite of information from several loci.

4.1. Single-Locus Frequencies

The first step is to show the frequencies at each of the loci in the matching profile, using a range of databases. For example, **Table 1** shows the frequencies at six of the VNTR loci in the profile said to match a bloodstain on a sock found in the defendant's bedroom and a blood sample from a victim in a recent court case.* The typing was performed by the California Department of Justice DNA Laboratory, and frequencies were found by seeking matching one-locus genotypes in four FBI databases. Presenting such a chart in court makes it very clear that matches at single loci are not common, regardless of whether frequencies based on African Americans, Caucasians, or Hispanics are used. At

*People of the State of California vs Orenthal James Simpson, Los Angeles County Case BA097211.

Table 2
Observed and Expected Genotypic Counts
in FBI Caucasian Database for Table 1 Profile

Locus	n	Frag. lengths	Floating bins			Fixed bins		
			Obs.	Exp.	'2p'	Obs.	Exp.	'2p'
D1S7	595	5319	6	3.5	91.0	2	2.1	71.0
D2S44	792	2638, 2528	3	4.4		2	2.7	
D4S139	594	5606	12	4.5	103.0	8	6.9	128.0
D5S110	511	4341, 2353	2	4.0		6	7.9	
D10S28	429	3597, 1449	4	2.5		4	2.2	
D14S13	751	6433	0	0.0	1.0	0	0.0	10.0

this point there have been no genetic assumptions. The statistical assumption is that the databases are sufficiently random with respect to genotypes at these loci that the observed proportions provide appropriate estimates of population frequencies.

4.2. Independence Within Loci

Another advantage of presenting **Table 1** is that it shows that a one-locus profile matching the one of interest may not be found in a database, even though alleles matching each of the two alleles at that locus are found in reasonable frequencies. It can be pointed out that a basic law in population genetics suggests that frequencies of genotypes are given by the products of frequencies of alleles. Although the biological conditions that lead to that law being true certainly do not hold in human populations, it is a simple matter to see whether the data are consistent with the law. **Table 2** is a very useful demonstration of the similarity between observed genotypic frequencies in the FBI databases and the frequencies expected under the law of independence.

Table 2 shows that similar conclusions are reached whether the observed or the expected genotype counts are used. For the profile in this example, three loci were single-banded, and the table also shows that doubling the single-allele frequency (the "$2p$" rule) rather than squaring it provides an overestimation of the genotype count. Although it will not be necessary in every case, it is also helpful to use tables like **Table 2** to answer questions about binning strategies. The counts in **Table 1** and the middle block of columns of **Table 2** use floating bins. Every fragment in a database that matches the profile fragment contributes to the estimated frequency of that fragment. The righthand block of columns in **Table 2** are for fixed bins, whereby a profile fragment is assigned to the fixed bins defined by Budowle et al. *(5)* and all database fragments falling into that same bin contribute to the estimated frequency of the fragment. Similar frequencies are obtained from both binning methods, although the six-locus product is larger (more conservative) for the fixed bins in this case.

Finally, although the comparison of observed and expected counts in such tables as **Table 2** make clear the consistency of databases with the law of independence of allele frequencies within loci, mention can be made of the many statistical tests for independence that have been conducted and published; recent publications include Maiste and Weir *(7)*, Evett et al. *(6)*, and Hamilton et al. *(8)*. Publications that challenge independence [e.g., Geisser and Johnson *(9)*] generally ignore the single-band issue and do not use conventional binning strategies.

4.3. Multilocus Frequencies

The discriminatory power of DNA profiles comes from the rarity of matching profiles when many loci are considered. Quantifying rarity is even more difficult than in the one-locus case because a specific profile is very unlikely to be seen in any database. To be 99% sure of seeing at least one copy of a profile that occurs once in a million people would require a sample of size 4.6 million. Once again, population genetics theory suggests that allele frequencies at different loci are independent and may be multiplied together to provide an estimate of the multilocus frequency.

Demonstrating independence follows the same path as for one locus, but only so far. Tables of observed and expected counts can be prepared, but beyond two loci most observed counts will be zero or one, and expected counts less than one. Reference can be made to published statistical tests, e.g., **ref. *10*** that show general consistency of forensic DNA databases with independence between loci. Indeed, it is difficult to imagine a biological reason for dependence among the frequencies for unlinked neutral markers, even in structured populations. Published accounts of linkage disequilibrium in human populations invariably refer to loci that are tightly linked. Once again, the best approach may be to demonstrate the effects of assuming independence by experiments on databases. The experiments conducted by Evett et al. *(2,6)* show that likelihood ratios for several loci calculated under this assumption are large when suspect and perpetrator are the same person and are small when they are different people.

There is the technical issue of bias in the product rule. Even at one locus, the genotype frequencies formed as the products of allele frequencies have an expected value (the average over many replicates of the same procedure) that is less than the true value for heterozygotes. If the true and sample allele frequencies for the ith allele are written as $\tilde{p}_i$ and $\hat{\tilde{p}}_i$, then the expected value of $2\hat{\tilde{p}}_i\hat{\tilde{p}}_j$ is $2\tilde{p}_i\tilde{p}_j(2n-1)/2n$ when the database is from n individuals and the population does satisfy the Hardy-Weinberg law of independent allele frequencies. For a database of size 250, the expected bias when the true allele frequencies are 0.1 is 0.0004 or 0.2%. There are also biases for products over several loci.

Balding *(11)* suggests increasing allele frequency estimates to compensate for this bias. If a database contains n_i copies of the *i*th allele among the $2n$ alleles listed, he would estimate $\tilde{p}_i$ as $(n_i +2)/(2n + 4)$ instead of the usual $n_i/2n$ (his equation 16). This can be thought of as the estimate resulting from adding the profiles of both suspect and perpetrator to the database. For a database of size $n = 250$ and true allele frequencies 0.1, this is expected to provide heterozygote frequency estimates of 0.0213, or 6.25% higher than the true value. An alternative procedure is to report confidence intervals for profile-frequency estimates.

4.4. Confidence Intervals

One of the most frequent questions raised of profile frequency estimates of the order of 1 in a million, or less is how they can be justified from databases of approx size 250. The response should point out that, because of independence of allele frequencies, there are effectively separate databases of size 500 for alleles at each locus and that this size gives allele-frequency estimates with low sampling variances. An effective response is to remind the questioner that statistical estimates can have levels of confidence attached to them, as is routine in public-opinion surveys. Such statements as: 47% of those surveyed support the President (based on a telephone survey of 1063 registered votes, margin of error ±3 percentage points) are common. Estimating *m*-locus profile frequencies amounts to asking a series of $2m$ questions, where each question can have many answers. Approximate formulas have been given to calculate confidence limits on products of frequencies *(12)* but a better procedure seems to be to use bootstrapping.

It can be explained that the profile-frequency estimate, or the likelihood ratio, is based on certain databases. If new databases were constructed for the same populations, different numerical values would result simply because different people were typed. It is not feasible to construct new databases, but the numerical resampling technique of bootstrapping allows new databases to be formed from the present one. The technique is standard and was described in a monograph by Efron *(13)* and a book by Efron and Tibshirani *(14)*. It is also treated by Weir *(15)*. For any particular case, a thousand bootstrap databases can be constructed and the value found that cuts off the most extreme 1% of the values. Only those values more favorable to the defense are likely to be of interest, and for the profile in **Table 2** the estimate and upper 99% confidence limit on the likelihood ratio are 8.7×10^{12} and 1.6×10^{12} for floating bins and 1.6×10^{12} and 0.5×10^{12} for fixed bins. There is 99% confidence in the statement that 1.6×10^{12} is less than the likelihood ratio for floating bins or that 0.5×10^{12} is less than the likelihood ratio for fixed bins. It is quite common for the confidence limit to be up to 10 times smaller than the original estimate and also common for the fixed- and floating-bin estimates to lie within the confidence intervals of the other.

A criticism of these confidence limits points to the discrepancy in orders of magnitude between confidence limits based on the most extreme 10^{-2} values and such estimates as 10^{-6}. A confidence limit of 99% is equivalent to acknowledging a probability of 1% of underestimating the frequency, which may have been estimated as 0.0001%. It may be better to have confidence limits of 99.9999% for such values. Bootstrapping would therefore require at least a million samples. A more practical approach would be based on confidence limits for each allele, as implicit in the approach of *(12)*, but exact limits should be used instead of appealing to normal theory *(15)*. Bootstrapping by sampling individuals repeatedly has the great advantage of preserving whatever dependencies there are among alleles within and between loci. Rather than attempting to answer such questions as "How large should a database be?" a forensic scientist can cite the confidence limits for the databases used and explain that these limits take into account the number of people sampled.

5. Population Structure

Simple calculation of profile frequencies is not sufficient when there are dependencies between the different people, suspect and perpetrator, featured in the defense explanation. The most common source of dependency is a result of membership in the same population and hence a common evolutionary history. The mere fact of populations being finite means that two people taken at random from a population have a nonzero chance of having relatively recent common ancestors.

The appropriate theory is well understood for pairs of alleles, as opposed to pairs of genotypes. The probability of choosing allele A from a population given that an A has already been chosen is

$$\Pr(A|A) = \theta + (1 - \theta)p_A \tag{12}$$

A full treatment was given by *(16)*, but here it is sufficient to note that p_A is the allelic frequency in the whole population, and θ is a measure of variability of allelic frequencies over subpopulations within that population. The quantity θ is often written as F_{ST}, and Balding and Nichols *(17)* have suggested that 0.05 is likely to be an upper bound for large populations. This conditional allelic frequency is greater than θ and p_A.

A theory for pairs of genotypes was given by *(18)*, but there are difficulties with estimating the three- and four-gene analogs of θ that this exact treatment requires. A good approximation appears to be that given by *(17)*:

$$\Pr(AA|AA) = \{[p_A + \theta(1 - p_A)][p_A + 3\theta(1 - p_A)]\}/[(1 + \theta)(1 + 2\theta)+] \tag{13}$$

$$\Pr(Aa|Aa) = \{2[p_A + \theta(1 - p_A)][p_a + \theta(1 - p_a)]\}/[(1 + \theta)(1 + 2\theta)] \tag{14}$$

It is useful to present a table showing the effects of different values of θ, from published estimates like 0.001 *(16)* to the likely upper bounds of 0.05. Such

Table 3
Likelihood Ratios for Table 1 Profile,
Using FBI Caucasian Database

θ	Likelihood ratio	99% Confidence limit
0.000	1.63×10^{12}	5.19×10^{11}
0.001	1.20×10^{12}	4.08×10^{11}
0.010	1.59×10^{11}	6.86×10^{10}
0.050	1.76×10^{9}	1.08×10^{9}

values are shown in **Table 3**. Clearly, allowing for a θ value of 0.05 can diminish the likelihood ratio by 1000. Just as clearly, however, likelihood ratios in the billions reflect the implausibility of finding the same nine-band VNTR profile in two different people.

6. Mixed Stains

The evidential value of a DNA profile can be substantially reduced when there is clearly more than one contributor to the profile and the prosecution explanation does not account for the whole profile. The issue has been dealt with by Evett et al. *(19)* and Aitken *(1)*, and a more detailed account is given by Weir et al. *(20)*. The problem should cause no difficulty, provided the likelihood-ratio framework is adopted.

In the language of Weir et al. *(20)*, suppose the evidence profile has a set of alleles $\{e\}$ at a locus. Under the prosecution explanation, there are some known contributors but they lack alleles $\{u\}$ in the profile. The probability that x unknown people have these alleles among them, and do not have any alleles not in the profile among them, is $P_x(\{u\}|\{e\})$. Likewise, the defense explanation may include certain known contributors, but leave alleles $\{v\}$ unaccounted for and the probability of finding this set among y unknowns is $P_y(\{v\}|\{e\})$. The likelihood ratio is

$$L = [P_x(\{u\}|\{e\})]/[P_y(\{v\}|\{e\})] \qquad (15)$$

As an example, suppose the evidentiary profile in a rape case has alleles *abcd* at a locus, the victim has alleles *ab*, and a suspect has alleles *cd*. The prosecution explanation for the evidence is that it is from the victim and the suspect. No alleles in the evidence need to be accounted for. If ϕ denotes the empty set, then the numerator of the likelihood ratio is $P_0(\phi|abcd) = 1$. The defense explanation may be that the evidence is from the victim and some unknown person. Alleles *cd* need to be accounted for, and $P_1(cd|abcd) = 2p_c p_d$. The likelihood ratio is the usual

$$L = 1/(2p_c p_d) \qquad (16)$$

If, however, the stain could not be positively said to contain the victim's DNA, the defense explanation may be that both contributors are unknown and there is a need to account for all four alleles *abcd*. The arrangement of these four alleles among two people is unknown, and the people could be *ab*, *cd* or *ac*, *bd* or *ad*, *bc*, and the order could be reversed in each case. The probability $P_2(abcd|abcd)$ is $24p_ap_bp_cp_d$. If the prosecution explanation is still that the victim and suspect were the contributors, then

$$L = 1/(24p_ap_bp_cp_d) \tag{17}$$

The strength of the evidence against the suspect is stronger (provided $12p_ap_b < 1$) because there are more unknown contributors in the defense explanation.

Suppose, instead, that the evidence is from a murder committed by two people, both of whom left blood at the crime scene. The evidence profile has alleles *abcd* but there is only one suspect, with profile *cd*. The prosecution explanation may be that the contributors were the suspect and an unknown person. The probability $P_1(ab|abcd)=2p_ap_b$ is needed. If the defense explanation is that both contributors are unknown, the likelihood ratio is

$$L = (2p_ap_b)/(24p_ap_bp_cp_d)$$
$$= 1/(12p_cp_d) \tag{18}$$

The strength of the evidence against the suspect is weakened because of unknown contributors in the prosecution explanation.

This discussion has assumed that the circumstances of the crime dictated there were exactly two contributors to the evidence. There will be cases when the number of contributors is unknown, and then a more extensive analysis is required. The simplest way is to give separate answers for each possible number of contributors, although a more satisfactory approach would allow the forensic scientist to express a professional opinion regarding to the number. This issue has been discussed in **ref. *6***.

The effect of the number of unknown contributors can quite substantial. In the case previously referred to, a stain on the steering wheel of the defendant's car had three or four VNTR alleles at each of three loci. The defense argued that the number of contributors was not known, and the court ordered calculations to be performed as though there were two, three, or four contributors *(21)*. The prosecution explanation was that the defendant (OS) and a victim (RG) were contributors. Using five FBI databases, the range of values for the likelihood ratios are shown in **Table 4** for a variety of pairs of explanations. The absence of likelihood ratios in the trial meant that only the frequencies with which two, three, or four contributors had the profile between them were presented. These frequencies correspond to the highest possible likelihood ratios. There is an additional issue of allowing for unknown contributors to have unseen VNTR alleles *(22)*.

Table 4
Likelihood Ratios for Interpreting Evidence[a]

Prosecution explanation	Defense explanation	Likelihood ratio
Two contributors		
OS+RG	OS+U1	65,000–150,000
OS+RG	RG+U1	38,000–73,000
OS+RG	U1+U2	100,000,000–220,000,000
OS+U1	U1+U2	720–20,000
RG+U1	U1+U2	1000–5800
Three contributors		
OS+RG+U1	OS+U1+U2	2000–5000
OS+RG+U1	RG+U1+U2	1200–3600
OS+RG+U1	U1+U2+U3	1,000,000–4,600,000
OS+U1+U2	U1+U2+U3	400–1100
RG+U1+U2	U1+U2+U3	650–2100
Four contributors		
OS+RG+U1+U2	OS+U1+U2+U3	880–2200
OS+RG+U1+U2	RG+U1+U2+U3	550–1500
OS+RG+U1+U2	U1+U2+U3+U4	260,000–1,300,000
OS+U1+U2+U3	U1+U2+U3+U4	110–680
RG+U1+U2+U3	U1+U2+U3+U4	410–1100

[a]U1,U2,U3,U4 are distinct unknown people.

7. Conclusion

Until such time as DNA profiles are of such detail and general acceptance that the fact of a match between two profiles can be presented without numerical qualification, the best approach in court rests on likelihood ratios. Even if the term "likelihood ratio" is not explicitly mentioned, forensic scientists are likely to convey the quantitative meaning of a match in the clearest possible way by casting their own thinking in this framework.

References

1. Aitken, C. G. G. (1995) *Statistics and the Evaluation of Evidence for Forensic Scientists.* Wiley, New York.
2. Evett, I. W., Scranage, J. K., and Pinchin, R. (1993) An illustration of the advantages of efficient statistical methods for RFLP analysis in forensic science. *Am. J. Hum. Genet.* **52,** 498–505.
3. Thompson, W. C. (1995) Subjective interpretation, laboratory error and the value of forensic evidence: three case studies. *Genetica* **96,** 153–168.
4. Lempert, R. (1995) The honest scientist's guide to DNA evidence. *Genetica* **96,** 119–124.

5. Budowle, B., Giusti, A. M., Waye, J. S., Baechtel, F. S., Fourney, R. M., Adams, D. E., Presley, L. A., Deadman, H. A., and Monson, K. L. (1991) Fixed-bin analysis for statistical evaluation of continuous distributions of allelic data from VNTR loci, for use in forensic comparisons. *Am. J. Hum. Genet.* **48,** 841–855.

6. Evett, I. W., Gill, P. D., Scranage, J. K., and Weir, B. S. (1996) Establishing the robustness of STR statistics for forensic applications. *Am. J. Hum. Genet.* **58,** 398–407.

7. Maiste, P. J. and Weir, B. S. (1995) A comparison of tests for independence in the FBI RFLP data bases. *Genetica* **96,** 125–138.

8. Hamilton, J. F., Starling, L., Cordiner, S. J., Monahan, D. L., Buckleton, J. S., Chambers, G. K., and Weir, B. S. (1996) New Zealand population data at five VNTR loci: validation as databases for forensic identity testing. *Sci. Justice* **36,** 109–117.

9. Geisser, S. and Johnson W. (1993) Testing independence of fragment lengths within VNTR loci. *Am. J. Hum. Genet.* **53,** 1103–1106.

10. Zaykin, D., Zhivotovsky, L., and Weir, B. S. (1995) Exact tests for association between alleles at arbitrary numbers of loci. *Genetica* **96,** 169–178.

11. Balding, D. A. (1995) Estimating products in forensic identification using DNA profiles. *J. Am. Stat. Assoc.* **90,** 839–844.

12. Chakraborty, R., Srinivasan, M. R., and Daiger, S. P. (1993) Evaluation of standard error and confidence interval of estimated multilocus genotype probabilities, and their implications in DNA forensics. *Am. J. Hum. Genet.* **52,** 60–70.

13. Efron, B. (1982) The jackknife, the bootstrap and other resampling plans. *CBMS-NSF Regional Conf. Series Appl. Math.*, Monograph 38. SIAM, Philadelphia, PA.

14. Efron, B. and Tibshirani, R. J. (1993) *An Introduction to the Bootstrap.* Chapman and Hall, New York.

15. Weir, B. S. (1996) *Genetic Data Analysis II.* Sinauer, Sunderland, MA.

16. Weir, B. S. (1994) Effects of inbreeding on forensic calculations. *Ann. Rev. Genet.* **28,** 597–621.

17. Balding, D. J. and Nichols, R. A. (1994) DNA profile match probability calculation: how to allow for population stratification, relatedness, database selection and single bands. *Forensic Sci. Int.* **64,** 125–140.

18. Cockerham, C. C. (1971) Higher order probability functions of identity of alleles by descent. *Genetics* **69,** 235–246.

19. Evett, I. W., Buffery, C., Wilott, G., and Stoney, D. (1991) A guide to interpreting single locus profiles of DNA mixtures in forensic cases. *J. Forensic Sci. Soc.* **31,** 41–47

20. Weir, B. S., Triggs, C. M., Starling, L., Stowell, L. I., Walsh, K. A. J., and Buckleton, J. S. (1997) Interpreting DNA mixtures. *J. Forensic Sci.* **42,** 213–222.

21. Weir, B. S. (1995) DNA statistics in the Simpson matter. *Nat. Genet.* **11,** 365–368.

22. Weir, B. S. and Buckleton, J. S. (1996) Statistical issues in DNA profiling. *Adv. Forensic Haemogenetics* **16,** 457–464.

11

Interpretation and Statistical Evaluation of Multilocus DNA Fingerprints in Paternity and Relationship Testing

Gillian Rysiecki and John F. Y. Brookfield

1. Introduction

Since its discovery by Prof. Sir Alec Jeffreys FRS in 1984, DNA finger-printing using multilocus VNTR probes (MLPs) has been widely used in the field of human identification *(1,2)*. Although multilocus probes were used in forensic investigations in the late 1980s *(3)*, forensic samples often yield too little undegraded DNA for the production of a complete DNA fingerprint. Therefore, single-locus VNTR probes (SLPs) are now more widely used in crime-stain analysis. The major application of MLPs is in the analysis of relationships, DNA fingerprinting remaining the most accurate method available for paternity testing *(4,5)*. Because MLPs detect multiple hypervariable loci simultaneously, the technology is especially powerful in the analysis of complex relationships and cases of incest. DNA fingerprinting is also widely used in establishing zygosity in twins and in relationship-testing in animals *(6,7)*.

The bands that make up a DNA fingerprint show Mendelian inheritance, such that all of the bands in a child's DNA fingerprint are inherited from either the mother or the father. The presence of several bands in a child's DNA fingerprint not matched by corresponding bands in those of the putative parents indicates that one or both of the "parents" must be incorrect. The more closely two individuals are related, the higher the proportion of bands that are shared between their DNA fingerprints (bandshare). Unrelated people have been found to share <25% of their bands, whereas at the other extreme, identical twins share 100% *(8,9)*. The complexity of DNA fingerprint patterns precludes accurate analysis using automated systems, and therefore manual analysis is recommended.

From: *Methods in Molecular Biology, Vol. 98: Forensic DNA Profiling Protocols*
Edited by: P. J. Lincoln and J. Thomson © Humana Press Inc., Totowa, NJ

2. Materials

1. High-quality MLP autoradiographs, labeled to identify individual cases and samples.
2. Transparent plastic sheets (e.g., overhead-projector transparencies).
3. Light box (e.g., Everything X-Ray Ltd).
4. Waterproof colored pens (e.g., Staedtler Lumocolor 313).
5. Ruler.

3. Method

1. Tape the autoradiograph onto the light box and overlay the case to be scored with a transparent sheet. Tape firmly in place. Using the waterproof pen, mark on the acetate the position of the tracks to be scored and the sample, case numbers, and the probe used.
2. Mark the positions of the bands in each sample on the transparent sheet using the colored pens and ruler. In a paternity case, different colors can be used to distinguish between maternal and paternal bands, e.g., red for bands in the mother's sample and matching bands in the child's sample, blue for bands in the putative father's sample and matching bands in the child's sample, green for bands in the child's sample that are not present in either the mother's or putative father's samples. A proportion of the bands in the mother's sample will match bands in the putative father's sample (up to 25% if they are unrelated), which should be indicated on the acetate (e.g., with a blue dot on the center of the band) (*see* **Fig. 1**).
3. Score all visible bands down to 3.5 kb (a suitable size marker should be run on all MLP gels to indicate this position). Bands below 3.5 kb are not scored because of poor resolution in this region of the gel (*see* **Fig. 2**).
4. Calculate the total number of bands that have been recorded in each sample. Calculate the level of bandsharing between samples using the following formula:

$$\% \text{ Bandshare} = [(2S)/(N)] \times 100 \tag{1}$$

   where S is the number of bands shared between samples a and b, and N is the total number of bands in sample a and sample b.
5. Interpretation of results.

In paternity cases in which mother, child, and putative father are all tested, the convention is to assume that maternity is correct. (In practice, this should always be verified by checking the percentage of mother/child bandsharing.) All nonmaternal bands in the child must therefore have been inherited from the father. If several (three or more) of the nonmaternal bands are not present in the sample from the putative father, then he can be excluded from being the true father.

If all or all but one of the child's nonmaternal bands are present in the sample from the putative father, then he *cannot* be excluded from being the true father. The significance of the results is usually calculated in the form of a likelihood ratio. This is the ratio of the likelihoods of seeing if the results the man tested

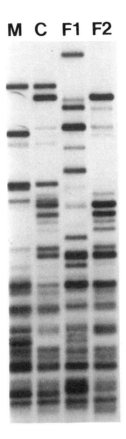

Fig. 1. Multilocus DNA fragments prepared from blood samples from a mother (m), her child (c), and two possible fathers, F1 who is excluded from paternity, and F2, who is not.

were the true father and seeing the results if the man tested were unrelated to the child. When all of the child's nonmaternal bands are present in the sample from the putative father then the probability of *a* is equal to 1 and the probability of *b* is equal to x^p, where x is the level of bandsharing expected between two unrelated individuals and p is the number of nonmaternal bands in the child's DNA fingerprint.

If all but one of the child's nonmaternal bands are present in the sample from the putative father (i.e., one band remains unassigned), then the most likely explanation is that paternity is correct but a mutation has occurred in one of the child's bands. It is impossible to tell whether the origin of the mutation is maternal or paternal. The probability of *a* is calculated as the probability that exactly one of the child's *c* bands is mutant, which is equal to

$$c\mu\,(1-\mu)^{c-1} \tag{2}$$

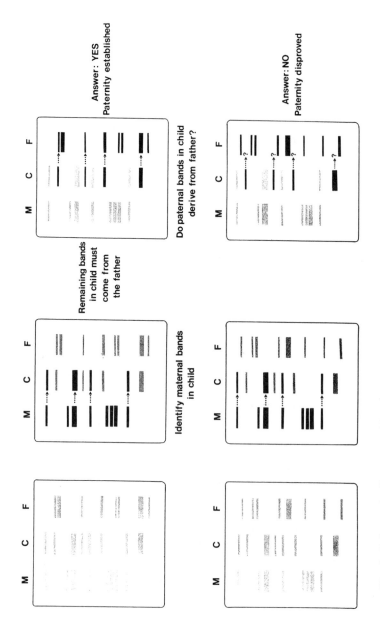

Fig. 2. Schematic diagram to illustrate scoring of multilocus DNA fingerprints in paternity analysis.

where μ is the mutation rate (mutation rate for multilocus probes 33-6 and 33-15 is 0.004/band/generation) and c is the total number of bands in the child. The probability of b is calculated assuming that nonmatching bands are all nonmutant and is

$$px^{p-1}(1-x) \tag{3}$$

where x is the level of bandsharing expected between two unrelated individuals and p is the number of nonmaternal bands in the child's DNA fingerprint.

In some cases, the relationship to be established is more distant (e.g., grandparent/grandchild or half-siblings). Here the level of bandsharing observed between the two individuals gives a good indication of the degree of their relationship. Since DNA-fingerprint bands are dominant genetic traits, the expected bandsharing values among various types of relatives are complex *(10)*. They are best calculated as functions, not of x, which is the probability that a band is present in a diploid individual, but of q, which is the probability that a band is present in a haploid gamete. The Hardy-Weinberg formula links these, since $x = 2q - q^2$. (In all formulas, it is assumed that x and q are constant for all bands.) The bandsharing between a parent and offspring is given by

$$(1 + q - q^2)/(2 - q) \tag{4}$$

and the bandsharing between siblings is given by

$$(4 + 5q - 6q^2 + q^3)/(8 - 4q) \tag{5}$$

The most commonly used value of x is 0.25, which gives $q = 0.134$. This value, when substituted into these formulae, gives 0.598 and 0.612 for the bandsharing between a parent and an offspring and between siblings, respectively.

Second-degree relatives include grandparents and their grandchildren, aunts or uncles and their nieces or nephews, and half-siblings. For second-degree relatives, the corresponding formula for bandsharing is

$$(1 + 5q - 5q^2 + q^3)/(4 - 2q) \tag{6}$$

This is 0.424 when $x = 0.25$ (and thus $q = 0.134$).

However, if a paternity test has been performed and the possibility exists that the father of the child, although not the man under test, is a close relative of this man, the probability that a paternal-specific band in the child will be in the child's uncle is not exactly the same as the bandsharing between siblings. (The difference is that, technically, the fact that the true father has passed on the band to the child makes it more likely that the band is homozygous in him and thus more likely that it will also be present in his brother.) In fact, the probability that a first-degree relative of the true father will possess a paternal-specific band seen in the child is just $(1 + x)/2$. Using $x = 0.25$, this gives a probability of 0.625. The corresponding probability that a second-degree relative of the father will possess a paternal-specific band is $(1 + 3x)/4$, which gives a probability of 0.4375, when x is 0.25.

4. Notes

1. Wherever possible, DNA fingerprints that are to be compared should be run on the same gel, side-by-side. Crossgel comparisons of these complex patterns are extremely difficult.
2. Prior to electrophoresis, sample DNA concentrations should be adjusted so that, within a case, all DNA fingerprints are of similar intensity. If a DNA fingerprint is very faint, some of the bands may not be visible. If it is too dark, faint bands may be obscured by darker ones.
3. It is essential that DNA samples be fully digested with restriction enzyme. Partial digestion can result in an altered DNA-fingerprint pattern, and therefore the digestion conditions and batch of enzyme used should be carefully monitored.
4. Because DNA-fingerprint patterns are highly complex, high-quality autoradiographs are essential if meaningful comparisons are to be made. Sample tracks should be straight, and bands should be sharp against clean backgrounds.
5. The value of x (bandshare between unrelated individuals) is a feature of the probe used, the way in which DNA fingerprints are prepared in the individual laboratory, and, in some cases, the population tested. This figure should be established by each laboratory for the relevant population.
6. A combination of MLP and SLP analysis maximizes the amount of information available for relationship analysis. For the most complex cases, two multilocus probes and several single-locus probes should be used.

References

1. Jeffreys, A. J., Wilson, V., and Thein, S. L. (1985) Hypervariable "minisatellite" regions in human DNA. *Nature* **314,** 67–73.
2. Jeffreys, A. J., Brookfield, J. F. Y., and Semeonoff, R. (1985) Positive identification of an immigration test-case using human DNA fingerprints. *Nature* **317,** 818,819.
3. Gill, P., Jeffreys, A. J., and Werrett, D. J. (1985) Forensic application of DNA "fingerprints". *Nature* **318,** 577–579.
4. Smith, J. C., Newton, C. R., Alves, A., Anwar, R., Jenner, D., and Markham, A. F. (1990) Highly polymorphic minisatellite DNA probes. Further evaluation for individual identification and paternity testing. *J. Forensic Sci. Soc.* **30,** 3–18.
5. Pena, S. D. J. and Chakraborty, R. (1994) Paternity testing in the DNA era. *Trends Genet.* **10,** 204–209.
6. Hill, A. V. S. and Jeffreys, A. J. (1985) Use of minisatellite DNA probes for determination of twin zygosity at birth. *Lancet* **ii,** 1394,1395.
7. Morton, D. B., Yaxley, R. E., Patel, I., Jeffreys, A. J., Howes, S. J., and Debenham, P. G. (1987) Use of DNA fingerprint analysis in identification of the sire. *Veterinary Rec.* **121,** 592–594.

8. Jeffreys, A. J., Wilson, V., Thein, S. L., Weatherall, D. J., and Ponder, B. A. J. (1986) DNA "fingerprints" and segregation analysis of multiple markers in human pedigrees. *Am. J. Hum. Genet.* **39,** 11–24.
9. Jeffreys, A. J., Turner, M., and Debenham, P. G. (1991) The efficiency of multilocus DNA fingerprint probes for individualisation and establishment of family relationships determined from extensive casework. *Am. J. Hum. Genet.* **48,** 824–840.
10. Bruford, M. W., Hanotte, O., Brookfield, J. F. Y., and Burke, T. (1992) Single locus and multilocus DNA fingerprinting, in *Molecular Genetic Analysis of Populations* (Hoelzel, A. R., ed.), IRL, Oxford, UK, pp. 225–269.

12

Overview of PCR-Based Systems in Identity Testing

Bernd Brinkmann

1. Introduction

PCR-VNTRs are important markers for questions of identification, individualization, and discrimination. They already form an integral part of forensic DNA analysis. They can be subdivided into the systems of short tandem repeats (STR) with fragment lengths ranging between approx 100 and 300 bp *(1–5)* and amplified fragment length polymorphisms (AmpFLP) with fragment lengths between 350–1000 bp *(6,7)*. Several hundred thousand STR loci are interspersed throughout the mammalian genome, and thousands have been further developed and mapped for the human genome *(8,9)*.

STR loci are useful to forensic science are characterized by their small range of alleles, their high sensitivity and suitability if the DNA is degraded. Furthermore, they show discontinuous allele distributions with consecutive alleles differing by 2–7 bp *(10)*. Among STRs, dinucleotide polymorphisms suffer from the disadvantage of quite intensive slippage, or stutter, bands leading to patterns that are difficult to interpret in stain work. Tetranucleotide polymorphisms are more suitable for two major reasons: They have negligible or no slippage effects, and consecutive alleles can be more easily resolved. Because of their polymorphic character and the aforementioned reasons, tetra- and pentanucleotide polymorphisms have revolutionized forensic case work.

2. Validation Studies

For their application in forensic casework, extensive validation studies have to be carried out. This would include population studies, investigation of a great number of meioses to determine mutation rates, and analysis of experimental stains. Studies of stains must include effects of the stain substrate, storage time, mixtures, and sensitivity studies.

From: *Methods in Molecular Biology, Vol. 98: Forensic DNA Profiling Protocols*
Edited by: P. J. Lincoln and J. Thomson © Humana Press Inc., Totowa, NJ

Table 1
STR Genetics

	System	Meioses (*n* =)	Mutations *n* %	DI Caucasian data	H (%)	HWE
Low	CD4	278	0	0.85	67	
	FES	525	0	0.83	65	Yes
	F13B	509	0 $<0.1\%$	0.85	67	
	TH01	725	0	0.91	78	
Intermed.	VWA	741	2 $<0.3\%$	0.91	81	Yes
High	D21	529	1	0.95	85	Yes
	ACTBP2	508	4 $<1.0\%$	0.99	99	

Overview of forensically relevant parameters of STRs with low, intermediate, and high microvariation. DI = discrimination index; H = heterozygosity; HWE = Hardy-Weinberg-equilibrium. comb. 1.5×10^8.

2.1. Population Studies

A minimum of 200 unrelated individuals and/or 500 meioses have to be investigated for each STR system to study allele frequencies, the mendelian inheritance, and whether significant numbers of mutations exist. Only STRs with a high forensic efficiency are suitable for application in forensic science **(Table 1)**; at present approx 30 such systems have been characterized. Besides the sensitivity, forensic efficiency values comprise a high discrimination index (DI) within populations and mutation rates, which should be <1% if the loci are to be used in parentage testing. The DI of individual STR systems varies in the range between 0.82 and 0.99. The combination of seven STR systems results in a combined DI of approx 1:100,000,000. The mutation rates are below 1% for all STR systems investigated so far *(11)*. If mutations are observed, insertions or deletions of single repeats are the predominant type *(12)*.

2.2. Stain Analysis

Before an STR system can be applied to casework, extensive validation is necessary. This must include the examination of a series of different stain substrates, a series of various storage conditions and times, and unequal mixtures of samples from different persons. It is necessary that the PCR system resolves DNA mixtures from different sources up to a proportion of 1:10 with no allelic dropout *(13)*. Microbloodstains and different sources of the stain material, e.g., mixtures of vaginal cells/semen, hair roots, saliva from stamps, and bones should be typeable. The success rate in stain analysis depends both on the quality and the quantity of DNA. Because of the heterogeneity of the stain material, it is necessary to establish different extraction procedures and make an indi-

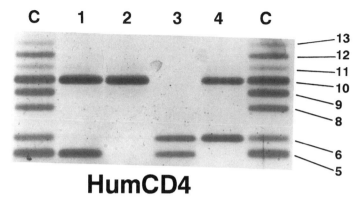

HumCD4

Fig. 1. The STR system HumCD4: amplification pattern in an identification case. DNA was extracted from a piece of femur of an unknown male cadaver according to *(18)*. Clarification of the relationship was achieved by comparing the amplification pattern of the putative parents and the putative grandmother. C = allelic ladder; 1 = putative father; 2 = putative grandmother; 3 = putative mother; 4 = unknown femur.

vidual decision for each case. Several extraction methods for different applications have been validated:

1. Organic extraction *(14,15)*,
2. Inorganic extraction *(16)*, and
3. Extraction by boiling *(17)*.

The most common method depends on the lysis of the cell membrane and the denaturation of proteins in the presence of proteinase K, EDTA, and detergents. Removal of the proteins and purification of the DNA is achieved by the organic phenol/chloroform extraction *(14)*. An alternative procedure is to use 6 M NaCl and to remove the proteins by precipitation instead of organic extraction. These procedures should be applied in cases where the co-extraction of PCR inhibitors is to be expected, e.g., tissue samples, hair shafts, and bone material *(18)* (**Fig. 1**).

A simpler and faster method of DNA extraction is the Chelex® 100 extraction *(5,17)*. Chelex 100 is a chelating resin and has the ability to bind metal ions. The basic procedure consists of boiling the sample in a 5% Chelex solution. The addition of proteinase K increases the efficiency of lysis *(19)*. This procedure has the disadvantage that PCR amplification of longer fragments (>500 bp) can fail. In consequence, the AmpFLPs with fragment lengths between 350 and 1000 bp cannot be analyzed. Nevertheless, this method can be applied in cases with bloodstains on different stain substrates, saliva on cigaret butts, and hair roots.

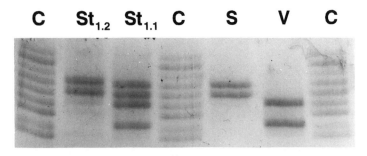

VWA

Fig. 2. Amplification pattern of the STR system HumVWA: DNA was extracted from a vaginal cell/sperm mixture using the preferential lysis method *(19)*. C = allelic ladder; $St_{1.1}$ = supernatant after the first lysis; $St_{1.2}$ = sediment after the first lysis; S = suspect; V = victim.

In the case of vaginal/sperm mixtures, a modification of the common proteinase K treatment in combination with the phenol/chloroform extraction is well established: preferential *(20)* or mild preferential *(19)* lysis. The basic principle is the lysis of the vaginal cells in the absence of DTT in a first step, and in a second step, lysis of the sperm cells in the presence of DTT **(Fig. 2)**.

The method of micromanipulation of single spermatozoa is another important application in forensic casework *(21)*, even in mixtures of several hundred vaginal cells and only very few spermatozoa (1–10). The spermatozoa are selectively scratched off from a microscope slide and after complete lysis are directly amplified **(Fig. 3)**.

3. Sequence Variability in STR Systems

According to the ISFH recommendations *(22,23,34)*, the allele nomenclature should be based on the sequence structure. Therefore, an essential part of research in legal medicine is the determination of the extent of sequence variability in each individual STR system, the exact allele definition according to the number of repeats, and the construction of system-specific allelic ladders. Several STR systems have been elaborated:

HumACTBP2 *(24)*; HumFES/FPS, HumVWA,
HumD21S11 *(25)*; HumF13B *(26)*
HumTH01 *(27)*; HumF13A1 *(28)*; HumCD4 *(11)*. The systems investigated so far show different extents of sequence variability and can accordingly be subdivided in three system categories **(Fig. 4)**.

TH01

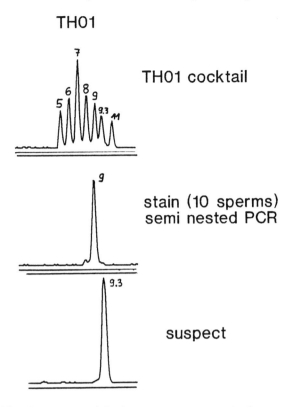

THO1 cocktail

stain (10 sperms)
semi nested PCR

suspect

Fig. 3. Amplification pattern of the STR system HumTH01 from a casework slide: DNA from single sperms were extracted using the method of micromanipulation *(21)* and amplified using seminested PCR for the TH01 locus. The pattern of the suspect did not match with the pattern of the sperms. In this case the suspect can be excluded.

4. Examples

4.1. Category (I)

HumTH01 *(27)*: Ten different alleles have been observed so far and verified by sequencing (**Fig. 5**). The repeat region is regularly spaced with the exception of the alleles 8.3, 9.3, and 10.3. In these alleles, the sixth and seventh repeats, respectively, are incomplete and have the trinucleotide unit ATG instead of the regular tetranucleotide repeat AATG. Sequencing data of these alleles from different unrelated individuals confirm that the position within the repeat region is constant. This system therefore constitutes an apparently regular STR polymorphism.

HumCD4 *(11)*: CD4 is a system of the first category with pentameric repeat units as basic motifs (**Fig. 6**). From allele 9 onward, the fourth repeat unit

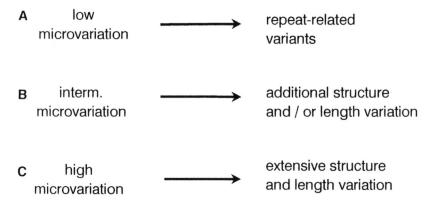

Fig. 4. Different types of microvariation. **(A)** STRs with low microvariation mainly consisting of regular repeated sequences (e.g., TH01; FES/FPS; F13B; CD4). **(B)** STRs with intermediate microvariation, where, in addition to the regular repeated sequences, structure and length variations exist. **(C)** STRs with high microvariation showing extensive structure and length variation within their repeated sequences.

shows a T to C transition, which is the only minor variation in this system. Eleven consecutive alleles have been observed.

4.2. Category (II)

HumVWA *(4,29)*: This system shows a more complex sequence structure with two tetranucleotide units (TCTA and TCTG) in combination. Two allele types were found that differ fundamentally from each other *(4)* **(Fig. 7)**. The consensus allele has a composite repetitive structure starting with one TCTA, continuing with three to five TCTGs and ending with the proper variable TCTA part **(Fig. 8)**. Allele 14 is unique and has a different structure at three sites leading to a slight electrophoretic shift under high-resolution/nondenaturing conditions **(Fig. 9)**.

4.3. Category (III)

HumACTBP2 *(24)*: HumACTBP2 (SE33) has a large number of alleles and is therefore the most informative, but is also the most complex STR system character-ized for forensic work to date. More than 250 alleles have been sequenced and ana-lyzed, and three different types of structure variation have been observed **(Fig. 10)**.

Sequence variability in the repeat region is further increased by the exist-ence of deletion/insertion mutations and/or point mutations in both flanking regions. Because of this high complexity there have been warnings regarding reproducibility, especially under different gel conditions: alleles with the same fragment length but different sequence structures can exhibit various electro-

Allele		Fragment length (bp)

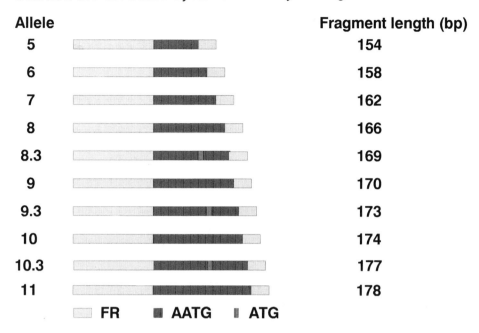

Allele	Fragment length (bp)
5	154
6	158
7	162
8	166
8.3	169
9	170
9.3	173
10	174
10.3	177
11	178

FR ■ AATG ▮ ATG

Fig. 5. Schematic representation of the repeat structure in the STR system HumTH01 (TC11). FR = flanking region.

phoretic mobilities in a high resolution/nondenaturing gel system **(Fig. 11)**. A better reproducibility was obtained using a denaturing gel system in combination with automatic fragment analysis and fluorescence detection **(Fig. 12)** *(24)*.

4. Methodologies for STR Analysis

For the demonstration of structure and/or length variation different methods have been established: discontinuous, horizontal, nondenaturing PAGE *(30)*; vertical, nondenaturing PAGE *(31)*; and vertical, denaturing PAGE *(27)*.

Visualization of the bands can be achieved either by silver staining *(6)* or by the much more sensitive fluorescence detection, which is often used in combination with automatic fragment analysis (e.g., Pharmacia, Uppsala, Sweden, Perkin–Elmer, Norwalk, CT).

For PCR systems with a low or an intermediate extent of microvariation (e.g., TH01, vWA, FES/FPS, etc.) the horizontal/nondenaturing gel system is the method of choice even in laboratories with a high sample throughput. The advantages of this gel system are ease of handling and low costs of reagents.

Microvariant alleles occurring in systems like FES/FPS and F13B can be easily detected using native gels **(Fig. 13)**. Their detection increases the extent of individualization and the chance of exclusion.

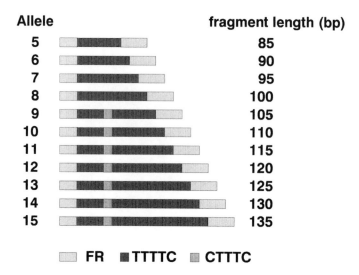

Fig. 6. Schematic representation of HumCD4 alleles 5–15. Nomenclature according to the number of repeats; FR = flanking region.

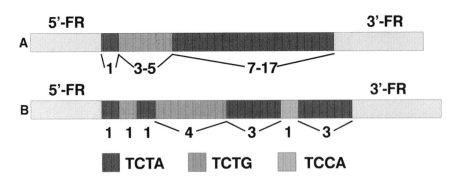

Fig. 7. Comparison between HumVWA **(A)** consensus allele and **(B)** variant allele 14. FR = flanking region, numbers below the diagram represent the number of repeats.

For the analysis of systems with an expressed microvariation such as ACTBP2 *(24)* it is not generally advisable to use native gels because of the great number of "interalleles." The ACTBP2 system needs a high level of standardization and good laboratory skill to achieve the necessary reproducibility. Therefore, especially in collaborative exercises, the use of a vertical, denaturing gel system in combination with fluorescence detection is the methodology of choice. The use of fluorescence labeled primers avoids the occurrence of double bands that could make the interpretation of amplification patterns diffi-

Allele

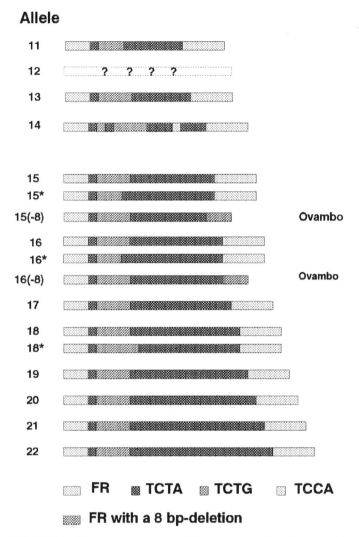

Fig. 8. HUMVWA. Sequence structure of the repeat region, fragment length, and allele designation of the STR system.

cult *(27)*. Extensive studies have been carried out within and between laboratories to validate this highly complex STR system. It could be shown that reproducible results were obtained when typing the sample DNA against an allelic ladder for the system and with fluorescence detection. "Interalleles," i.e., alleles with the same lengths but different mobilities, did not occur, because the electrophoretic mobilities depend only on the fragment length and not on the sequence structure *(24)*.

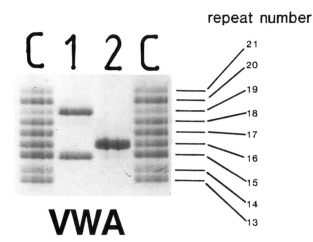

Fig. 9. HumVWA. Allelic ladder composed of nine sequenced alleles. Samples were separated on a high-resolution/nondenaturing gel and visualized by silver staining. C = allelic ladder; 1 = alleles 15, 19; 2 = allele 16.

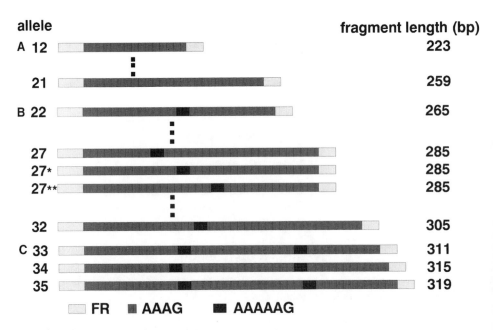

Fig. 10. Examples of three different classes of HumACTBP2 alleles. **(A)** The lower allelic range showed a common repeat structure with the regular 4 bp repeat AAAG. **(B)** The intermediate allelic range that contains an inserted hexanucleotide ARAAAG in addition to the regular 4 bp repeat. This irregular unit occurs only once in each fragment but at different positions. **(C)** The upper allelic range revealed two hexamer units at different sites within the repeat region leading to a highly complex sequence structure.

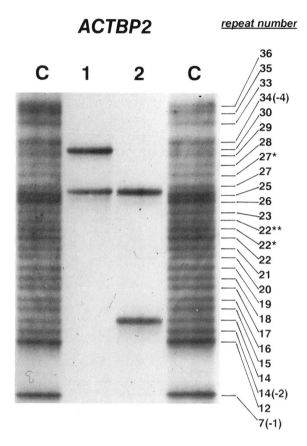

Fig. 11. ACTBP2 allelic ladder composed of 27 sequenced alleles. Samples were separated on a high-resolution/nondenaturing gel and visualized by silver staining.

The possibility of coamplification of several STR systems ("multiplexing") offers the possibility to achieve a high sample throughput in a very short time. The term "multiplexing" is not well-defined and there exist different approaches **(Fig. 14)**: independent amplification of each system and mixing before loading, coamplification of two systems ("duoplex") and mixing before loading, and coamplification of several systems and direct loading *(30,32,33)*.

Our experience shows that coamplification of several systems is disadvantageous in stain cases, especially if only a small stain quantity is available. Simultaneous amplification of several systems reduces the sensitivity (a minimum of 1 ng human genomic DNA is necessary) and increases the susceptibility to artefact bands. However, if enough template DNA is available, the simultaneous coamplification of several STR systems is an efficient method to

repeat number

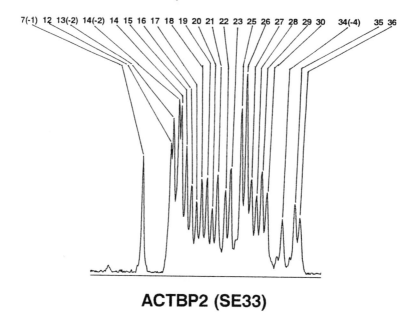

ACTBP2 (SE33)

Fig. 12. Separation of 23 ladder alleles in a vertical/denaturing gel system and automatic fragment analysis after fluorescence detection (373A Sequencer, ABI, Foster City, CA).

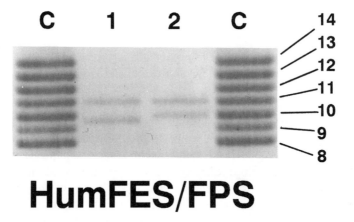

HumFES/FPS

Fig. 13. HumFES/FPS: Detection of microvariants in the FES/FPS system using the high-resolution/nondenaturing gel system. C = allelic ladder; 1 = alleles 10a, 11; 2 = alleles 10, 11.

quickly establish a convincing population database. However, it must be emphasized that the usage of internal standards *(33)* has some disadvantages: (1) the fragment sizes evaluated are not accurate, (2) there exists

Multiplexing

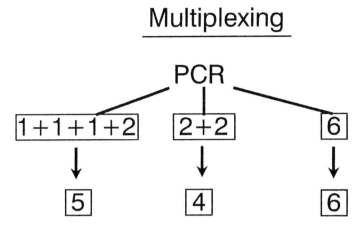

Fig. 14. "Multiplexing": examples of different multiplex approaches.

variation of the measurement error ("windows") up to or exceeding ±1.5 bp. In the case of microvariants that differ by 1 or 2 bp from the regular variant, these would be undetectable because of the window width. Therefore, the present approaches of coamplification, internal standards, and windows to allow for variation in size estimates are not acceptable. The usage of allele-specific internal ladders for each locus used should solve this problem.

References

1. Brinkmann, B. (1992) The use of STRs in stain analysis, in *Proc. Third Int. Symp. Human Identification.* Promega Corp., Madison, WI, 357–373.
2. Edwards, A., Hammond, H. A., Jin, L., Caskey, C. T., and Chakraborty, R. (1992) Genetic variation of five trimeric and tetrameric tandem repeat loci in four human population groups. *Genomics* **12,** 241–253.
3. Urquhart, A., Oldroyd, N. J., Kimpton, C. P., and Gill, P. (1995) Highly discriminating heptaplex short tandem repeat PCR system for forensic identification. *BioTechniques* **18,** 116–121.
4. Möller, A., Wiegand, P., Seuchter, S. A., Baur, M. P., and Brinkmann, B. (1994) Population data and forensic efficiency values for the STR systems HumVWA, HumMBP and HumFABP. *Int. J. Legal Med.* **106,** 183–189.
5. Wiegand, P., Budowle, B., Rand, S., and Brinkmann, B. (1993) Forensic validation of the STR systems SE33 and TC11. *Int. J. Legal Med.* **105,** 315–320.
6. Budowle, B., Chakraborty, R., Giusti, A. M., Eienberg, A. J., and Allen, R. C. (1991) Analysis of the variable number of tandem repeat locus D1S80 by the polymerase chain reaction followed by high resolution polyacrylamide gel electrophoresis. *Am. J. Hum. Genet.* **48,** 137–144.

7. Rand, S., Puers, C., Skowasch, K., Wiegand, P., Budowle, B., and Brinkmann, B. (1992) Population genetic and forensic efficiency data of 4 AmpFLPs. *Int. J. Legal Med.* **104,** 329–333.

8. Tautz, D. and Rentz, M. (1984) Simple sequences are ubiquitous repetitive components of eucaryotic genomes. *Nucleic Acids Res.* **12,** 4127–4138.

9. Weber, J. and Wong, C. (1993) Mutation of short tandem repeats. *Hum. Mol. Genet.* **8,** 1123–1128.

10. Edwards, A., Civitello, A., Hammond, H. A., and Caskey, C. T. (1991) DNA typing and genetic mapping with trimeric and tetrameric tandem repeats. *Am. J. Hum. Genet.* **49,** 746–756.

11. Brinkmann, B. (1996) The STR approach, in *Advances in Forensic Haemogenetics,* vol. 6, Springer-Verlag, Berlin, pp. 41–51.

12. Brinkmann, B., Möller, A., and Wiegand, P. (1995) Structure of new mutations in 2 STR systems. *Int. J. Legal Med.* **107,** 201–203.

13. Skowasch, K. and Brinkmann, B. (1992) Optimized amplification of the polymorphic system Col2A1. *Int. J. Legal Med.* **105,** 185–187.

14. Sambrock, J., Fritsch, E. F., and Maniatis, T. (1989) *Molecular Cloning: A Laboratory Manual,* 2nd ed. Cold Spring Harbor Laboratory, Cold Spring Harbor, NY.

15. Budowle, B. and Baechtel, F. S. (1990) Modifications to improve the effectiveness of restriction fragment length polymorphism typing. *Appl. Theor. Electrophoresis* **1,** 181–187.

16. Grimberg, J., Nawoschic, S., Bellvscio, L., McKee, R., Tirck, A., and Eisenberg, A. (1989) A simple and efficient nonorganic procedure for the isolation of genomic DNA from blood. *Nucleic Acids Res.* **17,** 8390.

17. Walsh, P. S., Metzger, D. A., and Higuchi, R. (1991) Chelex 100 as a medium for simple extraction of DNA for PCR-based typing of forensic material. *BioTechniques* **10,** 506–513.

18. Hochmeister, M. N., Budowle, B., Borer, U. V., Eggmann, U., Comey, C. T., and Dirnhofer, R. (1991) Typing of deoxyribonucleic acid (DNA) extracted from compact bone from human remains. *J. For. Sci.* **36,** 1649–1661.

19. Wiegand, P., Schürenkamp, M., and Schütte, U. (1992) DNA extraction from mixtures of body fluid using mild preferential lysis. *Int. J. Legal Med.* **104,** 359,360.

20. Gill, P., Jeffreys, A. J., and Werrett, D. J. (1985) Forensic applications of DNA "fingerprints." *Nature* **318,** 577–579.

21. Wiegand P, Madeja M and Brinkmann, B. (1994) Investigations on vaginal cell/sperm mixtures from microscopical slides, in *Advances in Forensic Haemogenetics,* vol. 5 (Bär, W., Fiori, A., and Rossi, U., eds.), Springer-Verlag, Berlin, pp. 256–258.

22. DNA recommendations—1992 report concerning recommendations of the DNA commission of the International Society of Forensic Haemogenetics relating the use of PCR-based polymorphisms. *Int. J. Legal Med.* **105,** 63,64.

23. DNA recommendations—1994 report concerning further recommendations of the DNA commission of the ISFH regarding PCR-based polymorphisms in STR (short tandem repeat) systems. *Int. J. Legal Med.* **107,** 159,160.

24. Möller, A. and Brinkmann, B. (1994) Locus ACTBP2 (SE33)—Sequencing data reveal considerable polymorphism. *Int. J. Legal Med.* **106,** 262–267.

25. Möller, A., Meyer, E., and Brinkmann, B. (1994) Different types of structural variation in STRs: HumFES/FPS, HumVWA and HumD21S11. *Int. J. Legal Med.* **106,** 319–323.

26. Alper, B., Meyer, E., Schürenkamp, M., and Brinkmann, B. (1995) HumFES/FPS and HumF13B: Turkish and German population data. *Int. J. Legal Med.* **108,** 93–95.

27. Puers, C., Hammond, H. A., Jin, L., Caskey, C. T., and Schumm, J. W. (1993) Identification of repeat sequence heterogeneity at the polymorphic short tandem repeat locus HUMTH01 $(AATG)_n$ and reassignment of alleles in population analysis by using a locus-specific allelic ladder. *Am. J. Hum. Genet.* **53,** 953–958.

28. Puers, C., Hammond, H. A., Caskey, C. T., Lins, A. M., Sprecher, C. J., Brinkmann, B., and Schumm, J. W. (1994) Allelic ladder characterization of the short tandem repeat polymorphism located in the 5' flanking region to the human coagulation factor XIII A subunit gene. *Genomics* **23,** 260–264.

29. Barber, M. D., Piercy, R. C., Andersen, J. F., and Parkin, B. H. (1995) Structural variation of novel alleles at the HumVWA and HumFES/FPS short tandem repeat loci. *Int. J. Legal Med.* **108,** 31–35.

30. Allen, R. C., Graves, G., and Budowle, B. (1989) Polymerase chain reaction amplification products separated on rehydratable polyacrylamide gels and stained with silver. *Biotechniques* **7,** 736–744.

31. Schneider, H. R. and Rand, S. (1995) High-resolution vertical PAGE: an alternative electrophoretic system with multiple forensic applications. *Int. J. Legal Med.* **108,** 276–279.

32. Kimpton, C. P., Gill, P., Walton, A., Urquhart, A., Millican, E. S., and Adams M. (1993) Automated DNA profiling employing multiplex amplification of short tandem repeat loci. *PCR Meth. Appl.* **3,** 13–22.

33. Kimpton, C. P., Fisher, D., Watson, S., Adams, M., Urquhart, A., Lygo, J., and Gill, P. (1994) Evaluation of an automated DNA profiling system employing multiplex amplification of four tetrameric STR loci. *Int. J. Legal Med.* **106,** 302–311.

34. Olaisen, B., Bar, W., Mayr, W., Lincoln, P., Carracedo, A., Brinkmann, B., Budowle, B., and Gill, P. (1997) DNA recommendations: further report of the DNA commission of the ISFH regarding the use of short tandem repeat systems. *Int. J. Legal. Med.*, in press.

13

An Introduction to PCR Primer Design and Optimization of Amplification Reactions

James M. Robertson and Jennifer Walsh-Weller

1. Introduction

PCR optimization is usually performed in order to obtain maximum *specificity* and *yield*. In some applications for which the amount of template may be limiting, or when there is a large amount of nontarget sequences, the *sensitivity* is maximized. Nonoptimized conditions promote artifactual bands resulting from primer dimerization and mispriming, broad bands containing a mixture of correct and incomplete products, and, generally speaking, a poor signal-to-noise ratio. Poorly designed PCR protocols also lack robustness. Such situations can occur in which the PCR is challenged with less than pristine DNAs or when the ingredients of the reaction are not added in the exact amounts as described by the authors of the protocol. Thus, a well-optimized PCR should be *reliable* in the hands of a coworker or in the laboratory of another institute.

Since the original description of the PCR *(1,2)*, many techniques have been described for the amplification of a wide variety of templates *(3–6)*. Ways were found to coamplify more than one target in a single reaction (i.e., multiplex PCR *[7]*), to analyze genes of a single cell *(8)*, to amplify several hundred samples at a time *(9)*, to rapidly cycle so that the PCR will take only approx 15 min *(10)*, and to amplify many-kilobase-long stretches of DNA in a reproducible manner *(11)*. Perhaps the widest use of the PCR is in genome screening. Obviously, a short chapter cannot cover all of these techniques. Instead, we shall focus on the optimization of two different PCR methods used in analyzing DNA polymorphism, which can be used to identify individuals *(12)* and to discover gene locations *(13,14)*. The principles described here can be adapted to the development of any PCR protocol.

From: *Methods in Molecular Biology, Vol. 98: Forensic DNA Profiling Protocols*
Edited by: P. J. Lincoln and J. Thomson © Humana Press Inc., Totowa, NJ

Conventional (target-specific) PCR uses locus-specific primers that flank the region of the polymorphism. This type of PCR requires prior knowledge of the target sequence, so that perfectly complementary primers can be designed. In addition, different techniques have been developed for situations in which the sequence around a gene or polymorphic site of interest is unknown. These methods are called variously, randomly amplified polymorphic DNA (RAPDs) *(14)*, arbitrarily primed PCR (AP-PCR), *(15)*, differentially amplified fragments (DAFs) *(16)*, and amplified fragment length polymorphisms (AFLPs) *(17)*. Although each of these techniques has unique aspects, all rely on a principle of genome-wide screening in an arbitrary, or nontargeted, fashion, and use non-locus-specific primers. The goal is to find a maximal number of useful polymorphisms that can be applied to the construction of a map, or if recombinant inbred or similar genetic stocks are being used, finding closely linked markers to a gene of interest *(18)*. Once the markers are found, they can be developed into sequence-confirmed amplified fragments (SCARs) *(19)*, which allow specific primers to be synthesized that can specify the analysis for a particular locus. A final refinement, bulked segregant analysis *(20)*, obviates the need for tedious generations of backcrossing in the field of plant molecular genetics, for example, or the need to develop specialized genetic materials in order to target the region around a gene of interest for marker discovery.

An important aspect of molecular-marker screening techniques is to perform many assays quickly, in order to allow only those systems that provide useful results to be retained for development. Thus, other goals of optimization of a PCR technique are *speed* and *simplicity*, and the aim is to provide these without sacrificing reliability.

1.1. Primer Selection

The first two partial reactions of the PCR involve denaturation of the template and hybridization of the primers to specific sites on the opposing strands of the matrix. Then, in the third partial reaction of the PCR, the DNA polymerase extends the primer from its 3' end. It is important to realize that only the 3' terminus of the primer needs to hybridize perfectly with the template for initiation of the polymerization to occur. This feature can result in the generation of a mixture of desired and unrelated PCR products if annealing conditions are permissive with respect to the entire primer sequence.

There are guidelines, as reported by numerous authors, that may be useful in designing effective primers. The suggestions for primer design presented in **Subheading 3.1.** are based on empirical observations from a limited number of primers and should not be considered as general rules. More important for a successful PCR are comparable T_ms for the primers.

Sometimes, the DNA sequence of the target allows little or no latitude in primer selection. However, visual inspection of the template annealing sites and careful selection of the primer sequences will often minimize the time required to optimize the PCR. For example, avoid potential hairpin-forming sequences in the primer binding site of the template that may limit the accessibility of the primer. Computer programs can also be used for predicting template secondary-structure effects on primer binding efficiency *(21)*. Despite a gallant attempt at optimization of the PCR and primer design, poor sensitivity might only be relieved when new primer pairs are tried *(22)*.

1.1.1. Primers Designed to Provide Specificity

When the target sequence is known, specific primers can be designed for amplifying the region between the annealing sites of the two oligonucleotides. An optimized, target-specific PCR has good yield with a minimum number of amplification cycles. When an adequate signal requires very many cycles, the noise will generally be high as well. With good primers, there is a better chance that the desired signal will appear early in the PCR relative to the undesired signals from misprimed sites. Thus, the goal is to have the highest yield of desired product without having nonspecific products.

1.1.2. Primers Designed to Exploit Target Degeneracy

When the sequence of the target is unknown, randomly primed or arbitrarily primed PCR (e.g., RAPDs) can be utilized to generate patterns of fragments at sufficiently high concentration to be readily visualized *(14,15)*. As in the case of conventional PCR, attention to the principles of primer design can lead to oligonucleotides, the majority of which will give satisfactory results under one or two PCR conditions. Here, in contrast to targeting a specific sequence, a reproducible pattern is the goal, rather than one unique product. The more complex the PCR fingerprint, the more reproducible it seems to be. Primers yielding fewer than ten distinct products should be avoided.

Another type of genome-wide arbitrary screen can be accomplished using the AFLP technology. Here, the genomic DNA is restricted with appropriate endonucleases, and the PCR target sequences are ligated to the sticky ends so generated. The design of these "adapters" should follow the guidelines for designing any good primer, including the lack of duplex formation between or within the adapters. Several adapter pairs have been optimized *(17)*, and there are commercial kits available as well. Following modification of the fragmented DNA by adapter ligation, subsets are amplified using primers complementary to the adapter sequence and the restriction endonuclease recognition site. The primers are designed at the 3' end so there can be complementarity to only one internal nucleotide of the genomic DNA, to obtain a reduction of the complexity of the fragment pattern (so-called "preselective" amplification). The process can be repeated with another specific primer to reduce the complexity of the product mixture still further ("selective" amplification). The final result is a highly reproducible fragment pattern, which may be used as fingerprints for identification or as the basis for a search for polymorphisms between related individuals in the construction of genetic maps.

Touchdown PCR *(23)* is often used when many sets of primers are being evaluated. Initially, a T_{anneal} about 10°C higher than the highest estimated T_m is selected and the T_{anneal} is reduced by 1–3°C in successive rounds of amplification. In this protocol, mispriming is reduced, because the primers anneal at the most stable site early in the amplification process. An example of touchdown PCR is given in **Subheading 3.1.** under the ALFP protocol.

1.2. Optimal Annealing Temperature

Since the second step of the PCR is duplex formation between the primers and template, it is obviously helpful to have an estimate of the temperature (T_m) at which 50% of the template-annealing site is in a duplex with the primer. Of the several methods available to estimate T_m, the one considered to be most reliable is the nearest neighbor algorithm *(10,24,25)*, which is tedious to calculate manually and is best used in commercial software. If software is unavailable, then it is recommended to compare the results obtained from using one of the empirical formulas for T_m that is based on AT and GC content of the primers. The PCR yield will be greatest when the T_m values of the primers match, and even a rough estimate is better than none at all. In addition, mispriming can occur if there is a significant difference between the T_ms of the two primers. For example, if one primer, *b*, has a higher T_m than the other primer, *a*, there may be mispriming by *b* at the lower temperature required for adequate annealing of *a*. This scenario may lead to a considerable fraction of the product being something other than the desired target. Thus, once the primer sequences are chosen, take the time to estimate the T_m values of all primers to be used in the PCR and change the sequences as necessary to match T_m values.

Once the primers are available, they must be tested in the PCR to empirically determine the optimal temperature for annealing, T_{anneal}. One tries to find the T_{anneal} that provides a balance between maximum yield and minimum mispriming. The dissociation rate of the primer from the template and the initial elongation rate will affect the value obtained *(26)*. The guidelines given in **Subheading 3.1.** are useful in establishing T_{anneal}.

Regard data from computer programs with healthy skepticism. One reason for this is that the T_m values may be calculated for conditions of membrane hybridization (i.e., equilibrium without elongation) and high salt—not exactly what one uses in the PCR, *see* **Note 2**). If the condition of 50 mM Na$^+$ (K$^+$) is included, the T_m values are lower than those obtained in 1 M salt. The algorithms also neglect the contribution of Mg^{2+} in stabilizing the duplex. The

Table 1
Comparison of Calculated T_m Values
with Experimentally Determined T_{anneal}[a]

Primer	T_m	T_p	T_a	Method	Reference
Upper	64.2			Nearest neighbor (1 M salt)	25
Lower	62.9				
Upper	64.5			% G + C	28
Lower	65.7				
Upper	54			2(A+T) + 4(G+C)	29
Lower	60				
Upper		61		22 + 1.46(2[G+C]+[A+T])	26
Lower		66			
Both			47	Calculated (50 mM salt)	
Both			64	Derived in the PCR	30

[a]The system compared in the table is for the D21S11 locus of Sharma and Litt (accession # M84567 [31]) with the upper primer (5')GTGAGTCAATTCCCCAAG and with the lower primer (5')GTTGTATTAGTCAATGTTCTCC. The calculations were performed with Oligo™ Version 4.0, Primer Analysis Software (National Biosciences, Inc., Plymouth, MN) except for the fourth calculation, which is from *(30)*, using the method described in *(26)*. T_a is the optimum annealing temperature and T_p is highest temperature at which optimal primer-directed amplification and specificity occur *(26)*. The GC content of the amplified product in this example is 33.3%, and the product length is 225 bp.

empirically derived T_{anneal} for a PCR is usually higher than the value calculated for 50 mM salt. In spite of the shortcomings of the algorithms, the experimentally derived, optimal annealing temperature can be very close to the estimated T_m values obtained by a variety of methods, as depicted in **Table 1**. If the values do match, it is probably by chance, because there are many factors left out of the calculations. Algorithms for obtaining T_m cannot substitute for an experimentally derived temperature optimum, but they do provide help in matching the primers during the design stage.

The optimal multiplex system will have primers with matching T_m values. However, there are cases when one wants to use sets of primers developed by someone else or by a commercial supplier that do not have matching T_ms. When coamplifying such sets of primers, it may be necessary to operate at less than the optimal T_{anneal} for some of them, if there are primers that have a lower optimal T_{anneal} present in the mixture. For example, the primers presented in **Table 1** have an empirically derived optimal T_{anneal} of 64°C when used alone, but work best in a multiplex at 60°C when combined with two other primer sets *(32)*. This conclusion was derived from a systematic test of a variety of reaction conditions.

When there is no sequence information available and arbitrary primers must be used, a low T_{anneal} is chosen. It is important to realize that increasing the stringency does not increase the reliability of a fingerprint in arbitrary PCR; indeed, the reverse has been observed. This result is probably caused by the competition among annealing sites that occurs in these complex PCRs *(33)*.

Once primers have been designed properly, it is also important to keep in mind adverse-reaction conditions that could lead to a poor PCR result. One situation that can occur is polymorphism in the primer annealing site. This will lead to a weaker interaction of the primer with the template, resulting in low yield in the PCR, and could even prevent primer extension, if it is in the very important 3' region of the primer binding site. Another factor to consider is the purity of the primers themselves. Unpurified primer preparations may contain sequences that are truncated, oligonucleotides with base modifications that prevent hybridization, and fragments that are incomplete chains. In a PCR that is not operating with stringency (i.e., not well optimized), these oligonucleotides could lead to weak signals and irreproducible banding patterns. The best primer preparations are those purified by reversed-phase HPLC, but this option may not always be available, *see* **Note 3**). If purity is suspect, perform a preparative polyacrylamide-gel purification. When hundreds of primers are to be screened, they will probably be purchased, and it is important to choose a vendor that provides pure primers. "Cheap" oligonucleotides cost a lot, if time and template are lost. Finally, it is good practice to personally measure the optical density of the primer preparation, *see* **Note 4**).

1.3. PCR Artifact Bands

Two types of artifactual banding patterns may be observed even from a PCR working at high stringency, when sites of sequence polymorphism are amplified. In one, there is a mixture of two bands that differ in length by the addition of a single nucleotide to the expected size of the single-stranded DNA fragment. This PCR artifact is not restricted to amplification of polymorphic loci and is dependent on the concentration of free Mg^{2+}. In the other pattern, observed only in the amplification of microsatellite loci, shadow, or stutter, bands are observed. These bands are usually less than the size of the main band by an integral of one or more repeat lengths (i.e., at -4, -8 nucleotides for tetranucleotide repeats). A stutter band may be observed as well at one repeat integral larger than the main band. The propensity to form stutter bands increases as the repeat size decreases (i.e., tetranucleotide < trinucleotide < dinucleotide). Artifactual bands resulting from nontemplated nucleotide addition can be eliminated, as suggested in **Subheading 3.1.**, whereas the stutter artifacts are more difficult to control.

One of the activities of DNA polymerases is to make nontemplated additions to the 3' end of a primer extension product, called terminal transferase-like activity *(36)*, thus producing a mixture of blunt and non-blunt-ended molecules. The efficiency of the addition seems to be a function of the sequence at the 5' end *(37)*. We have confirmed that when the 5' end of a primer terminates with a pyrimidine, there is much less addition to the 3' terminus of the complementary strand. However, it has seldom been possible to obtain 100% blunt ends by this tactic; there is always some nontemplated addition. We have been more successful in obtaining single bands by using a primer with a purine at the 5' terminus and aiming for complete addition. If the template sequence does not have a complementary pyrimidine base for the 5' terminal purine of the primer, one can add an artificial sequence to the 5' end of the primer. The added sequence will be nontemplated only in the first round of thermal cycling. As long as the primer has sufficient complementarity at the 3' end, there is a wide latitude for adding mismatched "tags" of different lengths at the 5' end *(38)*.

The process of nontemplate nucleotide addition can be considered to occur when "the engine idles" *(39)*. What is meant by this metaphor is that the reaction will proceed in a forward direction when the binding of the DNA polymerase to the blunt-ended PCR products is favored. The forward reaction is favored when the PCR protocol includes a long extension step at the end of thermal cycling. Also, since nucleotide addition does not occur when ends are opened by breathing, a low temperature and a high Mg^{2+} concentration will increase the yield of the extra nucleotide addition.

Stutter bands are more difficult to manage. In discussions among persons performing the PCR on microsatellites, one hears that this or that primer leads to shadow (or stutter) bands. However, it is not the primer selected, but rather the *locus* chosen for amplification that results in the observance of these bands (*see* **Note 5** for speculations on the cause of stuttering). Stutter bands are a cause for concern, if they have an intensity close to that of the main band, because homozygotes may be mistyped in family studies of linkage analysis using dinucleotide tandem repeats. In forensics, the evidence sample is often a mixture of DNA from more than one human source, and stutter bands will make the analysis more difficult. A few recommendations to reduce the intensity of stutter bands are presented in **Subheading 3.1.**

1.4. Template Preparation

The importance of a good template is often overlooked during the development of a PCR protocol: It is generally considered that organically extracted DNA is sufficiently clean. However, after establishing a protocol with a control DNA, there may be difficulties with DNA obtained from samples of origins different from the control. Examples of refractory amplifications have been reported for DNA obtained postmortem, from fresh blood and stains, from cells isolated from urine, degraded human tissue, bones, and plant tissue, to list just a few. Before optimizing a PCR protocol with a commercially available

DNA or a control DNA obtained from cells grown in culture, consider the sample preparation that will be used to isolate the template. Obviously, an optimized PCR protocol will not work satisfactorily, if the test sample contains an effective concentration of inhibitors or unwanted templates that will also bind the primers (e.g., AP-PCR). Another important feature is a good estimation of the input amount of the specific template. In this chapter, not all of the various methods of DNA isolation and quantification can be discussed, and the reader is referred to the literature **(Notes 6–9)**. Protocols for purifying the extracted DNA and for its quantification are presented in **Subheading 3.1.** Features specific for forensic and plant templates are worth considering in more detail.

1.4.1. Templates for Forensic DNA Analysis

A robust PCR protocol should work well with DNA isolated by either of the two popular extraction methods, since laboratories differ in their preference. Organically extracted DNA is double-stranded and can be used for analysis by both restriction digestion *(46)* and the PCR. The organically extracted DNA is obtained in a small volume. DNA extracted by the Chelex® method is single-stranded, so it can only be used for analysis by the PCR *(47)*. The DNA obtained from the Chelex extraction method may be recovered in a rather large volume.

DNA quantification should be carried out on templates for PCR analysis of DNA in forensic applications. The concern is that, although the PCR systems are specific for human templates, the signal obtained from the PCR may be too low if the extracted DNA contains templates from nonhuman sources. The amount of template estimated from spectroscopic measurements of DNA extracted from biological evidence could lead to little or no human DNA. Thus, the quantitation assay must be able to exclude DNA from bacteria, yeast, dog, rodent, and so on. Specificity for human DNA can be obtained by utilizing a primate-specific probe in a slot-blot assay, and commercial kits are available *(48,49)*. One compares the signal of the sample with that observed for a series of standards.

1.4.2. Templates for Plant-Genome Analysis

Plant DNA is too variable in its manifestations to give a comprehensive list of the purification protocols that have been successfully used in particular instances. A rigorous CTAB procedure **(Note 10)** with at least two rounds of chloroform extraction and three rounds of successive CTAB precipitation and selective resolubilization of the DNA, followed by a high-salt precipitation *(50)*, have given sufficient purity and yield of DNA for all of the PCR procedures *(51)*. A final RNase treatment with a phenol-chloroform extraction has been helpful in reducing PCR artifacts and in the quantification of yield.

In AP-PCR analysis of plant genomes, two independently purified DNA preparations must be used for each sample, in particular if one wishes to compare field samples. Independent samples are desirable, because other DNAs may be present from molds, mildews, and so on, and the copurified DNA may contribute a signal that can be mistaken for a polymorphism in the plant DNA (**Note 11**). It must be remembered that if DNA is purified from plant tissue, mitochondrial and chloroplast DNAs are present, and the short primers used in the assays for polymorphisms may well bind and give rise to a PCR product. Another important control in plant-genome analysis by AP-PCR is to use at least two concentrations of DNA template for each primer, in order to eliminate many false positives. For example, with barley DNA, we use 10 and 50 ng DNA per PCR analysis, of sample obtained from axenically grown plants (**Note 12**). In AFLP analysis, it is unnecessary to use two concentrations of DNA, but it is still necessary to consider impurities from field samples and organellar genomes.

1.5. Detection of PCR Products

PCR optimization is responsive to the means of detecting the reaction products: The method chosen must be simple, reliable, and sensitive. PCR products may be separated on either an agarose gel or an acrylamide minigel, denaturing or native, and detected after staining with either ethidium bromide or silver *(12)*. In the denaturing system, many more bands can be seen, and more products will appear to be polymorphic, thanks to strand separation (**Note 13**). Detection can be supplemented by photographic recording of the stained fragments, or the intensity can be digitized and recorded with the Eagle Eye™ (Stratagene, La Jolla, CA) followed by densitometry using an instrument such as the LKB Ultrascan XL and analysis with software such as the Gelscan XL (Pharmacia, Uppsala, Sweden). The primers can be radioactively labeled for autoradiography with subsequent densitometry or scanning with a Phosphor-Imager™ (Molecular Dynamics, Sunnyvale, CA), followed by analysis with the accompanying software. Automated methods for analysis of PCR products *(30)* with database-management possibilities exist for the DNA sequencer employing laser-induced fluorescence detection. Primers labeled with dyes at the 5'-end can be used in the PCR, since they do not inhibit the *Taq* DNA polymerase, and the sensitivity is greater than that obtained with silver staining of short amplicons (<400 nt). This approach allows automation of genome-wide screening *(52)*.

1.6. Optimizing the PCR

A set of primers with similar T_m values, a sensitive detection method, and an understanding of the template limitations, provide the tools for the optimization endeavor. Examination of the PCR literature reveals that many amplifications work well under a standard set of buffer conditions composed of 1.5 mM Mg^{2+}, low salt (50 mM KCl), 0.2 mM of each dNTP and 0.2 μM each primer. Optimization experiments can be started by looking for "windows" around these established values.

Many researchers regard optimization as the process, which defines conditions that give the maximum yield in a reaction. However, that concept is limited and can be compared with the act of only seeing the trees and not the forest. What one should strive for is finding those conditions under which similar yields are obtained even when the constituents vary somewhat. For example, in an optimized PCR, a 10% variation in Mg^{2+} concentration should not make much difference in the yields. The reason for this requirement is that inorganic analytical assays can have an inherent variation associated with them, generally about 10%. Variations in pipetting and weighing the reagents is generally only a few percent. Finally, as the careful analyst knows, material is often left clinging to the inside of a pipet tip after ejection of the buffer stock solution. Some people will rinse the tip by pulsing solution in and out of the pipet tip before they withdraw the tip from the mix in order to get all the material into the solution. This practice is impractical when hundreds of solutions are being pipetted. Thus, the goal should be to determine a concentration *range* at which similar results will be obtained in the PCR, rather than a concentration at which a maximum yield is obtained. The concentration range will be fairly flat, but when the limits are known, it will be possible to chose conditions within the plateau (i.e., toward one of the limits, for example) to compensate for performance variables. Examples of shifting along the plateau to satisfy specific PCR requirements are given in **Subheading 3.1.** at various locations.

1.6.1. Prevention of Pre-PCR Artifacts

Taq DNA polymerase has significant activity at 0–25°C *(53)* to give rise to multiple bands resulting from mispriming and especially primer dimers (primer oligomerization) in some PCR systems (**Note 14**). Even the practice of mixing the reagents and holding the tubes on ice until they are placed in the thermal cycler block may not prevent these PCR artifacts from emerging. There is a solution to the problem *(54)*. "Hot-start" PCR involves the practice of leaving a critical reaction component out of the master mix and then adding it when the solution reaches 70–94°C *(55)*. In this temperature range, primers involved in weak annealing interactions will have dissociated. The elimination of pre-PCR mispriming and primer dimerization may be the most important feature of the PCR protocol, especially when sets of primers are being coamplified. The yield will increase, and there will be fewer artifact bands. It is difficult to design multiple primer sets that exclude intermolecular interactions; whereas the hot start is fairly simple.

Wax can be used to separate an incomplete master mix from the missing component as described in **Subheading 3.1.** When the tubes are placed in the thermal cycler, the wax will melt at a temperature sufficiently high that weak primer interactions will no longer survive, and the aqueous layer above the wax will blend with the master mix by thermal convection. After the PCR, the wax acts as a barrier to prevent evaporation and contamination, but it must be penetrated to withdraw a sample for analysis. Commercially available, chemically modified wax beads used with small reaction volumes (50–100 µL) are easily penetrated.

Another way to perform hot-start PCR is to pipet the missing component through an oil layer once the tubes are in the thermal cycler and at the elevated temperature. This procedure is simple, but there are a number of precautions. Heated air in the pipet tip can push out the volume containing the missing component, so one must not linger in the oil layer too long, or the reaction mix might receive only a portion of the missing component (the rest will be above the oil layer). The template preparation room normally does not have a thermal cycler, so care must be taken not to introduce contaminants. Adding the missing component through an oil layer would be very tedious when screening hundreds of primers.

Advances in enzymology also provide solutions for eliminating pre-PCR artifacts. Recently, a modified version of recombinant AmpliTaq® DNA Polymerase, which is inactive until it reaches temperatures around the normal T_{den} (92–95°C), has become available for "invisible" hot-start. Since the polymerase is activated in a pre-PCR heating step at high temperature, mispriming events will have dissociated before the enzyme has become activated (*see* **Subheading 3.1.**). A similar stratagem utilizes a monoclonal antibody to *Taq* DNA polymerase, which deactivates the polymerase at 25°C (Clontech, Palo Alto, CA). When the PCR is started, the deactivation is reversed as the temperature rises during the denaturation step. By these two approaches, all of the reaction components can be added to one PCR solution, and there are no extra steps. The enzymatic-activation approach provides a method for hot-start in capillary PCR, since the tubes cannot be opened to add the missing component at high temperature.

1.6.2. Finding the Optimal Reaction Components of the PCR

1.6.2.1. DEFINING THE OPTIMAL MAGNESIUM ION CONCENTRATION

Magnesium ion stabilizes the DNA duplex and has been regarded as the single most important constituent in the PCR. Exhaustive experiments in variations of the salt, pH, primer, and dNTP concentrations often lead to very little effect on the yield of the PCR. However, high Mg^{2+} concentrations can promote incorrect annealing of the primer to its intended binding site, resulting in extraneous bands, and may make it more difficult to melt internal template hairpins, reducing the yield. A too low Mg^{2+} concentration will increase the stringency, but could decrease the activity of *Taq* DNA polymerase, resulting in lower yields. Commercially available "PCR optimization" kits can be used to help find the optimal Mg^{2+} concentration window (**Note 15**). These kits can save time required for optimization. In **Subheading 3.1.**, a protocol is given for investigating the Mg^{2+} concentration without a kit.

1.6.2.2. Adjusting Reaction Components Other than the Magnesium Ion

Instead of "binning" primers in sets to be coamplified according to their estimated T_ms, it is possible to adjust the ionic conditions so that a series of primers can work optimally at one temperature *(56)*. Different series of primers would require different buffers for optimal performance. Thus, the primers are "grouped" according to the optimal buffer, making primer screening amenable to automation. The pH of the buffer may effect the PCR efficiency, especially in multiplex PCR, when many primers are present. Other components to evaluate are the salts. KCl and $MgCl_2$ may be less effective than $(NH_4)_2SO_4$ and $MgCl_2$ *(57)*.

1.6.3. Optimization for Nonstringent PCR

Many descriptions exist for the optimization of RAPDs, including systematic variations in the magnesium ion and primer concentrations, the effect of additives such as detergents, formamide, glycerol, and so on. However, two of the advantages in using molecular marker screening techniques are simplicity and speed. Perhaps for that reason, researchers using the RAPD/AP-PCR/DAF techniques choose the standard PCR buffer containing 1.5 mM $MgCl_2$, Tris-HCl buffer, and 50 mM KCl. However, the recommended primer concentration depends on the specific technique being used.

The AFLP technique, which combines the accuracy and reliability of the RFLP technique with the power of PCR, needs much less optimization. The increase in reliability derives from the ligation of oligonucleotides to restricted DNA, allowing the use of longer PCR primers and thus more stringent annealing conditions. By adding "selective bases" to the 3' end of the PCR primer sites, the method can be "tuned" to amplify from one to thousands of restriction fragments simultaneously in a highly reproducible way, without having to carry out a lengthy optimization procedure.

1.7. Finding the Right Thermal Cycler Parameters

In the optimization process, it is important to examine the temperature windows, the duration of the holding times, and the ramping speed between the denaturation, annealing, and extension steps. These experiments should be done in order to arrive at consensus parameters. For example, a particular temperature for denaturation (T_{den}), or the time spent at T_{den}, might give the best yields in one thermal cycler, but a slightly different temperature, or duration, may be optimal for the same PCR with a different thermal cycler. Thus, if the PCR protocol is to be utilized in other laboratories or with other thermal cyclers within the same laboratory, then the best protocol would be to utilize a T_{den}, or duration at T_{den}, that lies in the middle of the range that gives satisfactory yields.

One paper often cited compares rapid cycling in capillaries with conventional cycling in a heat block *(10)*, where it is shown that clean reactions are obtained only in the rapid system. This comparison is a bit unfair, because the reactions were performed at 3 m*M* Mg²⁺ concentration: One will invariably get spurious bands at this magnesium-ion concentration in conventional thermal cyclers! A 2 min ramp-down time from 94 to 54°C at 1.5 m*M* Mg²⁺ concentration resulted in no spurious bands at 1.5 m*M* Mg²⁺ in a four-loci multiplex system *(7)*. In fact, the slow ramping results in greater yield than one can obtain with rapid ramping in this system. In arbitrarily primed PCR, the ramping time from T_{den} to T_{anneal} has been reported to be one of the factors leading to variability between laboratories *(58)*. The number of polymorphisms increased as the ramping time interval was prolonged, suggesting that differences among laboratories using the same primers and template might be because of the ramping-speed differences among the various thermal cyclers.

1.8. Amplifying Regions of High GC Content

When *Taq* DNA polymerase must amplify a segment having a high proportion of G + C nucleotides, it may encounter undissociated template, because the local T_m of these regions is high. These templates may be refractory to the PCR. To increase the efficiency of amplification through such regions, a number of techniques has been employed, which are presented in **Subheading 3.1.** Upon PCR optimization, the amplification of GC-rich regions will often be successful. An initial denaturation at alkaline pH, followed by neutralization and precipitation of the single-stranded DNA may provide the solution for refractive templates containing high GC content *(59)*.

1.9. Prevention of Contamination

Optimization will come to naught if one does not guard against contamination. In forensic DNA analysis, the issue is not whether such-and-such PCR technique can be used to analyze crime-scene mixtures containing template from more than one source. The issue that should concern forensic scientists, or any researcher who is contemplating the PCR for DNA analysis, is how to avoid contamination of samples by either the PCR products of previous experiments or by other templates at the workbench. The supreme PCR protocol will be worthless if the laboratory is not set up to avoid potential contamination. We cannot discuss in detail ways to control contamination in this chapter, but we shall give some suggestions for good laboratory practice.

2. Materials

The items listed in this section were used by the authors with repeated success. Similar materials from manufacturers or suppliers other than those listed (i.e., with a similar degree of quality control) may also work satisfactorily.

2.1. PCR Protocols

2.1.1. PCR with Primers to Provide Specificity

1. GeneAmp® PCR Reagent Kit with AmpliTaq DNA Polymerase containing buffer and dNTPs (Perkin-Elmer, Norwalk, CT, N801-0055).
2. Molecular-biology-grade water, 18 Megaohm (Sigma, St. Louis, MO, W-4502).
3. Light mineral oil (Sigma M-5904).

2.1.2. PCR with Primers to Exploit Target Degeneracy

2.1.2.1. RAPDs PROTOCOL

1. AmpliTaq DNA Polymerase, Stoffel Fragment (Perkin–Elmer, N808-0038), with 10X PCR buffer and 25 mM MgCl$_2$ (Perkin-Elmer, N808-0010).
2. Formamide (Gibco-BRL Life Technologies, 5515UA).
3. GENESCAN 2500-ROX Size Standard (Perkin–Elmer, 401100).
4. Long Ranger™ (AT Biochem, Malvern, PA).

2.1.2.2. AFLP PROTOCOL

1. AFLP™ Plant Mapping Kits: AFLP Amplification Core Mix Module, AFLP Ligation and Preselective Amplification Module, and AFLP Selective Amplification Start-Up Module (Perkin–Elmer, 402005, 402004, 402006, respectively).
2. TE (10^{-4}): 0.1 mM EDTA, 10 mM Tris-HCl, pH 8.3. Dissolve 121 mg Tris-HCl in 80 mL 0.1 mM EDTA solution, adjust pH to 8.3 with 0.2 N HCl and bring volume up to 100 mL. For 0.1 mM EDTA, make a 1 mM solution and dilute 1:10. For 1 mM EDTA, dissolve 93 mg EDTA, disodium salt, in 100 mL molecular-biology-grade water.

2.2. Establishing the T$_{anneal}$

Tubes for the PCR: 0.5 mL thin-walled tubes for the GeneAmp PCR System 9600 or DNA Thermal Cycler 480; 0.2 mL MicroAmp Reaction Tubes for the GeneAmp PCR System 9600.

2.3. PCR Artifact Bands

1. T4 DNA Polymerase, 3000 U/mL (New England Biolabs, Beverly, MA, 203).
2. 2 mM EDTA: dissolve 74 mg EDTA, disodium salt, in 90 mL molecular-biology-grade water, adjust the pH to 8.0, and bring the volume to 100 mL.

3. 7.5 M ammonium acetate: dissolve 58 g anhydrous salt in molecular-biology-grade water to give a final volume of 100 mL.

2.4. Template Preparation

2.4.1. Quick Pre-PCR Template Purification

Microcon™-100 ultrafiltration device (Amicon, Beverly, MA).

2.4.2. Rapid DNA Quantitation

1. PicoGreen™ dye (Molecular Probes, Eugene, OR, P-7581).
2. TE: 1 mM EDTA, 10 mM Tris·HCl, pH 8.3. Dissolve 121 mg Tris-HCl and 93 mg EDTA, disodium salt, in 80 mL water, adjust the pH to 8.3 with 0.2 N HCl and bring up the volume to 100 mL.

2.5. Detection of PCR Products

1. Loading buffer: make 2 mL blue dextran (Sigma, D-5751) solution, 50 mg/mL, filtered through a Gelman™ Acrodisc. Mix with 2 mL 10X TBE and 8.8 mL deionized water.
2. 10X TBE buffer (Gibco-BRL Life Technologies, Gaithersburg, MD, 15581-036).
3. Sizing ladder: GENESCAN 2500-ROX, or GENESCAN 350-ROX Size Standards (Perkin-Elmer, 401100 and 401735, respectively).

2.6. PCR Optimization

2.6.1. Prevention of Pre-PCR Artifacts

2.6.1.1. Using Wax as a Barrier in Hot-Start PCR

AmpliWax™ PCR Gem 100 (Perkin-Elmer, N808-0100).

2.6.1.2. Using DNA Polymerase Activation to Avoid Pre-PCR Artifacts

AmpliTaq® Gold, 5 U/µL (Perkin-Elmer, N808-0240).

2.6.2. Adjusting the Reaction Components of the PCR

2.6.2.1. Defining the Optimal Magnesium Ion Concentration

1. 25 mM MgCl$_2$ stock solution (Boehringer-Mannheim, Mannheim, Germany, 1 699 113).
2. dNTP stock solution: Set of dATP, dCTP, dGTP, dATP, 100 mM (Boehringer-Mannheim, 1 277 049). Add 12.5 µL of each dNTP to 450 µL molecular-biology-grade water.
3. Enzyme solution: use AmpliTaq DNA Polymerase, 5 U/µL (Perkin-Elmer N801-0055). Dilute 16 µL stock enzyme with 64 µL molecular-biology-grade water and dispense immediately.
4. PCR optimization kits are available from commercial sources: PCR Optimizer™ Kit (Invitrogen, San Diego, CA, K1220-01); Opti-Prime™ PCR Optimization Kit (Stratagene, La Jolla, CA, 200422).

2.7. Finding the Best Thermal Cycler Parameters

GeneAmp PCR System 9600 or DNA Thermal Cycler 480 (Perkin-Elmer).

2.8. Primers and Remedies
for Amplifying Regions of High GC Content

1. 7-Deaza-dGTP, Li-salt, 10 mmol/L, pH 7.0 (Boehringer-Mannheim, 988 537).
2. DMSO: methyl sulfoxide (Aldrich, Milwaukee, WI, 27, 685-5), stored with molecular sieves at room temperature in the dark.
3. UlTma™ DNA polymerase, 6U/µL (Perkin-Elmer, N808-0117).
4. Vent®(exo-) DNA Polymerase (New England Biolabs, Beverly, MA, 257S).

2.9. Guidelines for Prevention of Contamination

ART® Tips (Molecular Bio-Products, San Diego, CA).

3. Methods
3.1. Guidelines for Primer Design
3.1.1. Primers Designed to Provide Specificity

1. Target-specific PCR primers are generally from 17 to 25 nucleotides in length. Primers can be much longer or even shorter *(38)* and still provide specificity *(60)*. The length of the two primers does not need to be identical (*see* **Table 1** and *[31]*).
2. The G + C content is usually between 40 and 60% and is similar between the primers. When this procedure is limited by the template sequence, one of the primers is lengthened or shortened to compensate for the difference in G + C content. (For an example, refer to the two primers described in the legend of **Table 1**.)
3. Primers with a GC "clamp" at the 3' end may increase the sensitivity *(61)*, but mispriming may be observed in multiplex PCR with primers having runs of more than two GCs at the 3' end.
4. Pick 3' sequences that are noncomplementary with sequences in the other primer, since primer dimer amplification can occur and reduce the concentration of substrates available for producing the desired product. When incorporating sets of primers that worked well in singleplex PCR into a multiplex PCR protocol, re-examine the primer complementarity.
5. Avoid palindromes and hairpin loops that are sufficiently stable to compete with the target sequence and therefore reduce the sensitivity.
6. If the length and %GC of the region of primer complementarity are minimal, then the primer/primer duplex should not be a cause of concern with regard to primer/template formation.
7. Avoid stretches of sequences at the 3' end that are complementary to repetitive motifs, unless degeneracy is desired.

3.1.1.1. CONVENTIONAL PCR PROTOCOL

1. Program the thermal cycler. Start with 5 min at 94°C for one cycle. Link to a regime of 1 min at 94°C, 1 min at the T_m, 1 min at 72°C for 30 cycles. Link to

10 min at 72°C for one cycle and then to a holding step at either 4 or 25°C. The T_m can be estimated from one of the formulas in **Table 1**. The number of cycles necessary to obtain a signal will depend on the amount and quality of the template. If insufficient signal is obtained with 30 cycles, try 35 and 40 cycles.

2. On ice, prepare a cocktail containing the proportion of PCR buffer recommended by the supplier of the DNA polymerase. If Mg^{2+} is to be added to the buffer, choose 1.5 mM in the initial experiments. Add primers to a final concentration of 0.25 µM each, dNTPs to 0.2 mM each, and use 10–50 ng template. Bring the volume to 45 µL with molecular-biology-grade purified water. Add 5 µL solution containing 2.5 U thermal-stable DNA polymerase. Mix gently and pulse spin to bring the contents of the tube to the bottom.

3. Layer a drop of light mineral oil over the mixture if a thermal cycler without a heated lid is to be used.

4. Start the thermal cycler and when the temperature reaches 94°C, place the reaction tube in the block.

5. When analyzed, if the signal from amplified products is a smear, or if the product is produced in too high an amount, reduce the number of cycles or the amount of template.

3.1.2. Primers Designed to Exploit Target Degeneracy

1. DNA amplification fingerprinting in the arbitrarily primed mode uses primers of length 6–18 nt.

2. Most primers are minimally 50% GC, with no obvious dimer-forming characteristics. One can tailor the primers to the genome, if some information is available for the overall GC composition and on any structural motifs, such as di- and trinucleotide repeats *(62)* and Alu-repeats *(63)*.

3. It has been shown that primers having a "core" of arbitrary sequence and a stable mini-hairpin at the 5' terminus (GCGAAGC-) can increase detection of polymorphic DNA *(64)*. However, in some situations, it can be helpful to use pairs of unstructured primers, as they lead to products that do not have complementary ends, which may form poorly amplified "panhandles" (**Note 16**).

4. In detecting a "consensus" sequence or a group of related sequences, primers with deoxyinosine can be helpful *(65,66)*. Primers with deoxyinosine can pair with all four nucleotides at positions of complete degeneracy *(67)*. The deoxyinosine is placed in the primers at sites of potential sequence ambiguity.

5. The concentration of the arbitrary primers can be used to complement the structure in generating bands. For example, in the DAF approach, the primers are present in a high concentration (3 µM), which leads to a large number of bands *(16)*. However, the RAPD system uses a more conventional concentration of primers (0.3 µM), so that fewer products result *(14)*.

3.1.2.1. RAPDs Protocol

Use from 2–8 mM $MgCl_2$, 0.075–0.1 U/µL *Taq* DNA polymerase or 0.2–0.4 U/µL AmpliTaq DNA Polymerase, Stoffel Fragment (which is reported to give up to twice as many fragments, although of smaller size), and 0.1–0.2 m*M* dNTPs. The template concentration should be at least 5 ng/μL DNA for plasmid, 1 ng/μL DNA for low-complexity genomes and 0.1 ng/μL for high-complexity genomes. Restriction digestion of the template DNA, either before or after the PCR reaction, will also increase the detection of polymorphisms. It is probably easier to standardize the restriction of the post-PCR product than of the starting DNA sample, because the large number of cycles in these experiments causes a plateau to be reached and the same volume of PCR sample will give a very reproducible amount of signal. An example of a RAPD protocol follows.

1. Make a PCR cocktail (for 50 reactions, the volume will be 1 mL): 100 μL 10X PCR buffer (no MgCl$_2$), 30 μL MgCl$_2$ stock (25 m*M*), 3 μL 0.1 m*M* primer, 50 U *Taq* DNA polymerase, 607 μL molecular-biology-grade water.
2. Dispense 15 μL per tube.
3. Add 5 μL template DNA at 10 ng/μL.
4. Thermal cycling conditions vary, but the conditions recommended here are 72°C, 2 min (once); 94°C, 30 s; 36°C, 1 min; 72°C, 2 min (25 cycles). Ramping times are not faster than 2°C/s. There is a final incubation at 72°C for 7 min.
5. Analyze 10 μL of the PCR sample on an agarose gel or 5 μL on an acrylamide gel. If fluorescent-dye labeled oligonucleotides were used in the PCR, the products can be analyzed on the ABI Prism™ DNA Sequencer (Perkin-Elmer, Foster City, CA). Dilute the sample with formamide either to 1:50 (Model 373 DNA Sequencer), or to 1:150 (Model 377 DNA Sequencer), and load 1 μL of the dilution with a dye-labeled size standard after denaturation at 95°C for 3 min. With the 373 DNA Sequencer, analyze the samples on a 6% polyacrylamide gel (29:1 acrylamide/bis) containing 6 *M* urea in 1X TBE at 1800 V for 11 h. With the 377 DNA Sequencer, use a 5% Long Ranger gel in 1X TBE with 4 *M* urea and run for 7 h at 3 kV. Use of the internal-lane size standard and GeneScan™ Analysis software (Perkin-Elmer) allows a high level of sensitivity and precision in identifying the bands.

3.1.2.2. AFLP Protocol

Prior to the PCR, AFLP reactions require biochemical steps, which involve restriction of the genomic DNA with endonucleases followed by ligation of adapter molecules. This technology has been licensed by Keygene, N.V., the inventor of the technique, to Life Technologies and Perkin-Elmer (*see* **Note 1**). These companies supply AFLP kits that include the adapters, PCR primer sets, and protocols with detailed instructions. An example of an ALFP protocol follows.

1. Amplification of target sequences: Combine the following materials in a thin-walled 0.2 mL PCR tube: 4.0 μL diluted DNA from a restriction-ligation reaction, 1.0 μL preselective primer pairs (have two sets per DNA sample: one with

*Eco*RI + A and *Mse*I + C; one with *Eco*RI and *Mse*I + C), 15.0 µL AFLP mix (which includes 1X PCR buffer) 0.1 mM dNTPs, 0.5 U *Taq* DNA polymerase in amounts appropriate for 20 µL reaction volume. If using a thermal cycler without a heated lid, overlay the sample with 20 µL light mineral oil.

2. Cycling the preselective amplification reaction: Thermal cycling conditions are 65°C, 2 min (once); 94°C, 10 s; 56°C, 30 s; 72°C, 2 min (20 cycles). Ramping times should not be faster than 2°C/s. At the end, the reactions are brought to 4°C. If the reactions were successful, then there should be a smear of product from 150–1000 bp visible on a 1.5% agarose gel.

3. AFLP reaction: Combine the following in a thin-walled 0.2 mL PCR tube: 3 µL diluted preselective amplification reaction product (10 µL preselective amplification reaction product diluted with 190 µL TE 10^{-4} buffer), 1 µL *Mse*I[primer + CXX] at 5 µM, 1 µL *Eco*RI[labeled primer + AXX] at 1 µM, 15 µL AFLP PCR mix. If using a thermal cycler without a heated lid, add 20 µL light mineral oil to each tube.

4. Cycling the AFLP reactions: The method follows a touchdown regimen.

   94°C, 2 min: 1 cycle
   94°C (10 s) 66°C (30 s) 72°C (2 min): 1 cycle
   94°C (10 s) 65°C (30 s) 72°C (2 min): 1 cycle
   94°C (10 s) 64°C (30 s) 72°C (2 min): 1 cycle
   94°C (10 s) 63°C (30 s) 72°C (2 min): 1 cycle
   94°C (10 s) 62°C (30 s) 72°C (2 min): 1 cycle
   94°C (10 s) 61°C (30 s) 72°C (2 min): 1 cycle
   94°C (10 s) 60°C (30 s) 72°C (2 min): 1 cycle
   94°C (10 s) 59°C (30 s) 72°C (2 min): 1 cycle
   94°C (10 s) 58°C (30 s) 72°C (2 min): 1 cycle
   94°C (10 s) 57°C (30 s) 72°C (2 min): 1 cycle
   94°C (10 s) 56°C (30 s) 72°C (2 min): 23 cycles
   4°C Hold.

3.2. Guidelines for Establishing T$_{anneal}$

1. Test the PCR at various T_{anneal}, using 2°C intervals and starting 4°C below the estimated T_m, working up to 10°C above it. If the product yield does not significantly decrease at the higher temperatures, choose a higher temperature for the PCR. If the yield decreases, drop back to the plateau area. Stringency becomes important, when the target is contaminated with other genomic DNA, so operate close to the high value if possible.

2. To screen for the best T_{anneal} of a multiplex system, start 4°C lower than the lowest primer T_m and perform the PCR at various T_{anneal}, increasing the value by 2°C intervals to the point at which the first locus fails to amplify. Cut back to a temperature at which all loci yield strong signal-to-noise ratios.

3. When arbitrary primers are used, the choice of the temperature depends on whether the primers are structured or not. For example, use 35 cycles of 94°C (30 s), 30°C (30 s), 72°C (30 s) for primers with minihairpins at the 5'-terminus

and 45 cycles at 94°C (1 min), 35°C (1 min), 72°C (2 min) for unstructured primers (**Note 17**).

4. When adopting a PCR protocol optimized in another laboratory with a different thermal cycler, the optimal T_{anneal} or the time at the T_{anneal} may need to be adjusted. Amplify a low and medium amount of template at the temperatures described in the protocol, performing the assays in duplicate. If pattern changes are observed, raise the T_{anneal} in 1°C increments or the time at T_{anneal} by 10–15 s. If the yield drops, increasing the denaturation temperature or holding times may improve the yield. If the pattern is identical, but the yield is low, it may be helpful to consider the following: Try a tube with a thin wall, check the temperature at which the clock starts counting, and verify that the instrument is still calibrated using the manufacturer's recommended procedure.

3.3. Reduction or Elimination of PCR Artifact Bands

3.3.1. Removing a Nontemplated Nucleotide Addition with T4 DNA Polymerase

The nonuniform end of a primer-extended PCR product can be "polished" with either T4 DNA polymerase *(68)* or with Pfu DNA polymerase *(69)*. The following procedure is a simple and cheap way to polish the ends of PCR products. The treatment is a post-PCR one, so do not perform this assay in the template preparation area (refer to **Subheading 1.9.**). Even though using a thermal stable DNA polymerase that has 3' to 5' exonuclease activity in the PCR, such as UlTma (Perkin-Elmer) will help somewhat in removing nontemplated nucleotide additions, this type of polymerase is still sufficiently active in nontemplated nucleotide addition that one obtains a mixture of blunt and 3'-overhanging ends.

1. Generally, this procedure is performed as a group of a series of samples obtained from different PCRs that have been stored at –20°C. Thaw the samples, vortex, and pulse-spin the tubes to collect the solution before opening the tube.
2. Remove 10 µL PCR solution and place in a 0.5 mL microfuge tube with cap. Heat 2 min at 98°C.
3. Add 5 U of T4 DNA polymerase and incubate at 25°C in a thermal cycler block for 30 min. Additional dNTPs do not need to be added if the PCR was carried out with 0.2 mM of each dNTP.
4. Add 1.3 µL 2 mM EDTA to quench the reaction or heat-inactivate at 80°C for 10 min.
5. The amplicons (i.e., PCR products) may be precipitated after adding 0.5 vol of 7.5 M ammonium acetate and 2.5 vol 95% ethanol. The pellet is washed with 70% ethanol and dried.

3.3.2. Adjusting the Thermal Cycling Protocol to Effect Blunt-End Addition

After all the thermal cycles have been completed, there is a one-time incubation at 72°C in most protocols. Change this incubation step to 30 min at

60°C to convert most of the blunt ends to +1 additions at Mg^{2+} concentrations of 1.25 mM or above *(70)*.

3.3.3. Reducing Shadow (Stutter) Band Signals

1. Keep the template concentration as low as possible, preferably below 10 ng, and do not overamplify. As the main band reaches the plateau state, the stutter band appears to increase in proportion to it.
2. Since stutter tends to increase at higher Mg^{2+} concentration, using a concentration toward the low end of the Mg^{2+} window may be beneficial (refer to **Subheading 1.6.**).
3. Change the locus. If there is a large amount of stutter with the chosen locus, look for another one that provides a similar discriminatory power but does not produce intense stutter bands.
4. Perform the PCR with different DNA polymerases to find one that is more processive.

3.4. Template Preparation for the PCR

3.4.1. Quick Pre-PCR Purification of the Template

DNA extracted from forensic samples by the organic extraction method still may contain inhibitors of the PCR *(7)*. We have found that further purification by ultrafiltration may improve the success rate of amplification of compromised samples. Perform these operations in the template preparation room (if available) to avoid contamination with amplicons, but spin the samples in a different area to avoid problems associated with aerosols.

1. Label both the filter and reservoir with a marking pen.
2. Place 500 μL dH$_2$O on top of the membrane of a Microcon-100, being careful not to touch the membrane with the pipet tip. To see if it leaks and to remove the moisturizer from the membrane, let the apparatus sit vertically for 15 min, or pulse centrifuge at 500g. If the membrane seal is not tight, a large volume of water will pass quickly through the device. Throw away leaky spin columns.
3. Force the water through the membrane by centrifuging at 500g for 15 min, discard the ultrafiltrate from the reservoir, and reattach the filter.
4. Place a known volume of DNA sample without chloroform residue on the membrane. We use up to 100 μL and rinse the walls of the apparatus with another 100 μL dH$_2$O. Cap the column. Work with one uncapped column at a time in order to avoid crosscontamination.
5. Centrifuge according to the manufacturer's recommendations (500g), then remove and measure the volume of the ultrafiltrate (**Note 18**). Since the membrane was prewet, there should be approx 200 μL. If not, spin longer.
6. Rinse the walls of the filter with 200 μL dH$_2$O, working with one apparatus at a time to avoid cross-contamination, and spin again. This step may be repeated.
7. Finally, collect the retentate containing the purified template according to the manufacturer's recommendations.

8. Measure the volume: in our hands, it varies from about 10 to 50 μL. Reconstitute to the original volume of the sample applied to the membrane and store in a 0.5 mL capped tube.
9. If an inhibitor is suspected, include a positive control in the PCR (i.e., primers and template that give a characteristic band) to insure that the purification was successful.

3.4.2. Protocol for Rapid DNA Quantification

DNA can be quantified accurately with a fluorescence method that distinguishes between double- and single-stranded nucleic acids. An example that has worked well is presented here.

1. Make a 1:400 dilution of PicoGreen^{TM} in TE buffer.
2. Dispense 100 μL per well of a 96-well microtiter dish for reading on the fluorometer LS-50B (Perkin-Elmer, Norwalk, CT) or an equivalent instrument.
3. Add 1–10 μL of sample or standard to wells.
4. Read on the LS50B using the "WPR" software. Use the following settings for detection and quantification: Excitation Wavelength, 480 nm; Excitation Slit Width, 5 nm; Emission Wavelength, 520 nm; Emission Slit Width, 4 nm; Emission Filter, 515 nm.
5. Use the Generic WPR Macro and template for generating a spread sheet that displays the concentrations in μg/mL (**Note 19**).

3.5. Detection of PCR Products

3.5.1. Manual Detection Methods

1. For arbitrarily primed PCR: in the stage of doing the initial screen for useful primers, 10–15 μL PCR product commonly is run on a 1.5% agarose gel in 1X TBE buffer at 4 V/cm for 3–4 h, stained with ethidium bromide, and visualized on a UV transilluminator.
2. The products may also be separated on a 6% acrylamide minigel, denaturing or native, and stained with either ethidium bromide or silver *(12)*.
3. For AFLP, the samples must be radioactive *(17)*.

3.5.2. Automated Analysis on a DNA Sequencer with Fluorescence Detection

1. If converting from manual detection methods to automated fluorescence detection methods, initially try 0.1 μ*M* concentration of both primers in the PCR and cut the number of cycles of amplification by 2–3 rounds. Increase the concentration of primers in 0.5 μ*M* increments as necessary to obtain sufficient signal.
2. Dispense 1.5 μL PCR solution and 0.5 μL dye-labeled sizing ladder into a 1.5 mL microfuge tube, dry on a heat block at 90°C for 3 min, and add 6 μL loading buffer (**Subheading 2.**). Denature 2 min at 95°C and load onto the proper polyacrylamide gel containing urea as suggested in the chemistry guide of the manufacturer for the particular DNA sequencer being utilized. Perform electrophoresis as recommended by the manufacturer for the size plates selected.

3. For AFLP, mix 0.3 µL ROX sizing ladder, 0.2 µL loading buffer, 0.7 µL deionized formamide, 0.4 µL FAM- or JOE-labeled PCR products, and 0.8 µL TAMRA-labeled PCR products in a 0.5 mL tube with cap. Denature for 3 min at 95°C and snap-cool. Load immediately onto a 5% Long Ranger gel with 6% urea and 1X TBE. Perform the electrophoresis as recommended by the manufacturer (at 3 kV for 4 h or at 1.8 kV for 8 h with the Applied Biosystems DNA Sequencer models 377 or 373, respectively).

3.6. Procedures for PCR Optimization

3.6.1. Prevention of Pre-PCR Artifacts

3.6.1.1. USING WAX AS A BARRIER IN HOT-START PCR

Perform the following steps in the template preparation room or special area clear of amplicons (refer to **Subheading 1.9.**).

1. Dispense 20–40 µL reaction mix that lacks one or more components such as the DNA polymerase and the template in the PCR tube. Pulse spin to bring the liquid to the bottom of the tube.
2. Place a wax bead over the reaction mix, put the tube in a heat block set at 80°C for 5 min to melt the wax, then remove the tube so the wax will solidify and seal off the master mix. A large number of tubes can be prepared in advance and stored at 4°C, but do not store frozen.
3. Carefully dispense 30–60 µL of a solution containing the missing component, usually combined with the test sample, on the wax surface (without disturbing the barrier and), cap the tube.
4. Load the tubes in the thermal cycler and perform the amplification as normally done without the wax barrier.

3.6.1.2. USING DNA POLYMERASE ACTIVATION TO AVOID PRE-PCR ARTIFACTS

1. Program the thermal cycler with a one-time incubation at 95°C for 11 min as the initial step before initiation of the thermal-cycling regime and increase the normal number of cycles by one round.
2. When preparing the reaction cocktail, add 2.5 U AmpliTaq Gold™ for 1–50 ng template. Add oil to tubes if the thermal cycler does not have a heated lid.
3. Place the capped tubes in the thermal cycler and initiate the pre-PCR heating step.

3.6.2. Adjusting the Reaction Components of the PCR

3.6.2.1. DEFINING THE OPTIMAL MAGNESIUM ION CONCENTRATION

In the following protocol, the samples are assayed in duplicate.

1. Obtain mixes of (a) the buffer without Mg^{2+}, (b) a dilution of *Taq* DNA polymerase (**Note 20**), (c) $MgCl_2$, (d) dNTPs, and (e) primers (**Note 21**).
2. On ice, make cocktails from these reagents for the optimization experiments, instead of pipeting each constituent in each individual reaction. This practice will

Table 2
Cocktails for MgCl$_2$ Optimization

	dH$_2$O	10X Buffer	dNTP mix	Primer mix	*Taq* 1 U/μL	MgCl$_2$ 25 m*M*
1.0 m*M* MgCl$_2$	66	25	20	16.5	12.5	10
1.2 m*M* MgCl$_2$	64	25	20	16.5	12.5	12
1.5 m*M* MgCl$_2$	61	25	20	16.5	12.5	15
1.8 m*M* MgCl$_2$	58	25	20	16.5	12.5	18
2.2 m*M* MgCl$_2$	54	25	20	16.5	12.5	22
2.6 m*M* MgCl$_2$	50	25	20	16.5	12.5	26

The values are in microliters and make a total vol of 150 μL. This is sufficient for four aliquots (i.e., four reactions) of 30 μL each, which are added to 20 μL DNA, and an extra 30 μL to allow for pipetting errors. The stock dNTP solution is 2.5 m*M* in each dNTP, and the stock primer solution has a 5 μ*M* concentration of each primer in water.

save time and reduce variations that arise in pipetting concentrated solutions. Use molecular-biology-grade water (various suppliers) if there is no laboratory means of obtaining high-purity water lacking ions and organic substances. A scheme for making the cocktails is presented in **Table 2**.

3. Dispense 10–50 ng of two different DNA preparations (in water) in a total volume of 20 μL into the four PCR tubes on ice for each magnesium ion concentration (4 × 6 = 24 reactions). Keep all tubes capped except for the one you are working with in order to avoid crosscontamination. Immediately add 30 μL of the respective cocktail containing the buffer and other reaction components. The amount of DNA to choose depends on the sensitivity of your detection system.

4. Set the thermal cycler for 93°C, 3 min (once) and a three-step PCR format of 94°C, 30–60 s; T_m, 30-60 s; 72°C, 30–120 s (30 cycles), followed by 60°C, 30 min; and then 15°C (hold).

5. Repeat the experiment at 4°C below and 8°C above the estimated T_m (24 × 3 = 72 reactions per complete experiment).

6. Plot the results and note where the plateau begins. Choose a Mg^{2+} concentration for your PCR near the middle of the plateau. If the yield does not level off (as might occur under rapid PCR conditions in capillaries), continue the experiment up to 4 m*M* Mg^{2+} concentration.

3.6.2.2. Adjusting the Other Reaction Components of the PCR

1. Two components worth investigating in the optimization endeavor are the primer concentration and the optimal amount of *Taq* DNA polymerase. Perform the experiment with 0.1, 0.2, 0.5, and 0.8 μ*M* primer (i.e., both primers are at these concentrations) with 1.25, 5, and 10 U *Taq* DNA polymerase (12 reactions). If the primer is limiting, there will not be much effect from an increase in the amount of enzyme. Use that primer concentration at which there is a significant increase in yield when the enzyme concentration is higher. If stringency is required in the PCR,

then cut back the primer concentration to the level at which no extraneous bands are detected.

2. *Taq* DNA polymerase is inactive in buffers of high KCl *(53)*. To investigate the effect of KCl at the optimal Mg^{2+} concentration, run the PCR at 40, 50, 55, 60, and 70 mM concentrations.

3. Investigate the effect of pH on the system by increments of 0.2-pH unit from 8.1 to 8.9. Test at least ten DNA samples extracted by the same procedure.

3.7. Finding the Best Thermal Cycler Parameters

3.7.1. Establishing Windows for T_{den}, T_{anneal}, and T_{ext}

To establish the range, perform the thermal cycling with denaturation at 92, 93, 94, 95, and 96°C, keeping T_{anneal} constant. Perform the experiment with duplicate samples and use a 1 min duration at T_{den}. When the optimal T_{den} is established, investigate the variation around the $T_{optimum}$ for annealing in the same way (i.e., with ±1.5–2°C increments). The temperature of the extension step, T_{ext}, can be examined too, but it does not affect the results as critically as T_{den} and T_{anneal}. Repeat the testing at a duration of 30 s for T_{den}.

3.7.2. Adjusting Ramping Times for Stringency and Degeneracy

To test the effect of ramp-down time on the yield and stringency, examine 60-, 80-, 100-, 120-, and 160-s time intervals from T_{den} to T_{anneal}. Keep the default settings for the time intervals between T_{anneal} and T_{ext} and T_{ext} and T_{den}, and do not change T_{den}, T_{anneal}, and T_{ext}.

3.8. Primers and Remedies for Amplifying Regions of High GC Content

1. Extend the primer at a temperature high enough to reduce or eliminate the template GC hybridization. Design high-melting primers with T_m values between 70 and 74°C and use them with a combined annealing and extension step between 70 and 80°C as a two-step PCR to amplify regions of >70% GC *(71)*.

2. Partially replace the dG with either 7-deaza-2'-deoxy-guanosine *(72,73)*: try 150 μM 7-deaza-dGTP with 50 μM dGTP for long stretches of G/C nucleotides. Deoxyinosine has also been used with dGTP for amplifying recalcitrant GC-rich templates *(74)*. For detection, *see* **Note 22**.

3. Employ an additive in the PCR, such as formamide (up to 5%, *[76]*) or DMSO *(57)*, to reduce the melting temperature of the GC-rich regions. Test the effect of 2, 5, 10, and 15% DMSO and try the PCR at a T_{anneal} 5, 10, 15, and 20°C lower than that used without the additive. The DMSO should be the purest available and should be reserved for molecular-biology work only.

4. Perform the PCR at the highest T_{den} possible. Use a short time at very high T_{den}; only a few seconds are needed. Addition of glycerol to the PCR buffer (up to 20%) may help stabilize the *Taq* DNA polymerase at high temperatures and

improve the yield. Use a DNA polymerase such as UlTma or Vent(exo-) that has a higher thermal stability than *Taq*, which has a half-life of only 10 min at 97.5°C. Be aware that the hyperthermostable DNA polymerases may have an intrinsic 3'-to-5' exonuclease activity, which can shorten the primers from the 3' end.

3.9. Guidelines for Prevention of Contamination

1. Establish a separate room for template preparation and PCR setup if the laboratory budget permits. If a separate room is impossible, then devote a "clean" space in the laboratory for these two functions that is far removed from the bench where the thermal cycler is located and where amplicons are to be handled for analysis.
2. Purchase a set of pipetters for the clean-work area and leave them there. Use either positive-displacement pipeters (various suppliers) that work with a disposable barrel and plunger, or use standard pipetters with aerosol-resistant tips (ART Tips, Molecular Bio-Products, San Diego, CA). The object is to prevent template deposited in the pipetter via aerosols from the first sample from being injected into the second sample.
3. Work with one glove instead of two. Use the gloved hand to handle tubes with reagents and samples and the other hand to handle the pipeter, fixtures (i.e., door knobs, faucets, drawer pulls), utensils, and appliances. If the equipment becomes contaminated with template or amplicons, it will not be transferred to the reagents and samples so easily, since the gloved hand is used for these tubes. Change gloves often, since they can become contaminated too. The object of using gloves is not to protect your hands, but rather to control contamination.
4. Run a reagent-only control (i.e., template blank) in each PCR in order to detect contamination as early as possible. If a work area becomes contaminated with amplicons due to poor laboratory practice, it is important to discover the problem before it gets out of hand.
5. Dispense stock reagents and control template solutions in small portions into "one-use" tubes, and discard material not used in setting-up the PCR. This practice is time-consuming, but it will prevent contamination of the stock solutions.
6. Pulse-spin tubes containing reagents and template before opening them to control for aerosols. This operation should be performed outside of the clean area in order to avoid aerosols associated with centrifugation.
7. If many samples are to be treated, open only one tube at a time, keeping the other tubes capped (i.e., closed) when not in immediate use.
8. Dispose of solutions of amplicons and used gel-electrophoresis buffers with caution. Amplicon-containing liquids can almost be considered hazardous materials!
9. Clean the work areas and pipetter barrels with a weak solution of bleach on a regular basis, which is determined by the usage.

4. Notes

1. Zabeau, M. and Vos, P. (1993) European Patent Application, pub. no. EP 0534858.
2. A discussion of salt effects on the T_m can be found in *(27)*. Another factor of importance for the PCR is the dependency of the T_m on the concentration of the

primer. Primers are in a high concentration under the conditions of the PCR and do not change significantly during the PCR.

3. On a few occasions, HPLC-purified primers have been reported to yield "ladder" bands in the PCR of tetranucleotide repeats. In one case, this gross stuttering could be traced to an inhibitor of *Taq* that passed through the laboratory water-filtration unit. Commercially available purified, sterile water may be the best choice for a primer stock solution. For an introduction to HPLC purification of oligonucleotides, refer to *(34)* and the references therein.

4. Suppose the upper primer in **Table 1** ($A_5 C_5 G_4 T_4$, $n = 18$ nt) was obtained as a dry pellet from a core facility, and the amount was reported to be 3.2 OD units. It was resuspended in 0.5 mL molecular-biology-grade water with 20 mM Tris-HCl, pH 8.0, and 50 mM NaCl; 50 µL of the stock was diluted in 1 mL of the same for measuring the absorbance at 260 nm. An optical density of 0.26 was obtained. This is lower than expected, and the concentration of the stock is 27 µM instead of 33 µM. One may obtain the molar concentration of the stock solution from *(35)*:

$$(A_{260 \cdot \text{measured}} \times \text{dilution factor})/[A(15200) + C(7050) + G(12010) + T(8400)]$$

When the primer has one fluorescein or rhodamine derivative attached, one should add 26,662 to the denominator.

5. There are many speculations on the origin of stutter bands, but there is no proof. A currently popular idea is that stutter is thought to occur during pausing of the DNA polymerase because of the incorporation of a wrong nucleotide, the difficulty of extending the primer through a region of substantial single-stranded secondary structure, or encountering a region of the template that is not fully dissociated. After pausing, the polymerase could dissociate from the template, although there is experimental evidence against such a scenario *(40)*. The polymerization enzyme-template complex could become destabilized during the pause and enter into a secondary state that has a lower binding constant for the polymerase and the template and allows the less tightly bound strand to partially dissociate from the active site. When the strand rejoins, it may do so one repeat out of register (or by an integral multiple of repeats). This event would lead to a loss of a repeat (or a multiple of repeats) in primer extension if the template loops out. The mechanism is similar to slipped-strand mispairing *(41)*, and it has been suggested to be a likely cause of stutter bands *(40)*. We have found that stutter is not reduced when UlTma DNA polymerase, which has an inherent 3' to 5' exonuclease activity, is substituted for *Taq* DNA polymerase. This result appears to be inconsistent with the idea that stutter occurs when the polymerase is paused as a result of misincorporated nucleotides, because the 3' exonuclease activity should quickly correct such a situation. We also observed no reliable decrease in stutter band formation as a result of experiments with additives such as DMSO and formamide (i.e., denaturants that should melt regions of single-stranded secondary structure). Until the biochemical understanding of the PCR is available, predictions on the cause of stuttering will remain intuitive.

6. For example, in one paper, five different methods of pathogen DNA-sample preparation are compared, and the template isolation is presented as an important pre-PCR step *(42)*.

7. A number of methods were examined for removal of inhibitory substances from a stubborn substrate, and a chromatographic cellulose purification step proved to be a simple way to obtain amplifiable template at low copy number *(43)*.

8. A rapid method for removal of inhibitors of the PCR from urine involved the use of either polyethylene glycol precipitation or glass-powder extraction *(44)*.

9. Pathogens can be purified even from sewage for examination by the PCR *(45)*.

10. CTAB: hexadecyltrimethyl ammonium bromide.

11. Axenically grown plants will obviate the need for this precaution, if they are available.

12. Before screening a F2 population with a given primer, the PCR was repeated at least twice on different days and with the two DNA concentrations. If the polymorphic band still appeared (i.e., if it were consistently present), a test sample of DNA from the F2 generation was used to assess whether or not the marker segregated in a Mendelian fashion.

13. Depending upon the inherent level of polymorphism in the cross and the degree of marker saturation required, the researcher must decide whether to invest extra effort. The barley lines we examined had been back-crossed for seventeen generations, and the level of polymorphism was extremely low. Out of 1600 10-mer primers screened, only ten gave reproducible polymorphisms, and of those only half were detectable in a native gel system.

14. Primer dimers and mispriming are often observed in DNA analysis by multiplex PCR, long-range PCR *(11)*, and when the target is at low copy number and in the presence of a high amount of nontarget DNA, to mention a few cases. A good example of the latter situation is HIV-1 DNA (or cDNA) analysis from whole blood samples. Also, in forensics, the human-target template may be contaminated with an excess of bacterial DNA. Mispriming can occur to single-stranded DNA present in the preparation. It is not unusual to have single-stranded DNA if one uses a heat denaturation step during the template-preparation procedure. An example is heat denaturation of protease K. Chelex extraction utilizes a boiling step and leads to single-stranded DNA.

15. We suggest that before one invests a lot of time running a battery of tests, a control template should be assayed in duplicate. Use the kit components to run the PCR under the standard conditions of 1.5 mM Mg^{2+}, 50 mM KCl, and at a concentration of 0.2 mM each dNTP and 0.2 μM of each primer.

16. Panhandles: The 10-mers used as RAPD primers can anneal at the ends. They are as complementary as the primers and compete, so they are inefficiently amplified.

17. The PCR was performed in the GeneAmp® PCR System 9600 thermal cycler (Perkin–Elmer) with 10–50 ng DNA.

18. Spinning 200 μL in a microcentrifuge (Eppendorf model 5415 C) at 500g. Ten minutes is sufficient.

19. A good standard is salmon sperm DNA from Gibco-BRL (cat. no. 15632-011) supplied at a concentration of 10 mg/mL.

20. The *Taq* DNA polymerase should be diluted to 1 U/μL in order to avoid errors in pipetting viscous solutions. The enzyme may be diluted in 1X PCR buffer without Mg^{2+} if it is to be used immediately.

21. Make a solution containing each primer at a concentration of 5 μM. One may wish to screen primers for a number of loci by coamplification. Choose primers of similar T_m, ones that will yield amplicons in discrete size ranges, so they do not overlap. Check via computer or visually that the chosen primers will not anneal to each other, especially at the 3'-end.

22. Some precaution may be necessary, since the detection of 7-deazaguanine-containing PCR products with ethidium bromide is more difficult: The fluorescence is less with these duplex DNAs *(75)*. There is no reduction in the signal when deoxyinosine is used *(74)*.

Acknowledgments

The authors are grateful to Will Bloch for the critical reading of the manuscript and to colleagues Martin Buoncristiani and Sean Walsh for sharing ideas. The RAPD work was carried out in the laboratory of Dr. S. Somerville, Carnegie Institute, Stanford University, Stanford, CA. This chapter was completed when J. M. R. was an employee of Perkin-Elmer Applied Biosystems. The listing of commercial products is for identification purposes and does not imply endorsement by the Federal Bureau of Investigation.

References

1. Saiki, R. K., Scharf, S., Faloona, F., Mullis, K. B., Horn, G. T., Erlich, H. A., and Arnheim, N. (1985) Enzymatic amplification of β-globin genomic sequences and restriction site analysis for diagnosis of sickle cell anemia. *Science* **230,** 1350–1354.

2. Mullis, K. B. and Faloona, F. (1987) Specific synthesis of DNA in vitro via a polymerase-catalyzed chain reaction. *Meth. Enzymol.* **155,** 335–350.

3. Gibbs, R. A. (1990) DNA amplification by the polymerase chain reaction. *Anal. Chem.* **62,** 1202–1214.

4. Bloch, W. (1991) A biochemical prospective of the polymerase chain reaction. *Biochemistry* **30,** 2735–2747.

5. Arnheim, N. and Erlich, H. (1992) Polymerase chain reaction strategy. *Ann. Rev. Biochem.* **61,** 131–156.

6. Mullis, K., Ferré, F., and Gibbs, R. (1994) *The Polymerase Chain Reaction.* Birkhäuser, Berlin, Germany.

7. Robertson, J. M., Sgueglia, J. B., Badger, C. A., Juston, A. C., and Ballantyne, J. (1995) Forensic applications of a rapid, sensitive, and precise multiplex analysis of the four short tandem repeat loci HumvWF31A/1, HumTHO1, HumF13A1, and HumFES/FPS. *Electrophoresis* **16,** 1568–1576.

8. Zhang, L., Xiangfeng, C., Schmitt, K., Hubert, R., Naridi, W., and Arnheim, N. (1992) Whole genome amplification from a single cell: implications for genetic analysis. *Proc. Natl. Acad. Sci. USA* **89,** 5847–5851.

9. Garner, H. R., Armstrong, B., and Lininger, D. M. (1993) High-throughput PCR. *Biotechniques* **14**, 112–115.
10. Wittwer, C. T. and Garling, D. J. (1991) Rapid cycle DNA amplification: time and temperature optimization. *Biotechniques* **10**, 76–83.
11. Ohler, L. and Rose, E. A. (1992) Optimization of long distance PCR using a transposon-based model system. *PCR Meth. Applicat.* **2**, 51–59.
12. Budowle, B., Chakraborty, R., Giusti, A. M., Eisenberg, A. J., and Allen, R. C. (1991) Analysis of the VNTR locus D1S80 by the PCR followed by high-resolution PAGE. *Am. J. Hum. Genet.* **48**, 137–144.
13. Davies, J. L., Kawaguchi, Y., Bennett, S. T., Copeman, J. B., Cordell, H. J., Pritchard, L. E., Reed, P. W., Gough, S. C. L., Jenkins, S. C., Palmer, S. M., Balfour, K. M., Rowe, B. R., Farrall, M., Barnett, A. H., Bain, S. C., and Todd, J. A. (1994) A genome-wide search for human type 1 diabetes susceptibility genes. *Nature* **371**, 130–136.
14. Williams, J. G. K., Kubelik, A. R., Livak, K. J., Rafalski, J. A., and Tingey, S. V. (1990) DNA polymorphisms amplified by arbitrary primers are useful as genetic markers. *Nucleic Acids Res.* **18**, 6531–6535.
15. Welsh, J. and McClelland, M. (1991) Genomic fingerprints produced by PCR with consensus tRNA gene primers. *Nucleic Acids Res.* **19**, 861–866.
16. Caetano-Anolles, G., Bassam, B. J., and Gresshoff, P. M. (1991) DNA amplification fingerprinting using short arbitrary oligonucleotide primers. *Biotechniques* **9**, 553–557.
17. Vos, P., Hogers, R., Bleeker, M., Reijans, M., van de Lee, T., Hornes, M., Frijters, A., Pot, J., Peleman, J., Kuiper, M., and Zabean, M. (1995) AFLP: a new technique for DNA fingerprinting. *Nucleic Acids Res.* **23**, 4407–4414.
18. Young, N. D., Zamir, D., Ganal, M. W., and Tanksley, S. D. (1988) Use of isogenic lines abd simultaneous probing to identify DNA markers tightly linked to the Tm-2a gene in tomato. *Genetics* **120**, 579–585.
19. Paran, I. and Michelmore, R. W. (1993) Development of reliable PCR-based markers linked to downey mildew resistance genes in lettuce. *Theoret. Appl. Genet.* **85**, 985–990.
20. Michelmore, R. W., Paran, I., and Kesseli, R. V. (1991) Identification of markers linked to disease-resistance genes by bulked segregant analysis: a rapid method to detect markers in specific genomic regions by using segregating populations. *Proc. Natl. Acad. Sci. USA* **88**, 9828–9832.
21. Sakuma, Y. and Nishigaki, K. (1994) Computer prediction of general PCR products based on dynamical solution structures of DNA. *J. Biochem. (Tokyo)* **116**, 736–741.
22. He, Q., Marjamäki, M., Soini, H., Mertsola, J., and Viljanen, M. K. (1994) Primers are decisive for sensitivity of PCR. *Biotechniques* **17**, 82–87.
23. Don, R. H., Cox, P. T., Wainwright, B. J., Baker, K., and Mattick, J. S. (1991) Touchdown PCR to circumvent spurious priming during gene amplification. *Nucleic Acids Res.* **19**, 4008.
24. Rychlik, W. and Rhodes, R. E. (1989) A computer program for choosing of oligonucleotides for filter hybridization, sequencing and in vitro amplification of DNA. *Nucleic Acids Res.* **17**, 8543–8551.

25. Breslauer, K. J., Frank, R., Blocker, H., and Marky, L. A. (1986) Predicting DNA duplex stability from the base sequence. *Proc. Natl. Acad. Sci. USA* **83,** 3746–3750.

26. Wu, D. Y., Ugozzoli, L., Pal, B. K., Qian, J., and Wallace, R. B. (1991) The effect of temperature and oligonucleotide primer length on the specificity and efficiency of amplification by the polymerase chain reaction. *DNA Cell. Biol.* **10,** 233–238.

27. Schildkraut, C. and Lifson, S. (1965) Dependence of the melting temperature of DNA on salt concentrations. *Biopolymers* **3,** 195–208.

28. Baldino, F., Jr., Chesselet, M.-F., and Lewis, M. E. (1989) High-resolution in situ hybridization histochemistry. *Meth. Enzymol.* **168,** 761–777.

29. Suggs, S. V., Hirose, T., Miyake, E. H., Kawashima, M. J., Johnson, K. I., and Wallace, R. B. (1981) Using purified genes, in *ICN-UCLA Symposium on Developmental Biology,* vol. 23 (Brown, D. D., ed.), Academic, New York, pp. 683–693.

30. Frégeau, C. J. and Fourney, R. M. (1993) DNA typing with fluorescently tagged short tandem repeats: a sensitive and accurate approach to human identification. *Biotechniques* **15,** 100–119.

31. Sharma, V. and Litt, M. (1992) Tetranucleotide repeat polymorphism at the d21S11 locus. *Hum. Mol. Genet.* **1,** 67.

32. Frégeau, C. J., Bowen, K. L., Elliott, J. C., Robertson, J. M., and Fourney, R. M. (1994) PCR-based DNA identification: a transition in forensic science, in *Proc. Fourth Int. Symp. Hum. Ident.,* Promega Corp., Madison, WI, pp. 107–118.

33. McClelland, M. and Welsh, J. (1994) DNA fingerprinting by arbitrarily primed PCR. *PCR Meth. Applicat.* **4,** S59–S65.

34. Green, A. P., Burzynski, J., Helveston, N. M., Prior, G. M., Wunner, W. H., and Thompson, J. A. (1995) HPLC purification of synthetic oligodeoxyribonucleotides containing base- and backbone modified sequences. *Biotechniques* **19,** 836–841.

35. Wahl, G. M., Berger, S. L., and Kimmel, A. R. (1987) Synthesis and characterization of oligonucleotides. *Meth. Enzymol.* **152,** 399–407.

36. Clark, J. M. (1988) Novel non-templated nucleotide addition reactions catalyzed by procaryotic and eucaryotic DNA polymerases. *Nucleic Acids Res.* **16,** 9877–9686.

37. Hu, G. (1993) DNA polymerase-catalyzed addition of non-templated extra nucleotides to the 3' end of a DNA fragment. *DNA Cell Biol.* **12,** 763–770.

38. Rychlik, W. (1995) Priming efficiency in PCR. *Biotechniques* **18,** 84–89.

39. Gelfand, D. H., personal communication.

40. Hauge, X. Y. and Litt, M. (1993) A study of the origin of "shadow bands" seen when typing dinucleotide repeat polymorphisms by the PCR. *Hum. Mol. Genet.* **2,** 411–415.

41. Levinson, G. and Gutman, G. A. (1987) Slipped-strand mispairing: a major mechanism for DNA sequence evolution. *Mol. Biol. Evol.* **4,** 203–221.

42. Liedtke, W., Opalka, B., Zimmermann, C. W., and Schmid, E. (1994) Different methods of sample preparation influence sensitivity of *Mycobacterium tuberculosis* and *Borrelia burgdorferi* PCR. *PCR Meth. Applicat.* **3,** 301–304.

43. Wilde, J., Eiden, J., and Yolken, R. (1990) Removal of inhibitory substances from human fecal specimens for detection of group A Rotaviruses by reverse transcriptase and polymerase chain reactions. *J. Clin. Microbiol.* **28,** 1300–1307.

44. Yamaguchi, Y., Hironaka, T., Kajiwara, M., Tateno, E., Kita, H., and Hirai, K. (1992) Increased sensitivity for detection of human cytomegalovirus in urine by removal of inhibitors for the polymerase chain reaction. *J. Virol. Meth.* **37**, 209–218.

45. Tsai, Y.-L., Sobsey, M. D., Sangermano, L. R., and Palmer, C. J. (1993) Simple method of concentrating enteroviruses and hepatitis A virus from sewage and ocean water for rapid detection by reverse transcriptase-polymerase chain reaction. *Appl. Environ. Microbiol.* **59**, 3488–3491.

46. Budowle, B. and Baechtel, S. (1990) Modifications to improve the effectiveness of restriction fragment length polymorphism typing. *Anal. Biochem.* **92**, 497–500.

47. Walsh, P. S., Metzger, D. A., and Higuchi, R. (1991) Chelex® 100 as a medium for simple extraction of DNA for PCR-based typing from forensic material. *Biotechniques* **10**, 506–513.

48. Walsh, P. S., Varlaro, J., and Reynolds, R. (1992) A rapid chemiluminescent method for quantitation of human DNA. *Nucleic Acids Res.* **20**, 5061–5065.

49. Budowle, B., Baechtel, F. S., Comey, C. T., Giusti, A. M., and Klevan, L. (1995) Simple protocols for typing forensic biological evidence: chemiluminescent detection for human DNA quantitation and restriction fragment length polymorphism (RFLP) analyses and manual typing of polymerase chain reaction (PCR) amplified polymorphisms. *Electrophoresis* **16**, 1559–1567.

50. Fang, G., Hammar, S., and Grumet, R. (1992) A quick and inexpensive method for removing plant polysaccharides from plant genomic DNA. *Biotechniques* **13**, 52–54.

51. Kilby, N. J. and Furner, I. J. (1990) Another CTAB plant DNA extraction: isolation of high molecular weight DNA from small quantities of Arabidopsis tissue. http://weeds. mgh. harvard. edu menu item: Compleat Guide.

52. Reed, P. W., Davies, J. L., Copeman, J. B., Bennet, S. T., Palmer, S. M., Pritchard, L. E., Gough, S. C. L., Kawagucchi, Y., Cordell, H. J., Balfour, K. M., Jenkins, S. C., Powell, E. E., Vignal, A., and Todd, J. A. (1994) Chromosome-specific microsatellite sets for fluorescence-based, semi-automated genome mapping. *Nature Genet.* **7**, 390–395.

53. Gelfand, D. H. (1989) Taq DNA polymerase, in *PCR Technology: Principles and Applications for DNA Amplification* (Erlich, H. A., ed.), Stockton Press, New York, 17–22.

54. Mullis, K. B. (1991) The polymerase chain reaction in an anemic mode: how to avoid cold oligodeoxyribonuclear fusion. *PCR Meth. Applicat.* **1**, 1–14.

55. Chou, Q., Russell, M., Birch, D. E., Raymond, J., and Bloch, W. (1992) Prevention of pre-PCR mis-priming and primer dimerization improves low-copy-number amplifications. *Nucleic Acids Res.* **20**, 1717–1723.

56. Blanchard, M. M., Taillon-Miller, P., Nowotny, P., and Nowotny, V. (1993) PCR buffer optimization with uniform temperature regimen to facilitate automation. *PCR Meth. Applicat.* **2**, 234–240.

57. Chamberlin, J. S., Gibbs, R. A., Ranier, J., and Caskey, C. T. (1989) Multiplex PCR for the diagnosis of Duchenne muscular dystrophy, in *PCR Protocols: A Guide to Methods and Applications* (Innis, M., Gelfand, D., Sninsky, D., and White, I., eds.), pp. 272–281.

58. Schweder, M. E., Shatters, G., Jr., West, S. H., and Smith, R. L. (1995) Effect of transition interval between melting and annealing temperatures on RAPD analyses. *Biotechniques* **19**, 38–42.

59. Cusi, M. G., Cioé, L., and Rovera, G. (1992) PCR amplification of GC-rich templates containing palindromic sequences using initial alkali denaturation. *Biotechniques* **12**, 502–504.

60. Jeffreys, A. J., MacLeod, A., Tamaki, K., Neil, D. L., and Monckton, D. G. (1991) Minisatellite repeat coding as a digital approach to DNA typing. *Nature* **354**, 204–209.

61. Lowe, T. Shareifkin, J., Yang, S. Q., and Dieffenbach, C. W. (1990) A computer program for selection of oligonucleotide primers for the polymerase chain reaction. *Nucleic Acids Res.* **18**, 1757–1761.

62. Welsh, J., Petersen, C., and McClelland, M. (1991) Polymorphisms generated by arbitrarily primed PCR in the mouse: application to strain identification and genetic mapping. *Nucleic Acids Res.* **19**, 303–306.

63. Versalovic, J., Thearith, K., and Lupski, J. R. (1991) Distribution of repetitive DNA sequences in eubacteria and application to fingerprinting of bacterial genomes. *Nucleic Acids Res.* **19**, 6823–6831.

64. Caetano-Anolles, G. and Gresshoff, P. M. (1994) DNA amplification fingerprinting using arbitrary mini-hairpin oligonucleotide primers. *Biotechniques* **12**, 619–624.

65. Cassol, S., Rudnik, J., Salas, T., Montpetit, M., Pon, R. T., Sy, C. T., Read, S., Major, C., and O'Shaughnessy, M. V. (1992) Rapid DNA fingerprinting to control for specimen errors in HIV testing by the polymerase chain reaction. *Mol. Cell. Probes* **6**, 327–331.

66. Batzer, M. A., Carlton, J. E., and Deininger, P. L. (1991) Enhanced evolutionary PCR using oligonucleotides with inosine at the 3'-terminus. *Nucleic Acids Res.* **19**, 5081.

67. Bartl, S. and Weissman, I. L. (1994) PCR primers containing an inosine triplet to complement a variable codon within a conserved protein-coding region. *Biotechniques* **16**, 246–250.

68. Wang, K., Koop, B. F., and Hood, L. (1994) A simple method using T4 DNA polymerase to clone polymerase chain reaction products. *Biotechniques* **17**, 236–239.

69. Costa, G. L. and Weiner, M. P. (1994) Protocols for cloning and analysis of blunt-ended PCR-generated DNA fragments. *PCR Meth. Applicat.* **3**, S95–S106.

70. *AmpFISTR™ Blue User's Manual*, Perkin-Elmer Applied Biosystems, Foster City, CA, pp. 9–12.

71. Schuchard, M., Sarkar, G., Ruesink, T., and Spelsberg, T. C. (1993) Two-step "hot" PCR amplification of GC-rich Avian c-myc sequences. *Biotechniques* **14**, 390–394.

72. McConlogue, L., Brow, M. D., and Innis, M. A. (1988) Structure-independent DNA amplification by PCR using 7-deaza-2'-deoxy-guanosine. *Nucleic Acids Res.* **16**, 9869.

73. Ritchie, R. J., Knight, S. J. L., Hirst, M. C., Grewal, P. K., Bobrow, M., Cross, G. S., and Davies, K. E. (1994) The cloning of FRAXF: trinucleotide repeat expansion and methylation at a third fragile site in distal Xqter. *Hum. Mol. Genet.* **3**, 2115–2121.

74. Turner, S. L. and Jenkins, F. J. (1995) Use of deoxyinosine in PCR to improve amplification of GC-rich DNA. *Biotechniques* **19,** 47–52.

75. Latimer, L. J. P. and Lee, J. S. (1991) Ethidium bromide does not fluoresce when intercalated adjacent to 7-deazaguanine in duplex DNA. *J. Biol. Chem.* **266,** 13,849–13,851.

76. Sarkar, G., Kapelner, S., and Sommer, S. S. (1990) Formamide can dramatically improve the specificity of PCR. *Nucleic Acids Res.* **18,** 7465.

14

Analysis of Amplified Fragment-Length Polymorphisms (VNTR/STR Loci) for Human Identity Testing

Bruce Budowle and Robert C. Allen

1. Introduction

Highly polymorphic variable number tandem repeat (VNTR) loci have proven very useful for human DNA testing purposes *(1–3)*. Initially, these variants were characterized by restriction fragment-length polymorphism (RFLP) analysis. Subsequently, the polymerase chain reaction (PCR) provided significant methodological improvements, particularly enhanced sensitivity and specificity *(4)*. The PCR-based technology for detection of VNTR alleles has been termed amplified fragment-length polymorphism (AMP-FLP) analysis *(5)*. AMP-FLP analysis already has proven useful for the detection of multiallelic profiles *(6)*, D17S5 *(7)*, the 3' hypervariable region of the apolipoprotein B gene *(8,9)*, and particularly D1S80 *(5,10)*.

A subgroup of VNTR loci is the short tandem repeats (STR) loci. These loci are highly polymorphic and abundant in the human genome *(11,12)*. The STR loci are composed of tandemly arrayed repeat sequences, each 2–7 base pairs in length. Because the allele size is generally <350 base pairs, STRs are amenable to amplification by the PCR *(4,11)*. The alleles at a locus also are less likely to be subject to preferential amplification due to their small size. STR alleles are electrophoretically separated with high resolution on native or denatured polyacrylamide gels. Alleles differing in size by as little as 1 bp can be discriminated *(11,13–15)*.

The typing of STR loci for human identity testing has been facilitated by the ability to amplify two or more STR loci simultaneously in one amplification reaction by a procedure known as multiplex PCR *(11–13,16,17)*. The advan-

From: *Methods in Molecular Biology, Vol. 98: Forensic DNA Profiling Protocols*
Edited by: P. J. Lincoln and J. Thomson © Humana Press Inc., Totowa, NJ

tages of a multiplex system are that less sample DNA is required than when each locus is amplified independently, fewer reagents are consumed, and the time needed to perform validation studies, particularly population studies, on several loci, is greatly reduced. Multiple STR loci can be separated simultaneously by polyacrylamide-gel electrophoresis, and the amplicons are detected by silver staining or by a fluorescence detection system. The latter method includes a fluor attached to the 5' end of one primer in the amplification reaction and subsequent detection in real time (using, e.g., the ABI 373A or ABI 310, Perkin–Elmer, Norwalk, CT) or by using a decoupled detection system after gel electrophoresis (e.g., FluorImager SI or Hitachi FMBIO 100). Both detection schemes—manual silver staining and automated fluor detection—can be used to obtain reliable data for STR typing. The silver-staining approach is simple and does not require expensive equipment. However, in order to obtain unequivocal typing of the various loci in the multiplex, their sizes cannot overlap. In contrast, the loci in the fluor-labeled multiplex can be labeled with different colored fluors and thereby not necessarily be of different size.

The locus D1S80 is an AMP-FLP containing repeat sequences that generally are 16 bps in length; it has been well-defined in the human-identity typing literature *(10)*. The loci CSF1PO, TPOX, and HUMTHO1 are STRs containing tetranucleotide repeat sequences. These STR loci can be coamplified *(14,15)* and generally exhibit less stutter (or shadow) bands than STR loci containing smaller repeat units (i.e., tri- and dinucleotide repeat units) *(see* **Note 1**). Also, the size of the largest allele in the triplex (allele 15 in the CSF1PO locus allelic ladder) is <330 bp in size *(18)*. Thus, forensic biospecimens that contain substantially degraded DNA may be successfully typed. This chapter describes several approaches for typing the AMP-FLPs D1S80, CSF1PO, TPOX, and HUMTHO1 and can serve as a guideline for analyzing these, as well as other, VNTR/STR loci.

2. Materials

1. The primer sequences for amplifying the D1S80 locus *(10)* are:

   D1S80 5'GAA ACT GGC CTC CAA ACA CTG CCC GCC G3' (forward)

   D1S80 5'GTC TTG TTG GAG ATG CAC GTG CCC CTT GC3' (reverse)

2. The primer sequences for amplifying the three STR loci in a multiplex fashion *(15)* are:

   HUMTHO1 5'GTG GGC TGA AAA GCT CCC GAT TAT3' (forward)

   HUMTHO1 5'ATT CAA AGG GTA TCT GGG CTC TGG3' (reveaerse)

   TPOX 5'ACT GGC ACA GAA CAG GCA CTT AGG3' (forward)

   TPOX 5'GGA GGA ACT GGG AAC CAC ACA GGT3' (reverse)

CSF1PO 5'AAC CTG AGT CTG CCA AGG ACT AGC3' (forward)

CSF1PO 5'TTC CAC ACA CCA CTG GCC ATC TTC3' (reverse)

3. PCR buffers, dNTPs, and/or *TAQ* polymerase can be obtained commercially (Perkin-Elmer; Promega Corporation, Madison, WI; Gibco-BRL, Gaithersburg, MD; and so forth).
4. Bovine serum albumin (BSA) (*see* **Note 2**).
5. Tris-sulfate buffer, pH 9.0, is used to provide the leading sulfate ion. A stock solution of 0.09 M Tris-sulfate is prepared using 130 mL of 1 N H_2SO_4 and 89 g Tris diluted to 1 L.
6. Tris-formate buffer, pH 9.0, also can be used to provide a leading ion. A stock solution of 0.09 M Tris-formate is prepared using 3.51 mL concentrated formic acid and 67 g Tris diluted to 1 L.
7. Tris-borate, pH 9.0, is used to provide the trailing and counter ions. A stock solution of 0.42 M Tris-borate is prepared using 33.2 g boric acid and 183 g Tris diluted to 1 L.
8. Tris-serine, pH 9.0, also can be used to provide the trailing ion. A stock solution of 0.29 M Tris-serine is prepared by using 30.2 g serine and 74.4 g Tris diluted to 1 L.
9. Allelic ladders for the D1S80 locus and the STR triplex loci can be obtained commercially from Perkin–Elmer and Promega Corp, respectively.
10. Rehydratable polyacrylamide gels (crosslinker N,N'-methylenebisacrylamide [BIS]) (Micro-Map, Inc., Boca Raton, FL).
11. Acrylamide stock solution: acrylamide (Gibco-BRL), 29.4 g; piperazine diacrylamide (Bio-Rad, Richmond, CA), 0.6 g. Dissolve in 50 mL of distilled water, filter, and make up to a final volume of 100 mL with distilled water.
12. 20% glycerol.
13. Ammonium persulfate.
14. Tetraethylmethylene diamine (TEMED).
15. Whatmann paper strips.
16. Sponge strips (cat. no. 19-3664-01, Pharmacia-LKB, Piscataway, NJ).
17. 10% ethanol.
18. 1% nitric acid.
19. 0.012 M silver nitrate solution.
20. 0.28 M sodium carbonate (anhydrous), 0.019% formalin.
21. 10% glacial acetic acid.
22. Sample loading solution: 25 µg xylene cyanol, 25 µg bromophenol blue, and 4 g sucrose per 10 mL of 120 mM trisformate, pH 9.0.
23. EC Electrophoresis Apparatus 1001 (EC Corporation, St. Petersburg, FL).
24. SA 32 electrophoretic apparatus (Gibco-BRL).
25. FluorImager SI (Molecular Dynamics).

3. Methods

3.1. Extraction

Take aliquots of whole blood from EDTA vacutainer tubes (from venipuncture) or by fingerprick, place on cotton cloth, and air dry. The DNA is extracted by the phenol-chloroform method and washed using microcon 100 filters (Amicon) according to the method of Comey et al. *(19)*. (*see* Chapter 1).

3.2. DNA Quantitation (see Note 3)

The quantity of extracted DNA was estimated using the slot-blot procedure described by Waye et al. *(20)* using D17Z1, a human-specific alphoid probe, and chemiluminescent detection *(21)* (*see* Chapter 5).

3.3. PCR Amplification of the D1S80 Locus (22)

1. The PCR is carried out in 50 μL reaction volumes containing 0.4–5 ng template DNA, 10 mM Tris-HCl, pH 8.3, 50 mM KCl, 1.5 mM MgCl$_2$, 0.001% gelatin, 1 nmole of each of the four deoxyribonucleoside triphosphates, 12.5 pmol of each primer, and 2.5 units of Taq DNA polymerase (*see* **Note 4**).
2. Place the reaction tubes into a Perkin–Elmer 9600 thermal cycler and subject to 27 cycles of denaturation at 95°C for 10 s, primer annealing at 67°C for 10 s, and primer extension at 70°C for 30 s.

3.4. Electrophoretic Separation of Amplified D1S80 Products in Vertical, Native, Discontinuous Polyacrylamide Gels (22) (see Notes 5–7)

1. Prepare an acrylamide solution (7.5%T, 2.0%C; crosslinker was piperazine diacrylamide) and pour between two glass plates separated by 0.4 mm thick spacers. Affix Gel bond (FMC Corp., Rockland, ME) to one plate, such that the hydrophilic side was in contact with the gel. The gel dimensions are 17 × 33.5 cm. The gel buffer is trisformate, pH 9.0, which was 60 mM with respect to the formate ion. Use an 18-tooth comb (cat. no. 11092-095, Gibco-BRL).
2. Allow the gel to polymerize for 1 h at ambient temperature, followed by a minimum incubation of 30 min at 4°C, prior to use.
3. Place the gel in an SA 32 apparatus (Gibco-BRL) and add 250 mL of 28 mM Tris-borate, pH 9.0, (28 mM with respect to the borate ion) to the bottom reservoir. To the top buffer reservoir, add 300 mL of 60 mM trisformate.
4. Mix 4 μL of amplified DNA sample with 2.5 μL of sample loading solution. Apply the entire sample volume to a gel well at the cathodal end of the gel submersed in the Trisformate buffer. After all samples are loaded onto the gel, remove the upper reservoir buffer by draining and replace with 300 mL of 28 mM Tris-borate.
5. Electrophoresis is performed at ambient temperature with settings of 995 V, 200 mA, and 50 W. Allow to continue until the xylene cyanol tracking dye has migrated to the top of the lower reservoir buffer.

3.5. PCR Amplification of the STR Loci CSF1PO, TPOX, and HUMTHO1 (23)

The coamplification of HUMTHO1, TPOX, and CSF1PO is performed using the GenePrint kit (Promega) according to the following conditions.

1. PCR is carried out in 25 or 50 µL reaction volumes containing 0.1–5 ng template DNA, 10 mM Tris-HCl, pH 9.0, 50 mM KCl, 1.5 mM MgCl$_2$, 0.1% Triton X-100, 200 µM of each of the four deoxyribonucleoside triphosphates, 12.5 pmol of each primer, and 2.5 U of *Taq* DNA polymerase per 50 µL reaction.
2. Place the reaction tubes into a Perkin-Elmer 9600 thermal cycler and subject to denaturation at 95°C for 30 s, primer annealing at 67°C for 30 s, and primer extension at 70°C for 30 s, for a total of 28 or 30 cycles, depending on the initial quantity of template DNA.

3.6. Electrophoretic Separation of Amplified STR Products in Vertical, Denaturing, Discontinuous Polyacrylamide Gels (23) (see Note 7)

1. Prepare a polyacrylamide gel (6%T, 2%C; crosslinker, piperazine diacrylamide; 31 cm long and 0.4 mm thick) containing 7 M urea and 60 mM tris-formate, pH 9.0 (with respect to the formate ion). Allow the gel to polymerize for a minimum of 1 h at ambient temperature. Place the gel in a SA 32 apparatus (Gibco-BRL). The electrode buffer is 90 mM Tris-borate, pH 8.3 (90 mM with respect to the borate ion).
2. Mix 3 µL of sample loading dye with 3 µL PCR product. Denature the samples for 2 min in a Perkin-Elmer Model 480 DNA thermal cycler and load 5 µL onto the cathodal end of the gel.
3. Electrophoresis is performed initially at 80 W for approx 5 min, and then continued with settings of 25 W at ambient temperature. Allow the run to continue until the xylene cyanol tracking dye migrates to the top of the lower reservoir buffer (approx 3 h).

3.7. Rehydratable Gels (24–28) (see Note 8)

Alternatively, PCR-generated DNA fragments can be separated electrophoretically in precast rehydratable gels (Micro-Map, Inc., Boca Raton, FL).

1. The empty (i.e., devoid of polymerization byproducts and buffer), dried polyacrylamide gels, which are cast on a mylar support, are submerged, gel-side down, in a leading ion buffer of choice. Rehydration is accomplished in 1 h. After rehydration, the excess surface liquid is removed by using a piece of polyester film as a squeegee.
2. Gels of the appropriate pore size, e.g., 7%T, 3%C may be rehydrated in a variety of buffers containing the leading ions, such as chloride, formate, or sulfate ranging from 30 to 60 mM. Mobility modifiers *(27)*, such as glycerol or ribose, may be added when the trailing ion is borate. Alternatively, continuous buffer systems may also be employed.
3. Samples may be loaded directly onto the surface of the horizontal rehydrated gel by either of the following techniques:
   a. Schleicher and Schuell nylon DNA loading tabs (dimensions depend on the sample volume) are placed on the gel surface 1.0 cm anodal to the cathode

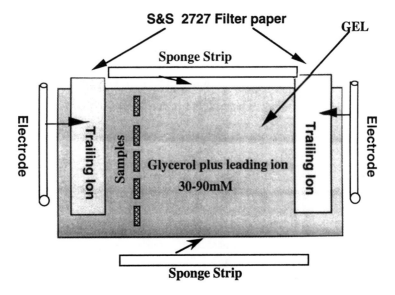

Fig. 1. Schematic setup of an open-faced horizontal polyacrylamide-gel system for the separation of single- and double-stranded nucleic acids. Sponge strips (cat. no. 19-3664-01, Pharmacia-LKB, Piscataway, NJ) are used along both sides of the gel to prevent drying at the edge, which can result in a "smile" effect.

wick edge. Between 0.25 and 10 µL of sample is applied to the tabs with a micropipet.
b. Samples may be loaded directly on the gel surface using a micropipet.
4. Gels are then run open-faced in a horizontal position as shown in **Fig. 1**. In this electrophoretic system, the temperature is controlled with Peltier cooling devices under the platen on which the rehydrated gel is placed. Separations of 20 cm are carried out for 2.5 h at 20°C at a constant current of 7 mA. Water-dampened sponge strips can be placed along the edges of the gel to prevent a "smile" effect from extended separation times, particularly when using high-ionic-strength buffers or 7 *M* urea in the gel.

3.8. Silver Staining

After electrophoresis, gels can be stained with silver according to the method of Budowle et al. *(5)* (*see* **Note 9**) and **Fig. 2**.

1. Oxidize the gel in 1% nitric acid solution for 3 min.
2. Briefly rinse the gel in distilled water.
3. Briefly rinse the gel in a 0.012 *M* silver-nitrate solution for 10–20 min.
4. Decant silver solution, rinse briefly in water, and reduce the gel in a 0.28 *M* sodium carbonate (anhydrous) and 0.019% formalin solution. Change solution when it turns

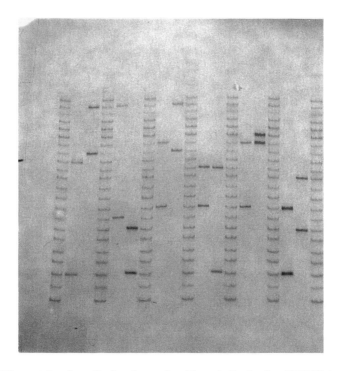

Fig. 2. Silver-stained vertical polyacrylamide gel displaying D1S80 types. From left to right the D1S80 types are: allelic ladder; 29-18; 36-30; allelic ladder; 36-23; 22-18; allelic ladder; 31-24; 36-30; allelic ladder; 28-24;28-18; allelic ladder; 31-24; 32-31; allelic ladder; 24-18; 27-22; allelic ladder. The cathode is at the top.

    brown. Continue to change every few minutes until the DNA bands show up well.
5. Stop reduction (and thus image development) by placing gel in 10% glacial acetic acid for 5 min.
6. Wash gel in distilled water for 10 min.
7. Place gel in 3% glycerol solution for 5 min.
8. Air dry or photograph the gel.

3.9. Fluor Detection

If one of the primer pairs was labeled at the 5' end with a fluorescent tag, the fluor-labeled amplicons can be detected using the FluorImager SI (Molecular Dynamics, Foster City, CA) with the PMT set at 1000 or the FMBIO 100 (Hitachi, San Bruno, CA).

4. Notes

1. Stutter bands: The tetranucleotide STR loci were selected for analysis because the alleles can be resolved by polyacrylamide gel electrophoresis *(11,12,14,15)*,

and these loci generally exhibit less stutter (or shadow) bands than STR loci containing smaller repeat-size sequences. The stutter bands are due to strand slippage and are much less intense than the true allelic product. In general, these stutter bands do not complicate interpretation of STR profiles, but should be evaluated carefully when there are mixed samples (when, for example, the two contributions in a mixed sample are not at equal concentration). The major component generally can be interpreted easily, but complications may arise when the minor component has bands of similar intensity as the stutter bands of the major component.

2. Inhibition of PCR may be overcome by the addition to the reaction of BSA (cat. no. A3350, Sigma, St. Louis, MO) at a concentration of 160 µg/mL *(29,30)*. The BSA may bind (a) soluble inhibitory factor(s) that copurify with DNA *(31)*, and/or BSA may stabilize the *Taq* polymerase. The source and quality of BSA can impact on effectivity.

3. The D1S80, as well as the STR, typing systems display a high sensitivity of detection. PCR samples containing from 4 ng to as little as 125 pg of template DNA can be typed readily. However, the slot-blot technique is a semiquantitative method for determining the quantity of DNA in a sample, and there is a need to avoid stochastic effects during PCR. Therefore, it is recommended that, for most forensic purposes, a minimum of approx 400 pg template DNA be used for the PCR.

4. The PCR conditions for amplifying the STR loci used in this study differ from those recommended by the manufacturer (Promega Corp). A higher annealing temperature of 67°C (i.e., a higher stringency) instead of 64°C during amplification did not compromise typing efficiency (**Fig. 3**). A fluorescently tagged quadplex kit also can be used (**Fig. 4**). The quadplex contains the same three loci as the triplex, with the addition of the locus VWF.

   In order to obtain amplification of all four STR loci, the annealing temperature in the PCR is reduced to 60°C. Thus, an annealing temperature of 60°C is recommended when attempting to multiplex VWA with the other three STR loci.

5. **Figure 2** shows a typical separation of amplified D1S80 alleles by high-resolution discontinuous vertical polyacrylamide-gel electrophoresis. With the vertical gel format, D1S80 variants tend to migrate to the nearest step in the allelic ladder. However, Budowle et al. *(22)* described the presence of anodal and cathodal D1S80 variants detected using a horizontal gel-electrophoresis system. The AMP-FLPs separated in vertical gels are in a warmer environment (45–50°C) compared with those separated in horizontal gels (15°C). Conformational differences in the DNA fragments resulting from sequence polymorphisms are less likely to manifest themselves during electrophoresis in a warmer gel because secondary structures are denatured. This observation suggests that most of the anodal and cathodal variants are caused by sequence polymorphisms.

6. Data obtained by typing the D1S80 locus can be useful for human-identity testing purposes. However, the method described here only enables typing D1S80 singly.

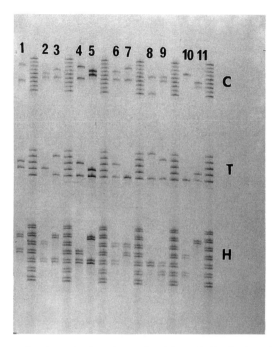

Fig. 3. Silver-stained vertical polyacrylamide gel displaying STR triplex loci—CSF1PO (C), TPOX (T), HUMTHO1 (H). The types from left to right for CSF1PO are: Sample 1,13-9; allelic ladder; Sample 2, 11-10; Sample 3, 12-10; allelic ladder; Sample 4, 13-10; Sample 5, 12-11; allelic ladder; Sample 6, 12-10; Sample 7, 13-10; allelic ladder; Sample 8, 11-7; Sample 9, 11-10; allelic ladder; Sample 10, 12-12; Sample 11, 10-9; allelic ladder; for TPOX are: Sample 1, 10-9; allelic ladder; Sample 2, 9-9; Sample 3, 11-8; allelic ladder; Sample 4, 10-8; Sample 5, 9-8; allelic ladder; Sample 6, 10-8; Sample 7, 8-8; allelic ladder; Sample 8, 12-8; Sample 9, 11-8; allelic ladder; Sample 10, 8-8; Sample 11, 9-8; allelic ladder; HUMTHO1 are: Sample 1, 9.3-8; allelic ladder; Sample 2, 9-7; Sample 3, 9.3-7; allelic ladder; Sample 4, 8-7; Sample 5, 9.3-7; allelic ladder; Sample 6, 9-7; Sample 7, 9-8; allelic ladder; Sample 8, 7-7; Sample 9, 7-6; allelic ladder; Sample 10, 8-6; Sample 11, 9.3-9.3; allelic ladder. The cathode is the top. The allelic ladders were obtained from the Promega Corp. (Madison, WI).

The advantages of a multiplex system are that less template DNA is consumed than when analyzing each locus independently, less reagents are consumed, and labor is reduced. As an example, Budowle et al. *(32)* demonstrated that the D1S80 locus can be analyzed simultaneously with the amelogenin gene. Typing the amelogenin gene enables determination of the sex of the contributor of a biological sample *(33,34)*. The only differences in the protocols for typing the D1S80 locus individually as compared with the multiplex fashion with the amelogenin

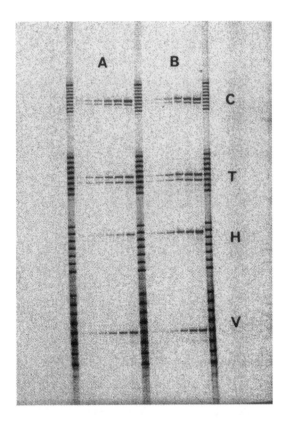

Fig. 4. Fluor-labeled STR quadplex loci—CSF1PO (C), TPOX (T), HUMTHO1 (H), and VWA (V)—on a vertical polyacrylamide gel. The samples are a dilution series using different amounts of template DNA in the PCR. From left to right the amounts of template DNA are: 0.15 ng; 0.30 ng; 0.60 ng; 1.25 ng; 2.5 ng; and 5.0 ng. A and B are the same dilution series, except samples in A were generated in a 50 μL PCR, and samples in B were generated in a 25 μL PCR.

locus are the addition of amelogenin primers to the D1S80 PCR, and the acrylamide concentration in the analytical gel is increased from 7.5%T to 8.5%T.

7. For a discussion on discontinuous gels and matrix modifiers, *see* **refs. *26,27*.**
8. Separations of DNA fragments on rehydratable gels are shown in **Figs. 5** and **6** and densitometric scans of **Fig. 6** are shown in **Figs. 7** and **8**. Horizontal rehydratable polyacrylamide gel electrophoresis systems for the separation of PCR-amplified products have been developed specifically to take advantage of the high resolution, discontinuous buffer techniques, and to provide flexibility of use that is not possible with conventionally cast gel systems. To provide the best possible electrophoretic conditions and reproducibility, these

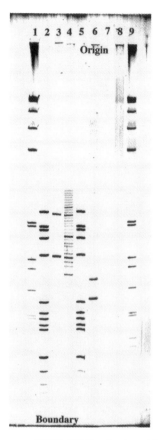

Fig. 5. A 20 cm separation on a 7%T, 3%C rehydratable polyacrylamide gel using 30 m*M* formate as the leading ion and serine as the trailing ion, with 1.34 *M* glycerol as moisturizing agent. The separation was carried out in 2 h of constant pulsed power at 20 pulses/s. Sample 1, 1 kb ladder; Samples 2 and 5, pBR322 ladder; Sample 3, D1S80 31-18; Sample 4, DIS80 allelic ladder; Sample 6, SE 33 (or ACTBP2) heterozygote; Samples 7 and 8, blank; Sample 9, 1 kb ladder. The gel was stained with silver.

gels are washed free of all polymerization byproducts following casting and then dried for storage at room temperature. These gels afford a number of significant advantages over conventionally cast gel systems. First, with commercially available rehydratable gels, the scientist is not exposed to potentially toxic acrylamide monomer. Second, a greater degree of reproducibility is possible; many gels can be made at one time from a single batch of reagents, and potentially deleterious polymerization byproducts are removed prior to initial dehydration. If not removed, these unwanted contaminants,

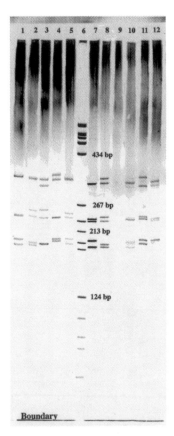

Fig. 6. A 20 cm separation of the STR Triplex loci on a 7%T, 3%C rehydratable polyacrylamide gel using 25 mM sulfate as the leading ion with borate as the trailing ion and 1.34 M glycerol as a mobility modifier. The electrophoretic separation was carried out at 7.5–10 mA constant current for 3.5 h.

most notably sulfate ion, derived from the breakdown of the common polymerization catalyst persulfate, can produce an unintentional multizonal electrophoresis system. This observation, in fact, led in part to the original "Disc" system described by Ornstein and Davis *(35)*. Perhaps the most important advantage is that the pre-prepared, empty, dried gels may be rehydrated in a variety of reagents that otherwise could interfere with the polymerization process.

Although the E-C Apparatus 1001 system is used here, any horizontal system designed for isoelectric focusing may be used, as long as the temperature is held constant during the electrophoretic run. Care must be taken not to use a platen temperature so low that moisture condenses on the surface of the gel. This can decrease resolution and result in surface smearing and band distortion.

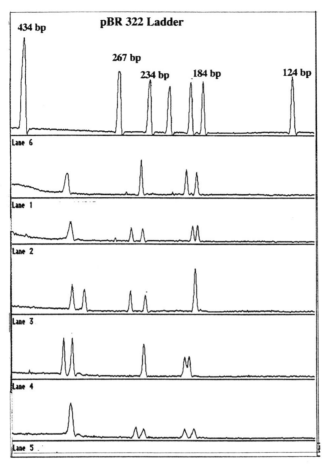

Fig. 7. Densitometric traces of Samples 1–6 from the gel shown in **Fig. 6**.

9. With silver staining, both fragments of the denatured STR products may be observed, depending on the conditions of the separation method. In the denaturing electrophoretic approach described in this chapter, the HUMTHO1-denatured, single-strand products are resolved, whereas the single strands for the CSF1PO and TPOX products are not separated. When one of the two primers for amplification of a locus is fluor-labeled, silver staining is not recommended as a detection method. The addition of a fluor molecule to the 5' end of one of the strands in the duplex (when denatured), will result in an altered migration of that strand, compared with the unlabeled complementary strand.

Acknowledgments

This is Publication Number 96-06 of the Laboratory Division of the Federal Bureau of Investigation. Names of commercial manufacturers are provided for

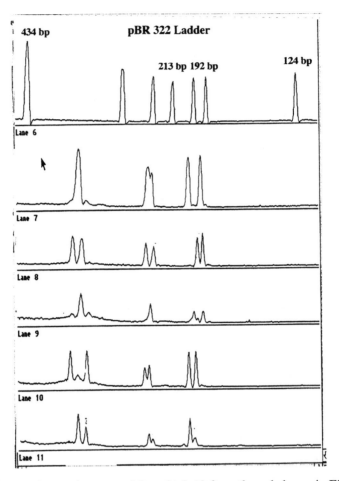

Fig. 8. Densitometric traces of Sample 6–12 from the gel shown in **Fig. 6**.

identification only, and inclusion does not imply endorsement by the Federal
Bureau of Investigation.

References

1. Wyman, A. R. and White, R. (1980) A high polymorphic locus in human DNA.
   Proc. Natl. Acad. Sci. USA **77,** 6754–6758.
2. Jeffreys, A. J., Wilson, V., and Thein, S. L. (1985) Hypervariable minisatellite
   regions in human DNA. *Nature* **314,** 67–73.
3. Jeffreys, A. J., Wilson, V., and Thein, S. L. (1985) Individual-specific fingerprints of human DNA. *Nature* **316,** 76–79.
4. Saiki, R. K., Scharf, S., Faloona, F., Mullis, K. B., Horn, G. T., Erlich, H. A.,
   and Arnheim, N. (1985) Enzymatic amplification of beta-globin genomic

sequences and restriction analysis for diagnosis of sickle cell anemia. *Science* **230,** 1350–1354.

5. Budowle, B., Chakraborty, R., Giusti, A. M., Eisenberg, A. J., and Allen, R. C. (1991) Analysis of the VNTR locus D1S80 by the PCR followed by high-resolution PAGE. *Am. J. Hum. Genet.* **48,** 137–144.

6. Jeffreys, A. J., Wilson, V., Neumann, R., and Keyte, J. (1988) Amplification of human minisatellites by the polymerase chain reaction: towards DNA fingerprinting of single cells. *Nucleic Acids Res.* **16,** 10,953–10,971.

7. Horn, G. T., Richards, B., and Klinger, K. W. (1989) Amplification of a highly polymorphic VNTR segment by the polymerase chain reaction. *Nucleic Acids Res.* **17,** 2140.

8. Boerwinkle, E., Xiong, W., Fourest, E., and Chan, L. (1989) Rapid typing of tandemly repeated hypervariable loci by the polymerase chain reaction: application to the apolipoprotein B 3' hypervariable region. *Proc. Natl. Acad. Sci. USA* **86,** 212–216.

9. Ludwig, E. H., Friedl, W., and McCarthy, B. J. (1989) High-resolution analysis of a hypervariable region in the human apolipoprotein B gene. *Am. J. Hum. Genet.* **48,** 458–464.

10. Kasai, K., Nakamura, Y., and White, R. (1990) Amplification of a variable number of tandem repeat (VNTR) locus (pMCT118) by the polymerase chain reaction (PCR) and its application to forensic science. *J. Forensic Sci.* **35,** 1196–1200.

11. Edwards, A., Civitello, A., Hammon, H. A., and Caskey, C. T. (1991) DNA typing and genetic mapping with trimeric and tetrameric tandem repeats. *Am. J. Hum. Genet.* **49,** 746–756.

12. Edwards, A., Hammond, H., Jin, L., Caskey, C. T., and Chakraborty, R. (1992) Genetic variation at five trimeric and tetrameric repeat loci in four human population groups. *Genomics* **12,** 241–253.

13. Hochmeister, M. N., Jung, M. M., Budowle, B., Borer, U. V., and Dirnhofer, R. (1994) Swiss population data on three tetrameric short tandem repeat loci—VWA, HUMTHO1, and F13A1—derived using multiplex PCR and laser fluorescence detection. *Int. J. Leg. Med.* **107,** 34–36.

14. Hochmeister, M. N., Budowle, B., Borer, U. V., and Dirnhofer, R. (1995) Swiss population data on three tetrameric short tandem repeat loci—HUMTHO1, TPOX, CSF1PO—derived using the GenePrint Triplex PCR Amplification Kit. *Int. J. Legal Med.* **107,** 246–249.

15. Huang, N. E., Schumm, J., and Budowle, B. (1995) Chinese population data on three tetrameric short tandem repeat loci—HUMTHO1, TPOX, AND CSF1PO—derived using multiplex PCR and manual typing. *Forensic Sci. Int.* **71,** 131–136.

16. Gill, P., Kimpton, C. P., and Sullivan, K. (1992) A rapid method for identifying fixed specimens by DNA profiling. *Electrophoresis* **13,** 173–175.

17. Sullivan, K. M., Pope, S., Gill, P., and Robertson, J. M. (1992) Automated DNA profiling by fluorescent labeling of PCR products. *PCR Meth. Appl.* **2,** 34–40.

18. *GenePrint Fluorescent STR Systems—Technical Manual.* (1994) Promega Corporation, Madison, WI.

19. Comey, C. T., Koons, B. W., Presley, K. W., Smerick, J. B., Sobieralski, C. A., Stanley, D. M., and Baechtel, F. S. (1994) DNA extraction strategies for amplified fragment length polymorphism analysis. *J. Forensic Sci.* **39,** 1254–1269.

20. Waye, J. S., Presley, L., Budowle, B., Shutler, G. G., and Fourney, R. M. (1989) A simple method for quantifying human genomic DNA in forensic specimen extracts. *Biotechniques* **7,** 852–855.

21. Giusti, A. M. and Budowle, B. (1995) Chemiluminescence-based detection system for human DNA quantitation and restriction fragment length polymorphism (RFLP) analysis. *Appl. Theoret. Electrophor.* **5,** 89–98.

22. Budowle, B., Baechtel, F. S., Smerick, J. B., Presley, K. W., Giusti, A. M., Parsons, G., Alevy, M., and Chakraborty, R. (1995) D1S80 population data in African Americans, Caucasians, Southeastern Hispanics, Southwestern Hispanics, and Orientals. *J. Forensic Sci.* **40,** 38–44.

23. Budowle, B., Koons, B. W., Keys, K. M., and Smerick, J. B. (1996) Methods for typing the STR triplex CSF1PO, TPOX, and HUMTHO1 that enable compatibility among DNA typing laboratories, in *Advances in Forensic Haemogenetics, vol. 6* (Carracedo, A., Brinkman, B., and Bar, W., eds.), Springer-Verlag, Berlin, pp. 107–114.

24. Allen, R. C., Graves, G., and Budowle, B. (1989) Polymerase chain reaction amplification products separated on rehydratable polyacrylamide gels and stained with silver. *Biotechniques* **7,** 736–744.

25. Allen, R. C. and Graves, G. M. (1990) Rehydratable gels: a potential reference standard support for electrophoresing PCR-amplified DNA. *Bio/Technol.* **8,** 1288–1290.

26. Budowle, B. and Allen, R. C. (1991) Discontinuous polyacrylamide gel electrophoresis of DNA fragments, in *Methods in Molecular Biology, vol. 9: Protocols in Human Molecular Genetics* (Mathew, C. G, ed.), Humana, Clifton, NJ, pp. 123–132.

27. Allen, R. C., Budowle, B., and Reeder, D. J. (1993) Resolution of DNA in the presence of mobility modifying polar and nonpolar compounds by discontinuous electrophoresis on rehydratable polyacrylamide gels. *Appl. Theoret. Electrophor.* **3,** 173–181.

28. Allen, R. C. and Goldberg, C. L. (1995) Forensic applications of gel electrophoresis. *Am. Biotechnol. Lab.* **13,** 8–12.

29. Hagelberg, E., Sykes, B., and Hedges, R. (1989) Ancient bone DNA amplified. *Nature* **342,** 485.

30. Hochmeister, M. N., Budowle, B., Borer, U. V., Eggmann, U. T., Comey, C. T., and Dirnhofer, R. (1991) Typing of DNA extracted from compact bone tissue from human remains. *J. Forensic Sci.* **36,** 1649–1661.

31. Rogan, P. K. and Salvo, J. (1990) Study of nucleic acids isolated from ancient remains. *Yearbook Phys. Anthropol.* **33,** 195–214.

32. Budowle, B., Koons, B. W., and Errera, J. D. (1996) Multiplex amplification and typing procedure for the loci D1S80 and amelogenin. *J. Forensic Sci.* **41,** 660–663.

33. Akane, A., Seki, A., Shiono, H., Nakamura, H., Hasegawa, M., Kagawa, M., Matsubara, K., Nakahori, Y., Nagafuchi, S., and Nakagome, Y. (1992) Sex determination of forensic samples by dual PCR amplification of an X-Y homologous gene. *Forensic Sci. Int.* **52**, 143–148.
34. Sullivan, K., Mannucci, A., Kimpton, C. P., and Gill, P. (1993) A rapid and quantitative DNA sex test: fluorescence based PCR analysis of X-Y homologous gene amelogenin. *Biotechniques* **15**, 636–642.
35. Ornstein, L. and Davis, B. J. (1962) *Disc Electrophoresis*, Parts 1 and 2. Distillation Ind., Rochester, NY.

15

Amplification of Short Tandem Repeat Loci Using PCR

Niels Morling

1. Introduction
1.1. Background

Short tandem repeat (STR) loci consist of repetitive elements of 3–7 nucleotides. The STR loci, which are numerous in the human genome, are highly polymorphic in length and may also vary in the sequences of the repetitive elements. The polymerase chain reaction (PCR) makes it possible to analyze very small amounts (nanograms) of DNA. STR loci from partly degraded DNA may be successfully amplified because the STR loci to be amplified are very short, only 100–400 bp. The PCR method is much faster than previous DNA methods. Investigations of the polymorphism of STR loci using PCR, thus, is of great value in identity and genetic testing.

In forensic work, STRs with four or five basepair repeats are most commonly used (1–6). The detection of the polymorphisms of STR loci is primarily based on the analysis of length polymorphisms by means of electrophoresis in polyacrylamide gels (PAGE) or, less frequently, in agarose gels. The electrophoretic migration of the STR DNA fragments is determined using silver or ethidium bromide staining, radioactivity, enzyme labeling, fluorescence detection, and others. Analysis of the sequence polymorphism may add further information, but sequencing is not commonly used in forensic investigations today.

1.2. General Strategy

The PCR makes it possible to amplify one or more STR loci simultaneously in one tube (multiplexing). Amplification of only one highly polymorphic STR locus (e.g. [5,6]), is mostly used if priority is given to sensitivity and robustness of the test. If two or more STR loci are to be amplified reliably together,

From: *Methods in Molecular Biology, Vol. 98: Forensic DNA Profiling Protocols*
Edited by: P. J. Lincoln and J. Thomson © Humana Press Inc., Totowa, NJ

Table 1
Examples of STR Loci Commonly Used in Forensic Genetics

STR locus	Chromosomal location	Related locus	Repeat sequence 5'-3'
HUMFES/FPS *(7)*	15q25-qter	c-fes/fps proto-oncogene	AAAT
HUMF13A1 *(8)*	6p24-25	Coagulation factor XIII a subunit gene	AAAG
HUMTH01 'TC11' *(9)*	11p15.5	Tyrosine hydroxylase gene, first intron	AATG
HUMVWA31A *(10)*	12p12-pter	von Willebrand factor gene, intron 40	AGAT
HUMCSF1PO *(4)*	5q33.3-34	c-fms proto-oncogene for CSF-1 receptor gene	AGAT
HUMTPOX *(11)*	2p23-2pter	Thyroid peroxidase gene	AATG
HUMACTBP2 'SE33' *(12)*	Chromosome 6	Beta-actin related pseudo-gene H-beta-Ac-psi-2	AAAG[a]
D21S11 (13)	21q11-21	None	TCTA[a]

[a]The main repeat unit. The sequences of these STR loci are complex.

the amplification parameters of the STR loci must be compatible. Especially the annealing temperature is important. Multiplexing of STRs, in general, decreases the sensitivity of the investigation to some extent compared to amplification of only one STR locus.

If STR loci are multiplexed, the DNA fragments of the different STR loci must be separated either by their electrophoretic migration or by other characteristics, such as the color of fluorescent markers. If the analysis depends on only the electrophoretic migration, it is important that the alleles of the different STR loci do not overlap each other in the gel *(4)* (*see* **Note 1**). If one has access to equipment with color-detection facilities, the primers of different STR loci may be labeled with dyes of different colors and, thus, two or more STR loci with overlapping allele sizes can be investigated *(1,2)*.

An example of four partly overlapping STR loci that can be amplified together and detected in a two- (or four-) color florescence system are HUMFES/FPS, HUMF13A1, HUMTH01, and HUMVWA31A *(1)* (*see* **Note 2** and **Table 1**). This quadruplex will be used as an example of amplification of STR loci using PCR. If one wants to amplify only one of these STR loci, basically the same method may be used.

Other STR loci may be PCR-amplified under very similar conditions. HUMCSF1PO, HUMTH01, and HUMTPOX *(3)* (*see* **Note 3** and **Table 1**) are examples of three nonoverlapping STR loci that can be amplified together and

Table 2
Primer Sequences of STR Loci Commonly Used in Forensic Genetics

STR Locus	Primers
HUMFES/FPS *(7)*	5'- -GGG ATT TCC CTA TGG ATT GG-3'[a]
	5'FAM-GCG AAA GAA TGA GAC TAC AT-3'[b,c]
HUMF13A1 *(8)*	5'- -GAG GTT GCA CTC CAG CCT TT-3'
	5'HEX-ATG CCA TGC AGA TTA GAA A.-3'[c]
HUMTH01 *(9)*	5'- -GTG ATT CCC ATT GGC CTG TTC CTC-3'
	5'FAM-GTG GGC TGA AAA GCT CCC GAT TAT 3'[c]
HUMVWA31A	5'- -GGA CAG ATG ATA AAT ACA TAG GAT GGA TGG 3'
(10)	5'HEX-CCC TAG TGG ATG ATA AGA ATA ATC AGT ATG 3'[c]
HUMCSF1PO *(4)*	5'-AAC CTG AGT CTG CCA AGG ACT AGC-3'
	5'-TTC CAC ACA CCA CTG GCC ATC TTC-3'
HUMTPOX *(11)*	5'-ACT CGC ACA GAA CAG GCA CTT AGG-3'
	5'-GGA GGA ACT GGG AAC CAC ACA GGT-3'
HUMACTBP2	5'-AAT CTG GGC GAC AAG AGT GA-3'
(12)	5'-ACA TCT CCC CTA CCG CTA TA-3'
D21S11 *(13)*	5'-ATA TGT GAG TCA ATT CCC CAA G-3'
	5'-TGT ATT AGT CAA TGT TCT CCA G-3'

[a]Forward primer corresponding to the AAAG strand.
[b]Reverse primer corresponding to the CTTT strand.
[c]As an example, one primer of each STR locus has been labeled with a fluorescent marker (*see* **Note 18**).

detected in, e.g., silver-stained polyacrylamide gels. ACTBP2 *(5)* and D21S11 *(6)* (**Table 1**) are examples of highly polymorphic STRs that are commonly amplified alone.

2. Materials
2.1. Equipment
1. Laminar air flow sterile bench (*see* **Note 4**).
2. Thermocycler (*see* **Note 5**).
3. Thermal block, 100°C.

2.2. Reagents
1. HPLC purified primers, 10 μM each. Store at –20°C. *See* **Table 2** for examples of primer sequences and labeling with fluorescent dyes for multiplexing (*see* **Note 6**).
2. Nucleotides (dATP, dCTP, dGTP, and dTTP), 10 mM each. Store at –20°C.
3. Thermostable *Taq* DNA polymerase 5 U/µL. Store at –20°C (*see* **Note 7**).
4. Nuclease-free, sterile water (*see* **Note 8**).
5. Mineral oil.

2.3. Solutions

1. 10X STR-buffer: 100 mM Tris-HCl, pH 8.3, 500 mM KCl, 15 mM MgCl$_2$, 10% Triton X-100. Autoclave and store at –20°C.
2. Primer-mix: 25 mM Tris-HCl, pH 8.3, 125 mM KCl, 3.75 mM MgCl$_2$, 2.5% Triton X-100, 0.5 mM dNTP (each), 500 nM primer (each). Store at –20°C.
3. PCR-mix: 10 mM Tris-HCl, pH 8.3, 50 mM KCl, 1.5 mM MgCl$_2$, 1% Triton X-100, 0.2 mM dNTP (each), 200 nM primer (each), 0.025 U/µL *Taq* DNA polymerase.

2.4. Preparation of Primer-Mix

1. Primer-mix for a quadruplex, one reaction: 3 µL sterile water, 5 µL 10X STR-buffer, 1 µL of each dNTP (10 mM), 1 µL of each primer (10 µM) (*see* **Note 9**). Total volume 20 µL per reaction.
2. Primer-mix for one STR locus, one reaction: 9 µL sterile water, 5 µL 10X STR-buffer, 1 µL of each dNTP (10 mM), 1 µL of each primer (10 µM). Total volume 20 µL per reaction.

If other numbers of STR loci are to be amplified together, use the same amount of the relevant primer pairs and adjust the total volume by adding more or less sterile water.

2.5. Additional Items

1. Disposable, aerosol-resistant pipet tips.
2. Eppendorf tubes (1.5 mL).
3. PCR reaction tubes (0.5 mL) with cap, e.g., GeneAmp Thin Walled Reaction Tubes (Perkin–Elmer, Norwalk, CT).
4. Powder-free, disposable gloves.
5. Ice.

3. Methods

3.1. Setting Up the PCR

Precautions should be taken to avoid crosscontaminations (*see* **Note 10**).

1. Calculate the amounts of reagents to be used by multiplying the amounts needed for one test by the number of tests and add, e.g., three tests to compensate for loss during pipetting. For each reaction, use the amounts of reagents stated below.
2. Thaw the primer-mix at room temperature.
3. Pipet the required volume of primer-mix in a reaction tube (20 µL per reaction).
4. Add the required volume of sterile water (27.75 µL per reaction).
5. Place the reaction tube in a thermal block at 100°C for 2 min.
6. Place the reaction tube on ice for 5 min.
7. Add 0.25 µL *Taq* DNA polymerase 5 U/µL, mix, and keep on ice. This is the PCR-mix.
8. Label the PCR reaction tubes (0.5 mL).

9. Place the PCR reaction tubes in a rack on ice for 10 min.
10. Pipet 48 µL of the PCR-mix into each tube.
11. Add 2 µL DNA extract (*see* **Note 11**). Total volume 50 µL.
12. Add 2 µL sterile water (instead of template DNA) to the reaction tube of the negative control.
13. Add 2 µL DNA from a positive control (instead of test DNA) to the reaction tube of the positive control (*see* **Note 12**).
14. Add 1 drop (approx 30 µL) mineral oil.
15. Close the tubes.
16. Perform the amplification procedure in a thermocycler.

3.2. PCR Amplification Protocol

3.2.1. Programming the Thermocycler

Program the thermocycler with the following parameters:

1. Incubate at 95°C for 60 s (*see* **Note 13**).
2. Ramp from 95°C to 54°C for 120 s.
3. Incubate at 54°C for 60 s (*see* **Note 14**).
4. Incubate at 72°C for 60 s.
5. A total of 28 cycles (*see* **Note 15**).
6. Extend the last incubation at 72°C to 10 min (*see* **Note 16**).
7. Lower the temperature to 4°C, which is kept until removal of the tubes.

This program can also be used for amplification of single STR loci of the quadruplex. ACTBP2 and D21S11 can be amplified with this program if the annealing temperature (**step 3**) is set to 60°C, and the triplex with HUMCSF1PO, HUMTH01, and HUMTPOX can be amplified at an annealing temperature of 64°C. **Step 2** can be omitted for the amplification of ACTBP2, D21S11, and the triplex with HUMCSF1PO, HUMTH01, and HUMTPOX.

3.2.2. Performing the Thermocycling

1. Select and run the temperature program.
2. When the thermocycler has reached 95°C, place the tubes into the thermocycler and continue the thermal cycling process.
3. After the completion of thermal cycling process, store the samples at 4°C (*see* **Note 17**).

4. Notes

1. If the sizes of the DNA fragments of two STR loci overlap, the size range of the alleles of the STR loci may be adjusted by changing the positions of the primers.
2. A number of multiplex systems that are detectable with multicolor detection systems have been described, (*1,2*) and some are commercially available.
3. A number of multiplex systems that can be detected in one-color detection systems, e.g., silver staining, have been described, (*3,4*) and some are commercially available.

4. The sterile bench can be decontaminated with hypochlorite-containing cleaning substances and ultraviolet light (254 nm) during nonactive periods.

5. Thermocyclers from a number of manufacturers are available. Cycling parameters should be optimized for individual thermocyclers. The amplification parameters presented here were optimized for use with the Perkin–Elmer GeneAmp PCR System 9600 and analysis at an ABD DNA Sequencer 373 equipped with GENESCAN software.

6. The use of STR loci for identity and genetic testing is the subject of patents in some countries.

7. The PCR and *Taq* DNA polymerase are the subject of patents in a number of countries.

8. The water should be purified by ultrafiltration or double distillation and autoclaved.

9. When two or more STR loci are amplified together (multiplexed), the PCR reactions of each STR locus may need to be balanced in order to obtain equal detection signals from the different loci *(1,4)*. This can be done by testing DNA from individuals who are heterozygous at all STR loci investigated. If the signal from a locus is too strong, the amplification of the STR locus may be reduced by decreasing the concentrations of both primers for the STR locus. Similarly, increased concentrations of the primers of a STR locus may increase the signal. For the quadruplex, we presently use 260 n*M* of HUMF13A1 and HUMVWA31A primers and 130 n*M* of HUMFES/FPS and HUMTH01 primers.

10. To minimize the risk of contamination of pre-PCR material with amplified DNA, a number of precautions are recommended. Pre-PCR and post-PCR activities should be performed in different areas and with different equipment. Measures to prevent crosscontamination resulting from manipulation or DNA-containing aerosols should be taken by using disposable gloves, aerosol-resistant pipets, and so forth. The setup of the PCR is recommended to be performed in a laminar airflow sterile bench using disposable or autoclaved equipment and DNA-free reagents, buffers, and solutions. When dealing with forensic samples, it is recommended to separate the work into three areas for handling of samples, and so forth, set up of PCR, and amplification and analysis of the PCR product. A negative control with all reagents except template DNA should be included in each amplification procedure.

11. Approximately 1–25 ng DNA may be amplified with these protocols. For the analysis with the ABD 373 and ABD 377, only 1–2 ng target DNA is necessary. Different preparations of DNA may be used for PCR amplification. In our experience, phenol-chloroform extraction leads most consistently to good results. In crime case work, we routinely use Chelex preparations of template DNA. We have found it important to standardize the amount of DNA to 1 ng per PCR for both single STR and quadruplex amplification, for fluorescence detection on an ABD DNA Sequencer. We routinely estimate the amount of DNA by a semiquantitative slot-blot method. The preparation of DNA from stains may result in low concentrations of DNA in a large volumes. This may be compensated by using a corresponding lower amount of sterile water in the PCR-mix. DNA preparations in TE

buffer (10 m*M* Tris-HCl, 1 m*M* EDTA, pH 7.5) should not exceed 20% of the total reaction volume, because the amplification efficiency and quality can be greatly altered by changes in pH and the available magnesium ion concentration. In paternity testing, we routinely use 2 µL supernatant of a Chelex preparation of whole blood corresponding to 1–4 ng DNA without measuring the amount of DNA. If the amount of PCR product is not sufficient for reliably detection in silver-stained polyacrylamide gels, the amount of target DNA may be increased.

12. Each PCR amplification should include a positive-control sample. DNA from the cell line K562 is widely used in the forensic community and is commercially available as purified DNA.

13. The temperature of the denaturing step may be decreased to 90°C after 10 cycles in order to save enzyme.

14. The annealing temperature of the quadruplex is a compromise for the four pairs of primers. It is often necessary to adjust the annealing temperature when implementing a new PCR reaction in the laboratory.

15. The number of cycles should be adjusted for each STR locus. If the amount of PCR product is too small for detection, the amplification factor may be increased by increasing the number of PCR cycles. However, it should be kept in mind that by doing so, the risk of unspecific amplification will increase. The number of cycles can usually be adjusted up to 30–32 cycles. Higher numbers of cycles are generally not recommended.

16. The Taq DNA polymerase tends to attach an extra nucleotide to the synthesized strands. The extension time in the last PCR cycle is increased to 10 min to allow the attachment of the extra nucleotide.

17. If the PCR product is to be stored for a longer period, the PCR product should be stored at –20°C and protected from light. Storage of the PCR product at 4°C or higher temperatures for a longer period may result in degradation of the DNA.

18. The primers of the quadruplex were labeled with 6-carboxyfluorescein (FAM, blue) and carboxy-2',4',7',4,6-hexachlorofluorescein (HEX, green).

References

1. Kimpton, C. P., Gill, P., Walton, A., Urquhart, A., Millican, E. S., and Adams, M. (1993) Automated DNA profiling employing multiplex amplification of short tandem repeat loci. *PCR Meth. Appl.* **3,** 13–22.

2. Oldroyd, N. J., Urquhart, A. J., Kimpton, C. P., Millican, E. S., Watson, S. K., Downes, T., and Gill, P. D. (1995) A highly discriminating octoplex short tandem repeat polymerase chain reaction system suitable for human individual identification. *Electrophoresis* **16,** 334–337.

3. Huang, N. E., Schumm, J., and Budowle, B. (1995) Chinese population data on three tetrameric short tandem repeat loci—HUMTHO1, TPOX, and CSF1PO—derived using multiplex PCR and manual typing. *Forensic Sci. Int.* **71,** 131–136.

4. Hammond, H. A., Jin, L., Zhong, Y., Caskey, C. T., and Chakraborty, R. (1994) Evaluation of 13 short tandem repeat loci for use in personal identification applications. *Am. J. Hum. Genet.* **55,** 175–189.

5. Wiegand, P., Budowle, B., Rand, S., and Brinkmann, B. (1993) Forensic validation of the STR systems SE 33 and TC 11. *Int. J. Legal Med.* **105,** 315–320.
6. Möller, A., Meyer, E., and Brinkmann, B. (1994) Different types of structural variation in STRs: HumFES/FPS, HumVWA and HumD2S11. *Int. J. Legal Med.* **106,** 319–323.
7. Polymeropoulos, M. H., Rath, D. S., Xiao, H., and Merril, C. R. (1991) Tetranucleotide repeat polymorphism at the human c-fes/fps proto-oncogene (FES). *Nucleic Acids Res.* **19,** 4018.
8. Polymeropoulos, M. H., Rath, D. S., Xiao, H., and Merril, C. R. (1991) Tetranucleotide repeat polymorphism at the human coagulation factor XIII A subunit gene (F13A1). *Nucleic Acids. Res.* **19,** 4306.
9. Edwards, A., Civitello, A., Hammond, H. A., and Caskey, C. T. (1991) DNA typing and genetic mapping with trimeric and tetrameric tandem repeats. *Am. J. Hum. Genet.* **49,** 746–756.
10. Kimpton, C., Walton, A., and Gill, P. (1992) A further tetranucleotide repeat polymorphism in the vWF gene. *Hum. Mol. Genet.* **1,** 287.
11. Anker, R., Steinbrueck, T., and Donis Keller, H. (1992) Tetranucleotide repeat polymorphism at the human thyroid peroxidase (hTPO) locus. *Hum. Mol. Genet.* **1,** 137.
12. Polymeropoulos, M. H., Rath, D. S., Xiao, H., and Merril, C. R. (1992) Tetranucleotide repeat polymorphism at the human beta-actin related pseudogene H-beta-Ac-psi-2 (ACTBP2). *Nucleic Acids Res.* **20,** 1432.
13. Sharma, V. and Litt, M. (1992) Tetranucleotide repeat polymorphism at the D21S11 locus. *Hum. Mol. Genet.* **1,** 67.

16

Manual Electrophoretic Methods for Genotyping Amplified STR Loci

Chris P. Phillips, Angel Carracedo, and Maviky V. Lareu

1. Introduction

This chapter describes the manual typing of STR markers, detailing the electrophoretic techniques used to separate amplified products, and the relevant methods for their detection.

While there is a general trend toward automated fluorescent detection of PCR products, many laboratories cannot afford the specialized equipment needed for this technique. For this reason, manual typing is presently a widely adopted approach in forensic laboratories. Genotyping amplified STR loci requires relatively simple analytical techniques; two of the most commonly used methods are outlined here. Each technique uses quite different gel media, formats, and detection systems, giving them distinct characteristics that offer the user a choice of approaches when analyzing STR markers. Electrophoresis in agarose gels can be accurately used for the separation of PCR products differing by 4 bp *(1)* and electrophoresis in polyacrylamide gels allows the detection of STR alleles differing by only a single base pair *(2)*. Here we describe appropriate protocols to achieve such results.

In the last ten years, hundreds of relatively simple protocols for silver staining polyacrylamide gels have been published. This detection technique was first introduced in forensic science for the detection of protein polymorphisms *(3)*. We have detailed in this chapter a method based on the procedure described by Budowle et al. *(4)*, based on the classical method of Merril *(5)*, with modifications that allow the entire procedure to be completed in only 25 min.

The two main advantages of using an agarose electrophoresis technique are the simplicity of the procedure and the ability to perform further analyses on the DNA after its electrophoretic separation. Because samples can go straight

From: *Methods in Molecular Biology, Vol. 98: Forensic DNA Profiling Protocols*
Edited by: P. J. Lincoln and J. Thomson © Humana Press Inc., Totowa, NJ

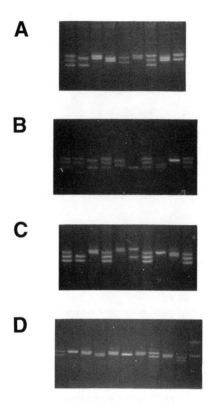

Fig. 1. **(A)** HumFES phenotypes obtained by high-sieving agarose electrophoresis. From left to right: lane (1) 8-10-12 control; lane (2) 8,11; lane (3) 11,12; lane (4) 10,11; lane (5) 9,11; lane (6) 11,12; lane (7) control; lane (8) 10,11; lane (9) 10,12. **(B)** HumTH01 phenotypes. From left to right: lane (1) 6-8-9.3 control; lane (2) 8,9.3; lane (3) 6,9; lane (4) control; lane (5) 7,9; lane (6) 6; lane (7) control; lane (8) 6,7; lane (9) 9; lane (10) control. **(C)** HumvWA1 phenotypes. From left to right: lane (1) 14-16-18 control; lane (2) 14,16, lane (3) 18,19; lane (4) control; lane (5) 19,20; lane (6) 16,20; lane (7) control; lane (8) 18; lane (9) 17,18; lane (10) control. **(D)** HumF13A1 phenotypes. From left to right: lane (1) 5,7 control; lane (2) 7; lane (3) 6,7; lane (4) 5,6; lane (5) 6,7; lane (6) 6; lane (7) 6,7; lane (8) 5,7 control; lane (9) 5,6; lane (10) 3.2, 5; lane (11) 6,15.

from the amplification reaction to the gel, and the staining of bands is an integral part of the electrophoresis procedure, it is possible to obtain a succession of results very quickly and easily, providing rapid feedback on the effects of changes in PCR conditions or on the quality of amplified samples. Typical agarose results are illustrated in **Fig. 1**. These show that, provided simple, tetrameric repeat STRs *(6)* are analyzed, the band patterns obtained with high-sieving agarose are clearly resolved and easy to read.

Multiplexes (i.e., multiple primer-set PCR) can be used in manual typing with the only extra requirement that the STRs combined in this way do not overlap in their fragment-size ranges. Many triplex (three different loci) and quadruplex (four loci) protocols have been published *(7)*, and there are commercial kits available specifically for multiplexing loci for forensic analysis using polyacrylamide electrophoresis *(8,9)*.

2. Materials

2.1. Polyacrylamide Gel Electrophoresis

2.1.1. Equipment for Electrophoresis

1. Horizontal electrophoresis cell.
2. Cooling system for the above cell. The system should be capable of maintaining a temperature of at least 15°C.
3. Power supply capable of delivering 1000 V, 50 W, and 50 A.
4. Casting system for horizontal formal: gels. A practical system can be homemade with a metacrilate plate ($30 \times 20 \times 1$ cm^3) and two layers of 9 mm wide embossing tape (Esselte Dymo, #5247.09) to act as a mold for the gel. Thirty single-thickness Dymo inserts (5×1.5 mm^2) are positioned across the mold 3 cm in from one end to create shallow depressions in the gel surface for inserting the samples.

2.1.2. Reagents for Electrophoresis

1. Gelbond PAG films (FMC Bioproducts, Rockland, ME) of suitable dimensions for use with the gel mould.
2. Chromatography paper blotting pads (e.g., Whatman Chr 17, Whatman, Clifton, NJ) 12×1 cm^2.
3. Loading dye (for denaturing gels): 10 mM NaOH, 95% formamide, 0.05% bromophenol blue, 0.05% xylene cyanol.
4. Monomer stock solution (30%T): 29 g acrylamide, 1 g piperazine diacrylamide (PDA), 100 mL H$_2$O. Fresh solutions should be prepared every few weeks. Store the solution in dark bottles at room temperature. Acrylamide is a strong neurotoxin and a potential carcinogen. It is absorbed through the skin and the effects are cumulative. Wear gloves and mask when weighing acrylamide powder and PDA. Wear gloves when handling all solutions containing these chemicals.
5. Gel buffer: 0.75 M Tris, 0.28 M formate, pH 9.0. Store in glass bottles at room temperature and discard any batches developing a precipitate.
6. Urea in denaturing gels. The gel mix should have a final concentration of 7 M urea.
7. 10% ammonium persulfate. This solution can be stored at 4°C for up to 4 wk.
8. TEMED (N,N,N',N' ethylenediamine). Store at room temperature.
9. Borate electrode buffer stock solution: 1.0 M Tris, 0.28 M boric acid, pH 9.0. Store in glass bottles at room temperature for up to 4 wk and discard any batches that develop a precipitate.

2.1.3. Reagents for Silver Staining

1. 1% nitric acid (stock solution).
2. 0.2% silver nitrate.
3. 0.28 M sodium carbonate solution.
4. 37% formaldehyde (stock solution).
5. 10% glacial acetic acid (stock solution).
6. 5% glycerol (stock solution).

2.2. Agarose Gel Electrophoresis

2.2.1. Equipment for Electrophoresis

1. Submarine electrophoresis tank allowing a minimum gel length of 10 cm, preferably with a buffer capacity of at least 1 L and, if possible, with ports that allow buffer recirculation with a pump and cooling system.
2. Power supply capable of delivering a constant voltage between 70 and 200 V.
3. Gel mold designed for use with the above electrophoresis tank. Molds that can be disassembled from the tank and allow gel pouring in a separate temperature-controlled environment are preferable.
4. Comb for use with the above that allows a minimum load volume of 15 mL. This can be homemade from a piece of 1 mm thick perspex cut to give the desired number and spacing of inserts (The agarose gel is robust enough to allow wells spaced 1–2 mm apart without splitting), which can be attached to the usual comb frame.
5. Long-wavelength ultraviolet transilluminator (300 nm) and appropriate eye protection (e.g., visors or an acrylic lid). Kodak red filter for UV photography.

2.2.2. Reagents for Electrophoresis

1. MetaPhor high-sieving agarose powder (FMC BioProducts # 50181).
2. 0.5 M EDTA, pH 8.0: 93.05 g. EDTA, 8 g. NaOH in 500 mL H_2O.
3. 20X TAE buffer stock: 96.8 g Tris, 222.8 mL acetic acid, 40 mL 0.5 M EDTA 8.0, H_2O q.v.
4. 1X TAE gel buffer, 0.5X TAE tank buffer: 20X stock is diluted appropriately and 50 mL of ethidium bromide/L of buffers added prior to use in the gel mix and electrophoresis tank. Ethidium bromide is a dye that intercalates with the DNA molecule and fluoresces when excited by long-wavelength ultraviolet light. Optimum fluorescence occurs at 300 nm—the strength of fluorescence being directly proportional to the amount of DNA present in the band. Ethidium bromide is a powerful teratogen, and all solutions containing this chemical should be handled with gloves. As it gradually loses activity once diluted, ethidium bromide should only be added to working buffer solutions just before use. The stock buffer can be kept at room temperature for several weeks, but laboratories performing electrophoresis without buffer cooling and recirculation will need to keep about 3 L of tank buffer/ethidium bromide equilibrated at 4°C prior to an electrophoresis run.

5. 5X loading buffer and tracking dye: 2.5 g Ficoll, 25 mg bromophenol blue, 20 mL. H_2O. Store at room temperature. Xylene cyanol tracking dye can also be added at the same concentration as bromophenol blue as an extra point of reference for monitoring electrophoresis.

3. Methods
3.1. Polyacrylamide Gel Electrophoresis

1. Dilute the monomer stock solution 5X with gel buffer to make a 6%T, 3.3%CPDA gel solution (*see* **Note 1**). For denaturing gels, add sufficient urea to give a final concentration in the gel mix of 7 *M* (*see* **Note 2**).
2. Initiate polymerization of the gel by adding 10 mL/mL of 10% ammonium persulphate and 1 mL/mL of TEMED (*see* **Note 3**). The gel mix soon starts to become viscous as polymerization progresses, so care should be taken that all the materials are in place and ready before adding the above two agents.
3. Cast the 500 μm gels (i.e., the thickness of two Dymo tapes) directly onto GelBond PAG films (cat. no. 54731, FMC Bioproducts) for ease of handling. The gel polymerizes anaerobically between a glass plate with attached GelBond film (fixed to the glass, hydrophilic side up, by adding a few drops of water between the two surfaces and pressing together) and the metacrilate plate. Pour most of the gel solution onto the metacrilate base plate and gently lower the glass top plate down, hinging at one end (*see* **Note 4**). Polymerization is carried out at room temperature for 25 min.
4. Set the cooling plate of the horizontal electrophoresis cell at 15°C. Carefully disassemble the glass/metacrilate sandwich by inserting a scalpel blade into a corner and twisting. Place the gel, now bonded to the GelBond support film, on the cooling plate with a few drops of water between them to achieve a good contact. Avoid air bubbles between plate and gel so that an even cooling effect can be produced across the whole area of the gel.
5. Soak paper blotting pads in the electrode buffer solution and position at each end of the gel (*see* **Note 5**). Set a distance between the electrodes of 22 cm for standard 240 × 120 × 0.5 mm^2 gels.
6. Mix together 2 μL of amplified product and 2 μL of loading buffer and pipet directly into the shallow troughs in the gel created by the small Dymo pieces of the mold. Place appropriate STR allelic ladders every five lanes (*see* **Note 6**).
7. The electrophoresis power settings for a gel of the dimensions described above are: 500 V, 25 A, and 20 W for 3 h. A short prerun (i.e., prior to loading the samples) of 30 min at the above settings is recommended.
8. Up to 30 V/cm can be used in 50 cm gels depending on the efficiency of the cooling system used. We recommend the use of 25 V/cm. Some protocols prefer to use constant wattage rather than constant voltage *(10)*.

3.2. Silver Staining

1. Remove the gel/support film from the cooling plate and place in 1% nitric acid until the bromophenol blue line turns yellow.

2. After rinsing briefly in water, stain the gel in 500 mL freshly prepared 0.2% (w/v) silver nitrate solution for 20 min.
3. Rinse the gel again in water and reduce in a solution containing 0.28 *M* sodium carbonate with formaldehyde (200 mL of 37% formaldehyde per 500 mL sodium carbonate solution). Initially add a small quantity of the above solution and rinse the gel briefly until the solution turns black, then discard this and add the rest of the solution, gently swirling around the gel until the brown bands become visible.
4. Give the gel a final rinse in water, then fix the bands in 10% glacial acetic acid for 10 min.
5. The gels can be dried at room temperature and protected by placing another plastic film on top of the dried gel. After taping the two edges of the films together, the gels are stored in the dark at room temperature (*see* **Note 7**).

3.3. Agarose-Gel Electrophoresis

1. A gel measuring 10 × 13.5 cm requires a volume of 40 mL buffer plus 1.8 g of agarose powder (*see* **Note 8**). Place the gel mold in an incubator at a temperature between 37 and 50°C before preparing the gel mix. Add the agarose powder slowly to the gel buffer/ethidium bromide (*see* **Note 9**) at room temperature in a 250 mL Pyrex Buckner flask (with a side arm) containing a magnetic stirring bar, then mark the top level with a pen. Stir the contents vigorously for 2–3 min to disrupt dry powder clumps; it is important to continue this process until even small clumps are no longer visible.
2. Microwave the gel mix at full power for 30 s, then swirl the contents vigorously. Continue heating the mix for 15 s periods at full power until it is near boiling out of the side arm (a small quantity of gel mix can be lost at this stage without effect). Stir the contents for a further 15 s, add water up to the original level, then briefly place in a water bath at 65°C until the flask has equilibrated to this temperature and degas with a vacuum line for 10 s before pouring immediately into the prewarmed gel mold and comb.
3. Once set, place the gel for a minimum of 30 min at 4°C, covered in a small volume of chilled tank buffer in order to complete the gelling process.
4. Add the cooled gel to the tank containing TAE buffer, previously cooled to 4°C, and just covering the gel to a depth of 3–4 mm. Position with the origin on the cathode side, remove the comb with care, and add 15 μL of PCR product mixed with 3 μL of load buffer. More sample can be loaded if required, but the bands become more diffuse, and the above volume represents a realistic amount to give clear bands for a PCR product derived from an optimized amplification reaction.
5. Perform an initial 5 min run of 20 V/cm (i.e., volts reading from power supply divided by interelectrode length) to stack in the samples. Follow this with a 2.5–3 h run of 8–10 V/cm.
6. Without recirculation, it is essential to change the buffer every hour with fresh stocks equilibrated to 4°C. This prevents the gel from overheating and the sieving effect of the agarose matrix progressively deteriorating. If buffer recirculation is used the temperature of the cooling system is set at 10–15°C, allowing settings up to 15 V/cm should a faster run be required.

7. After 2–3 h (*see* **Note 6**), view the gel on a UV transilluminator and record the results with a photographic or gel-documentation system.
8. Several critical factors may need to be adjusted to improve the resolution obtained in MetaPhor agarose electrophoresis. These are discussed in detail in the notes section (*see* **Notes 10–18**).

4. Notes

1. In polyacrylamide gels, the total monomer concentration is given by the percent value T, while the amount of crosslinker is given by the percent value C, with the polymerizer written as a subscript to C. Acrylamide and *N-N'* methylenebisacrylamide (Bis) are still the more commonly used copolymers. New crosslinkers are being introduced *(11)*, and of these, PDA *(12)* is the most widely used of the alternatives to Bis. The advantage of PDA over Bis is that considerably less background is observed after silver staining. The pore size of polyacrylamide gels can be changed by varying the total monomer concentration (%T) and the crosslinker concentration (%C). T values of 5–10% and C values of 2–5% are commonly used. We recommend a 6%T and 3.3%C for typing the commonly used STRs (ranging in size from 120 to 300 bp).
2. Although there are some studies using nondenaturing conditions, we strongly recommend using denaturing conditions for all STR typing. The advantage of nondenaturing gels is that additional variation can be detected in the majority of the STR systems and most notably in complex hypervariable STR loci. However, there are two principal problems with this approach. First, some STRs, especially those that have A-T rich sequences, are prone to anomalous electrophoretic behavior *(13)*. This is probably because of differing degrees of helical-axis distortion in each of the individual alleles in any one STR system, affecting their relative mobility *(14)*. Two STRs that particularly show this anomalous electrophoretic behavior are humACTBP2 *(15)* and humF13A1. The second problem is that when nondenaturing conditions are used and intra-allelic variation is detected, the electrophoretic conditions must be strictly standardized to achieve comparable results between different laboratories—an important factor in forensic analysis—which is most readily obtained by the adoption of denaturing electrophoresis in STR typing.
3. Polymerization is usually achieved with ammonium persulfate and TEMED. The time of polymerization can be reduced by heating the gel at 50°C. Photopolymerization can be performed using a riboflavin solution (0.3%) and an ultraviolet light source. The polymerization with riboflavin is slower, allowing an unhurried approach if practice is initially needed to pour the gels.
4. An alternative technique for pouring polyacrylamide gels, particularly longer gels, is to make a sandwiched assembly of both plates and arrange the top plate about 1 cm to one side, add the gel mix slowly into the assembly with a syringe, and when capillarity has brought the gel mix across the plates, gently slide the top plate into position to seal the assembly. Both this method and the "flap technique" described in the methods section may need clips to seal the two plates together more securely. The flap technique is prone to bubbles and requires practice.

5. Alternative electrodes can be constructed from 2% agarose-gel plugs (i.e., 2 g agarose/100 mL electrode buffer, gelled and cut into strips $12 \times 1 \times 1$ cm^3).

6. Reference ladders can be constructed as phenotyping controls from a combination of samples with different alleles. Polyacrylamide electrophoresis is more sensitive than agarose electrophoresis, so complete ladders can be constructed, whereas in the latter case typically there is a limit of 3–5 bands comprising a ladder. Normally, a 3 allele ladder can be used for STR systems, such as humTHO1, with five common alleles *(1)* and a 4 allele ladder for systems, such as humVWA, with seven common alleles *(16)*. The easiest way to produce a reference ladder for a particular STR system is to combine the DNA from the appropriate heterozygous and homozygous samples prior to PCR. Controls loaded into every third lane allow unknown samples to be directly compared with reference bands in an adjacent lane.

7. STR fragments can be detected using alternatives to silver staining. Laser excitation of a fluorogen (attached to the 5' end of one primer from each pair) can be used, but without the specialized detection system, it is still possible to develop the bands in this instance by using a decoupling detection system such as FluorImager SI *(8)* or Hitachi FM BIO *(3)*. Sensitivity is enhanced compared to normal silver staining with both of the decoupling systems but is not as good as automated fluorescent detection using a dedicated machine (as detailed in another chapter).

8. The concentration of MetaPhor agarose used to prepare a gel can be adjusted to give optimum resolution of fragments within a set basepair range. The majority of STR systems so far investigated and used by forensic laboratories give fragment sizes between 120–300 base pairs and the protocol outlined here involves 4.5% gels, i.e., 4.5 g agarose/100 mL gel buffer. If TBE (see other chapters for composition) is used as the buffer, then slightly less agarose is required. TBE gives faster run times and uses less of the expensive MetaPhor agarose, but in our experience bands are not so clearly defined. If different fragment sizes are to be investigated, then the following table should be used as a guide to achieving the best resolution of the PCR products.

Fragment size range (bp)	% Agarose using TAE	% Agarose using TBE	Equivalent PAGE %T
500–1000	2.0	1.8	3.5
150–700	3.0	2.0	4.0
100–450	4.0	3.0	6.0
70–300	5.0	4.0	8.0
40–130	—	5.0	10.0

9. The advantage of ethidium bromide in the buffer is that the gel can be periodically viewed on the UV transilluminator to check the progress of the bands in the STR system being studied: Experience will soon indicate the optimum time to read the band patterns for each STR. Care should be taken not to expose the gel for too long to the UV transilluminator while monitoring the run and before

recording the results, since fluorescence diminishes as the DNA is broken up by the ultraviolet light. In 4.5% MetaPhor gels, the bromophenol blue tracking dye corresponds in mobility to a 35 bp fragment. Xylene cyanol tracking dye can also be added as a further aid to monitoring a run; it should be noted that it corresponds in mobility to a 100 bp fragment and may overlap with STR bands in the gel. If required, the gel can be reloaded with new samples, or alternatively the gel can be remelted with about 10% of the volume being replaced as distilled water (reusing the flask with the original level marked). However, the sieving effects of the gel appear to diminish after two repeated runs or remelting of the gel.

10. One of the most important criteria in DNA electrophoresis is the resolution—i.e., the band sharpness and the separation of bands from similarly sized fragments *(17)*. The resolution obtained when phenotyping STRs using MetaPhor agarose is affected by several factors, including quantity of DNA loaded, comb dimensions, buffer used, percent of agarose, voltage gradient, temperature of buffer during run, and gel length *(17)*. Each of these factors is discussed in the following sections.

11. The first two factors are connected to some extent, in that a thin comb will give sharp, well-defined bands, but as the wells it creates are consequently smaller in volume, this requires samples with a high concentration of good-quality amplified DNA. So the PCR reaction from which such samples are derived needs to be optimized and amplified from mainly undegraded target DNA. In forensic analysis, the age and quality of target DNA cannot be guaranteed, and this represents a major limitation with agarose electrophoresis and ethidium-bromide detection— it needs a greater amount of DNA to be loaded compared to other manual electrophoresis formats *(1)*. Thin combs give sharp bands because the DNA stacks into the gel quickly and progresses through the gel in a narrow zone *(17)*. If the loaded volume required to give detectable bands is still large, a short initial period of electrophoresis at a higher voltage (5–10 min at twice normal V/cm) can be used. This stacks the DNA into the gel more quickly, improving band resolution. Overall, the thinnest comb and the smallest load volume that give clearly visible bands should be used.

12. Both TAE and TBE buffer should be compared when developing an agarose technique. Generally TBE gives faster runs, uses less agarose, and produces slightly sharper bands, but it may not be the best choice for the STR system being examined.

13. All STR loci so far analyzed by the authors separate well in 4.5% gels, but resolution can be further fine-tuned for any one STR system by adjusting the agarose concentration slightly. The smaller the molecular weight range of the alleles, the better the separation obtained, as four base pairs represent a larger percent of the total fragment length.

14. Voltage and buffer temperature are also related since both will directly affect the integrity of the gel matrix during a run. MetaPhor agarose is particularly sensitive to temperatures above 20°C—as the gel heats above this point, it loses its enhanced ability to resolve DNA fragments with similar sizes and the DNA diffuses at a faster rate, also progressively reducing the resolution. The voltage used

should be enough to keep the DNA-zone narrow but not so high that local heating within the gel affects resolution. Cooling the buffer with recirculation or replenishment with chilled stocks helps ensure that the gel does not overheat.

15. Gel length is an important factor when the difference between the fragment sizes in an STR system is a small percentage (<2%) of the total size of the fragments in base pairs. In this case, the best resolution is obtained with gels of 15–20 cm in length, and the longer the run, the better the separation between the fragments. The gel should be read when the fragments have migrated to 60% of the gel length—further electrophoresis does improve separation, as stated, but the bands begin to become increasingly diffuse as DNA diffusion accelerates and the sieving effect of the gel diminishes.

16. MetaPhor agarose can be used as the gel medium in several alternative electrophoresis formats for the analysis of STR loci. There is evidence that the resolution is improved when MetaPhor gels are run in a vertical format. Detailed protocols for pouring and running vertical gels have been published *(18–20)*. MetaPhor can be used as the gel medium in automated sequencers, but as there is no evidence for better resolution in this format *(21–23)*, it appears not to be worth the extra effort involved in casting the gels.

    MetaPhor gels can be cast directly onto a supporting plastic film such as GelBond (FMC Bioproducts, 54731) in the same way as that described in **Subheading 3.1.** Gels fixed to GelBond are stronger and easier to handle; they can also be poured to thinner depths, further improving the resolution.

17. Alternatives to ethidium-bromide stains are now becoming widely available. Of these, SYBR Green (FMC Bioproducts, 50512) offers improved sensitivity because of a greater affinity for DNA and a reduced background fluorescence in the gel *(24)*. Finally, it is possible to use silver staining to detect the separated DNA in MetaPhor gels using the same protocols outlined in the polyacrylamide section.

18. A particular advantage of agarose electrophoresis is the ability to use this separation technique in conjunction with other tests. A 20 µL PCR volume yields enough sample for a MetaPhor electrophoresis run as a check, before committing the same samples to more costly and time-consuming automated fluorescent detection. In this case, using fluorescently labeled primers does not affect the band quality of the PCR products. The use of MetaPhor offers the opportunity to perform further analyses on the separated fragments. The most advantageous feature of agarose electrophoresis is the ability to recover the DNA intact from a band and use it in sequencing reactions. Thus fragments differing by only 4 bp—the usual form of single repeats in many STRs—can be individually isolated from the MetaPhor gel for sequence analysis. Unlike agaroses such as NuSieve GTG, MetaPhor requires the DNA to be separated out before sequence analysis begins, but a variety of techniques have been described in detail to do this *(25,26)* and NuSieve would be unable to provide the 4 bp resolution obtained with the procedures outlined in this chapter. It is possible to perform Southern-blot analysis of MetaPhor gels in the same fashion as other agarose gels using the techniques outlined in another chapter.

References

1. Lareu, M. V., Phillips, C. P., Carracedo, A., Lincoln, P. J., Syndercombe-Court, D., and Thomson, J. A. (1994) Investigation of the STR locus HUMTHO 1 using PCR and two electrophoretic formats: UK and Galician Caucasian population surveys and usefulness in paternity investigations. *Forensic Sci. Int.* **66,** 41–52.
2. Allen, R. C., Graves, G., and Budowle, B. (1989) Polymerase chain reaction amplification products separated on rehydratable polyacrylamide gels and stained with silver. *Biotechniques* **7,** 736–744.
3. Carracedo, A., Concheiro, L., Requena, I., and Lopez-Rivadulla, M. (1983) A silver staining method for the detection of polymorphic proteins in minute bloodstains after isoelectric focusing. *Forensic Sci. Int.* **23,** 241–248.
4. Budowle, B., Chakraborty, R., Giusti, A. M., Eisenberg, A. J., and Allen, R. C. (1991) Analysis of the variable number of tandem repeats locus D1S80 by the polymerase chain reaction followed by high resolution polyacrylamide gel electrophoresis. *Am. J. Hum. Genet.* **48,** 137–144.
5. Merril, C., Dunau, M. L., and Goldman, D. (1981) A rapid sensitive silver stain for polypeptides in polyacrylamide gels. *Anal. Biochem.* **110,** 201–207.
6. Lygo, J. E., Johnson, P. E., Holdaway, D. J., Woodruffe, S., Whitaker, J. P., Clayton, T. M., Kimpton, C. P., and Gill, P. (1994) Validation of short tandem repeat loci for use in forensic casework. *Int. J. Legal Med.* **107,** 77–89.
7. Kimpton, C., Fisher, D., Watson, S., Adams, M., Urquhart, A., Lygo, J., and Gill, P. (1994) Evaluation of an automated DNA profiling system employing multiplex amplification of four tetrameric STR loci. *Int. J. Legal Med.* **107,** 77–89.
8. Budowle, B., Koons, B. W., Keys, K. M., and Smerick, J. B. (1996) Methods for typing the STR triplex CSFlPO, TPOX, and HUMTH01 that enable compatibility among DNA typing laboratories, in *Advances in Forensic Haemogenetics 6* (Carracedo, A., Brinkmann, B., and Bär, W., eds.), Springer-Verlag, Berlin, pp. 107–115.
9. Schumm, J., Sprecher, C., Lins, A., and Micka, K. (1996) Development and applications of high throughput multiplex STR systems, in *Advances in Forensic Haemogenetics 6* (Carracedo, A., Brinkmann, B., and Bär, W., eds.), Springer-Verlag, Berlin, pp. 145–148.
10. Barros, F., Lareu, M. V., and Carracedo, A. (1992) Detection of polymorphisms of human DNA after polymerase chain reaction by miniaturized SDS-PAGE. *Forensic Sci. Int.* **55,** 27–36.
11. Kozulic, B. (1995) Models of gel electrophoresis. *Anal. Biochem.* **231,** 1–12.
12. Hochstrasser, D., Patchornik, A., and Merril, C. (1988) Development of polyacrylamide gels that improve the separation of proteins and their detection by silver staining. *Anal. Biochem.* **173,** 412–423.
13. Stellwagen, N. C. (1983) Anomalous electrophoresis of deoxyribonucleic acid restriction fragments on polyacrylamide gels. *Biochemistry* **22,** 6186–6193.
14. Lareu, M. V., Pestoni, C., Phillips, C., Barros, F., SynderCombe-Court, D., Lincoln, P., and Carracedo, A. (1997) Normal and anomalous electrophoretic behaviour of PCR-based DNA polymorphisms (STRs and AMP-FLPs) in polyacrylamide gels. *Theoret. Appl. Electrophor.,* in press.

15. Lareu, M. V., Phillips, C. P., Pestoni, C., Barros, F., Munoz, J., and Carracedo, A. (1994) Anomalous electrophoretic behaviour of HUMACI BP2 (SE33), in *Advances in Forensic Haemogenetics 5* (Bär, W., Fiori, A. T., and Rossi, U., eds.), Springer-Verlag, Berlin, pp. 121–123.

16. Drozd, M. et al. (1994) An investigation of the HUMVWA31 A locus in British Caucasians. *Forensic Sci. Int.* **69,** 161–170.

17. Andrews, A. (1987) *Electrophoresis, Theory, Techniques and Biochemical and Clinical Applications,* 2nd ed. Clarenden, Oxford, UK.

18. Fast running protocols for high resolution in MetaPhor agarose. (1993) *FMC Tech. Guide T19.*

19. MetaPhor Agarose. (1994) *FMC Technical Guide T9.*

20. Vertical MetaPhor agarose gels. (1993) *Resolutions* **9.**

21. Vandenplas, S. et al. (1984) Blot hybridisation of genomic DNA. *J. Med. Genet.* **21,** 164–172.

22. Scharf, S. (1993) MetaPhor agarose gels used in the ABI genescanner to monitor bone marrow transplantation. *Resolutions* **9.**

23. Sullivan, K. Personal communication.

24. SYBR Green nucleic acid gel stains. (1995) *FMC Tech. Bull. P31.*

25. Sambrook, J., Fritsch, E. F., and Maniatis, T. (1989) *Molecular Cloning, A Laboratory Manual,* 2nd ed. Cold Spring Harbor Laboratories, Cold Spring Harbor, NY, pp. 6.24–6.27.

26. Vogelstein, B. and Gillespie, D. (1979) Preparative and analytical purification of DNA from agarose. *Proc. Natl. Acad. Sci. USA* **79,** 616–619.

17

Genotyping STR Loci Using an Automated DNA Sequencer

David Watts

1. Introduction

This introduction is intended to inform the forensic scientist of the historical development of STR typing on automated sequencers, as these precedents will have a bearing on the techniques and equipment used in the present-day laboratory. The section then concludes with a brief outline of the principle of a typical automated STR assay, which applies to all of the machines described. Equipment from other manufacturers operates on a similar basis, with the exception of the multi-fluor technique: other means are adopted for the provision of a suitable size standard on single-dye sequencers.

1.1. Development of the Automated DNA Sequencer

The Applied Biosystems 370A automated DNA sequencer was introduced in 1987. This machine was based on the principle that a DNA-sequence ladder could be recorded in real time as it separated on an electrophoretic gel by labeling the fragments with fluorescent dyes, to be excited and detected by a suitable scanner at the lower end of the gel (1). The use of four dyes, one for each nucleotide type, ensured the minimization of electrophoretic variability for any given template, as well as greatly increasing sample throughput per gel. The basic principle of the scanning laser excitation and photomultiplier detection system is shown in **Fig. 1**. This equipment was used primarily for research in the first few years, but wider applications of the basic technology soon became apparent.

Besides sequencing, the other major use of electrophoresis in DNA analysis is the sizing of DNA fragments, normally achieved by running unknown fragments and known-size standards in separate lanes of a gel followed by postrun

From: *Methods in Molecular Biology, Vol. 98: Forensic DNA Profiling Protocols*
Edited by: P. J. Lincoln and J. Thomson © Humana Press Inc., Totowa, NJ

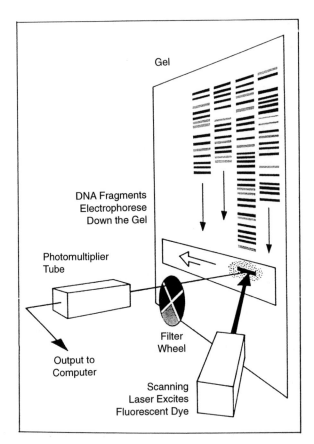

Fig. 1. Fluorescent DNA analysis technology (FDAT). FDAT as implemented on the 373A DNA sequencer. As fluorophore-tagged DNA fragments electrophorese (on a polyacrylamide gel) past a scanning laser beam, fluorescence measurements are taken and stored as a function of scan number (time) at four wavelengths using four bandpass filters. Each bandpass filter is used to detect one of the four fluorophores. Stored data are analyzed by the system's software to provide either primary DNA sequence information or, alternatively, sizes (in base pairs), colors, and concentration of fluorescent molecules in each sample relative to an internal fluorescent standard.

visualization. The advantages of a multi-fluor system, allowing internal-lane standards and multiplex loading, together with real-time detection, led to the development of the 362 Genescanner *(2)*. This machine used a similar optical arrangement to the sequencer, but mounted it under an agarose submarine gel. This choice of separation medium was made to satisfy the anticipated applications of the 362: PCR-RFLP analysis and the sizing of VNTR minisatellite markers, generally in the size range of a few hundred to several thousand bases.

Around the time of the 362's commercial release in 1990, attention began to center on the potential of STR microsatellite markers as more convenient, abundant markers for DNA studies *(3)*, including genotyping for forensic purposes. To achieve the resolution needed for typing of repeat units as small as two bases, experimental fragments had to be run on sequencing gels, which led to the Genescanner 362 software being adapted to the then current model of the sequencer, the 373A, and named Genescan. The attraction of a dual-purpose machine and the overwhelming move to microsatellite markers for automated typing meant that the 362 would become obsolete in 1992. Some early development work on DNA typing for forensic evidence had been done on the 362 at the British Forensic Science Service Laboratories and the Royal Canadian Mounted Police *(4)*, but these pioneering laboratories moved on to sequencer-based systems. A number of laboratories proceeded to develop STR-based typing assays using the 373 sequencer and Genescan software *(5)*, culminating in the introduction of fluorescent-based typing to police casework in British laboratories in 1994 *(6,7)*. This routine work was first performed with the four locus STR multiplex described in **ref. *18***.

In 1995, the Forensic Science Service in British laboratories began the compilation of a national DNA database of profiles from convicted criminals, using a new STR multiplex covering six microsatellite loci and the amelogenin gene as a sex determinant (*see* Chapter 21). The development of this program coincided with the release of a new, higher speed DNA sequencer for multi-fluor detection, the ABI 377. This machine has replaced the 373 and is being adopted for the database work, alongside the existing panel of 373 units. Any forensic laboratory setting up to do automated genotyping at present is likely to evaluate the 377 as a possible workhorse. The updated CCD camera detection system of the 377 is illustrated in **Fig. 2**. In late 1995, a low-throughput sequencer, the ABI 310, was introduced, employing the same CCD unit as the 377, but separating one sample at a time by electrophoresis through an automatically loaded capillary. This machine has been evaluated for forensic applications by some US laboratories and may well present a useful option for smaller, self-contained regional laboratories (*see* **Fig. 3** for a diagram of the 310 capillary-separation system).

1.2. The Principle of an Automated STR Assay

For detection on the sequencer, PCR-generated fragments must be labeled with a fluorescent dye. The family of dyes used is described in **Subheading 2.**; up to three different color channels can be used for the experimental material, retaining the fourth (usually longest wavelength) channel for the internal-lane standard. Different loci labeled with the same dye can be multiplexed by size in the same lane, provided that their respective size ranges are well separated, as was done in the initial FSS quadruplex. The dye is attached to the 5' end of

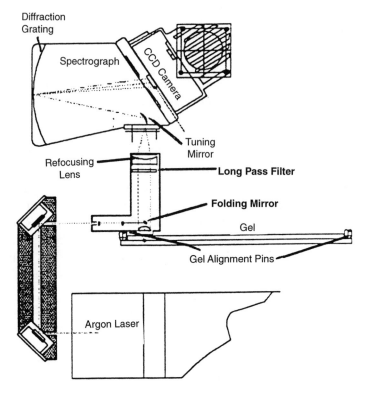

Fig. 2. 377 Detection system.

one of the PCR primers during synthesis, so that subsequent use of the primer in the PCR reaction incorporates one fluor per molecule, on one strand only. This is important for resolution because when the fragments are run on a denaturing gel, only one strand is visualized, which gives the clearest possible results. Labeling of fragments during PCR is feasible with dye dNTP analogs, but will label both strands to the detriment of clear band resolution, and is therefore not suitable for forensic applications.

The PCR amplification can include all the primers in one multiplex, or a number of single locus reactions can be pooled. The internal-lane standard is added to the sample together with formamide, and the samples are denatured by heating prior to loading. As the fragments migrate through the gel (*see* **Fig. 1**), they are detected by their fluorescent label, and a record of the retention times of all the fragments is built up as a computer file. After the run, the data are processed to remove any cross-talk between channels and brought down to a common baseline for all four dyes. A standard curve is created for each lane according to the known size of the standard ladder, and the unknown frag-

Block diagram of the optical system:

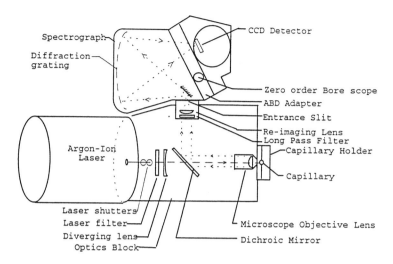

Fig. 3. 310 Detection system.

ments are sized against it. Results can be presented as a table of fragment sizes through the Genescan-analysis software or put into a further program, Genotyper, to assign allele designations rather than sizes and collate the results in a spreadsheet format. This can be used as a final data output.

2. Materials

The operation of the available automated genotypers is fully covered by the supporting documentation supplied by the manufacturers, as are the recommended sources for reagents and ancillary equipment needed for their successful use. A summary of the items needed, with important points to consider, is given here (*see* **Note 1**).

1. Water: for all reagents, gels and buffers, use 18 MΩ MilliQ water or equivalent quality.
2. Acrylamide: BioRad (Hercules, CA) or Amresco is recommended, as powders or as a premixed 40% solution (BioRad). Stock solution should not be kept for more than 1 mo, at 4°C.
3. Urea: Analar grade from most suppliers is suitable.
4. Tris/borate/EDTA buffer: Use Analar-grade reagents and make up according to the directions in the manual. The EDTA must be the disodium salt. The pH must be 8.3 ± 0.1 U; if not, discard the buffer and check the quality of the water and reagents.
5. TEMED: BioRad, IBI Technologies (Irvine, CA) (Kodak) or Amresco. Store according to instructions.

6. Ammonium persulfate: Analar grade; make up the 10% w/v stock fresh every day.
7. Formamide: BioRad, IBI (Kodak) or Amresco. Deionize with mixed bed resin and store frozen; start a fresh stock every few weeks.
8. PCR primer sets should be made or purchased to provide one dye labeled primer per locus. They should be HPLC-purified, and stored at –20°C. in appropriate aliquots to prevent excessive freeze-thaw damage. Three fluorescein-based dyes (trade names FAM, TET, and HEX) are available for labelling dye primers during synthesis. *See* **Note 2** for details of emission wavelengths for use on automated sequencers.

3. Methods

Since other chapters in this book deal comprehensively with the extraction and PCR amplification of appropriate material for forensic typing and the statistical handling of data obtained, this section covers machine setup, sample loading and running, and immediate downstream data analysis in isolation. Full training in the basic operation of these systems should be undertaken with the services available from the manufacturers; this chapter is intended to build on that basis by adding a further layer of information and advice that is specifically relevant to the forensic area.

The process divides into four main areas:

1. Preparation of samples and machine setup,
2. Loading the samples and running the machine;
3. Analyzing and reanalyzing data, and
4. Downstream processing of data, e.g., Genotyper software.

3.1. Preparation of Samples and Machine Setup

3.1.1. Preparation of Samples

1. Samples will be run on a denaturing gel for maximum resolution and so must be denatured before loading. The denaturing agent usually recommended is formamide, together with a heating step to melt out the double-stranded sample followed by snap-chilling on ice to prevent renaturation. The formamide, with a density of 1.13 g/mL, also acts as a weighting agent to assist easy loading.
2. Take an aliquot of PCR product from the original PCR reaction tube, which is likely to contain 25 or 50 μL of aqueous product. Unless the PCR machine used for the assay has a heated lid to minimize sample evaporation, the reaction will be overlaid with mineral oil. If pipetting PCR product from under oil, take care not to recover oil along with the aqueous solution, and wipe the pipet tip carefully to prevent carryover of oil to the loading buffer.
3. The sample mix to be analyzed consists of amplified PCR product, internal-lane standard and formamide. The formamide and size standard can be mixed on the day of use to form a stock reagent, for convenience and to minimize pipetting errors. The reagents should be stored separately before use. To ensure full dena-

turation and adequate density of the mixture, it is advisable to keep the final concentration of the formamide at 50% (v/v) or above.

4. If the PCR product is from a single multiplex reaction, it is possible to use the following mixture for loading the 373 *(9)*: Mix, per lane; 1.5 µL multiplex PCR product, 0.5 µL Genescan size standard (*see* **Note 3**), 2 µL deionized formamide (*see* **Note 4**).

5. Heat tubes at 90°C for 2 min, then immediately chill on ice. The denatured samples should be run on the 373 within 1 h.

6. If more than one PCR reaction must be pooled to produce a complete sample, the aqueous portion of the sample may be large enough to cause problems in loading. The volume should be reduced, if necessary, with a vacuum centrifuge (e.g., Savant Speed-Vac) or by a standard ethanol precipitation. It is not possible to load more than 5 µL on the 373, and it is better to limit the volume to 4 µL or less, especially when using the 36-lane comb.

7. The 377 sequencer has a much thinner gel than the older 373 model (0.2 vs 0.4 mm), which gives a lower maximum load volume, approximately half that of the 373. However, the 377 is more than twice as sensitive as the 373, so that reducing the volume of the loading mix by a factor of two will give the required performance.

8. The 310 capillary electrophoresis system uses electrokinetic injection of samples, so volume is not an issue. To give adequate signal strength, similar sample and standard quantities to the 377 loading should be used. These are then made up with around 12 µL of formamide, which allows them to remain denatured at room temperature for many hours on the 310 autosampler before running.

3.1.2. Machine Setup

3.1.2.1. GEL PREPARATION

1. The performance of an automated sequencer, for sequencing or fragment analysis, is critically dependant on the quality of the gel used for electrophoresis and on the cleanliness of the glass plates. (These form part of the optical system and so must be cleaned and cared for with this requirement in mind, as well as the usual standards needed for pouring even, bubble-free gels). It is vital that the manufacturer's recommendations be followed regarding reagents, plate cleaning, and gel preparation and pouring. If any departure from these guidelines appears to be forced by operational requirements, the performance of any replacement reagents or procedures would have to be rigorously validated (*see* **Note 5**).

2. Plates should be cleaned by soaking and gentle scrubbing with a weak solution of Alconox detergent in tap water. This is then removed by thorough rinsing in tap water followed by deionized water. Rinsing under a running stream of water is best, as is air-drying of the washed plates, to avoid the buildup of dust and fibers by drying with paper towels. When dry, plates and spacers should be carefully assembled to ensure correct alignment of all components, especially the well-casting comb. Always keep the same faces of the plates on the inside of the gel "sandwich" to ensure reproducibility of performance.

3. Acrylamide-gel solutions should be made up on the day of use, and deionized, filtered through a 0.2-μm filter and degassed to remove dissolved oxygen before pouring and polymerization. Keep to the recommended levels of TEMED and ammonium persulfate for catalyzing the gel polymerization, again to ensure reproducibility.
4. The 310 machine uses a liquid-polymer solution in a fused-silica capillary for electrophoretic separation. There is a much smaller likelihood of any problems analogous to those associated with poor-quality polyacrylamide slab gels.

3.1.2.2. COMPUTER AND SOFTWARE SETUP

The older 373 sequencer range was equipped with a series of Apple Macintosh computers, from the earliest IIcx to the latest (end-of-line) Quadra 650. All of these computers run the successive iterations of the version 1 series Genescan (672) software for fragment analysis. The following advice applies to all of these models:

1. Always restart the computer before setting up a run. This ensures the correct shutdown of any unwanted applications and the launch of the appropriate Genescan collection and analysis programs.
2. Restrict the non-ABD software on the Macintosh to those programs that are known not to cause any conflicts. Ideally, the fewer extraneous applications, the better.
3. If the Macintosh is on a network, it must be isolated from it prior to and during data collection. Any attempt to communicate or transfer data via a network or external drive while collecting data from a current 373 run can cause serious problems.
4. Always ensure that a generous amount of free harddisk space is available to collect data to before starting a run. A good rule of thumb is to have at least 3 MB of disk space for every hour of proposed run time. The more free space available, the more secure the system. A rigorous approach to data management and backup is vital.
5. The Macintosh computer requires careful operation and regular maintenance for reliable performance. All recommended operating procedures must be followed, and routine harddisk defragmentation and file repair should be carried out with a suitable commercial maintenance program such as Norton Utilities or Silverlining (**Note 6**).
6. The newer 310 and 377 sequencers use the Power Macintosh range of computers, which must run the version 2 series of Genescan analysis software. Unlike the earlier system, these computers can be accessed during data collection without detrimental effect, so long as they are not used continuously. The same observations about data management and disk maintenance apply to the Power Macintosh.

3.1.2.3. GEL MOUNTING AND PRERUN

1. In the slab-gel machines, it is vital to align the assembled gel with the laser read window. Any obstruction to the contact fit of the inner glass plate to the surround of the laser window, such as dried gel, urea crystals, and so forth, must be cleaned away prior to mounting the gel assembly. The exterior surfaces of the gel plates must be scrupulously clean for the same reason. Before running the machine and loading samples, a "plate check" should be carried out to ensure that the gel gives a clean, low

baseline for the fluorescent detector. The plate-check test is a dry run of the machine that runs the scanning detector without activating gel electrophoresis. If the baseline is not satisfactory, the gel assembly can be taken out and the surfaces cleaned again.

2. When the baseline is good, the electrophoresis buffer can be added to the system, the wells of the gel flushed out (by syringe) with running buffer to remove urea that leaches out of the gel, and the gel pre-electrophoresed for 10–15 min to bring it up to operating temperature.

3.2. Loading Samples, Running the Machine, and Collecting Data

3.2.1. Sample Loading

1. Standard instructions in the Genescan manuals recommend the use of cast-in wells produced by the use of a square-tooth former in casting the gel to avoid lane-to-lane leakage. However, it is common in many laboratories (e.g., Forensic Science Service, UK) to use the DNA sequencing sharks-tooth combs, which are easier to set up, flush, and load. If using a shark-tooth comb, follow the staggered-loading approach described in the manuals.

2. The loading pattern of the gels should avoid placing any case-related samples in adjacent wells to obviate queries about lane-to-lane contamination. It is good practice, if throughput is high enough, to run suspect and scene of crime samples on separate gels, or even different days, to remove any suggestion of sample-identity error. Prior to any sample loading, the wells of the gel must be flushed again to remove urea that would interfere with effective loading.

3. The gel is most effectively loaded with flat "duck-bill" tips expressly designed for this purpose (*see* **Note 7** for suppliers/part numbers).

3.2.2. Running the Machine

1. For optimum resolution on the 373 and 377, it is best to select DNA sequencing gels and run conditions, as described below.

Machine	Dates of availability	Gel type	Run conditions[a]
373 Classic	Jan. 1987 to Dec. 1992	24 cm 6%	2500 V/40 mA/30 W
373 Leon	Jan. 1993 to Jun. 1994	24 cm 6%	2500 V/40 mA/30 W
373 Stretch	Jul. 1994 to Dec. 1994	34 cm 4.75%	2500 V/40 mA/32 W
377	Jan. 1995 onward	36 cm 4%	3000 V/60 mA/200 W

[a]377 voltage-limiting, 373 power-limiting.

The 310 should be run with the Genescan format and the long denaturing reagents and run conditions.

2. Ensure that the correct filter set is selected for the dyes used in the assay. *See* Genescan manual and **Note 8**.

3.2.3. Collecting Data

1. For the 373, the run time set on the Macintosh computer must match or slightly exceed the time set on the scanning unit, to ensure that data output stops before

data collection does. Start computer collection some minutes after electrophoresis to help ensure this, noting that after starting electrophoresis on the 373, data collection must also be started separately on the computer to record the data to disk. While data collection is in progress, the computer must not be actively connected to a network nor can it be used for other tasks.

2. The 377 and 310 machines are operated via the Power PC Macintosh, so that data storage is automatically triggered by the start of the electrophoresis run. These computers can be used for other tasks during data collection, provided the use is not intensive. The 310 analyzes one result at a time after each sample has run. The 377 analyzes all sample data as a batch after the run finishes. During the run, the gel baseline and the progress of fragments passing the detector can be monitored by use of the scan and map windows. An overview of the entire run so far can be gained from the gel window. On the Power Macintosh, this gel window can be left open throughout the run; on the earlier Macintosh models (373), it should be left closed after viewing.

3.3. Analyzing and Reanalyzing Data

1. Once a routine assay procedure (PCR parameters, sample preparation, loading, and running conditions) has been established, data analysis on all ABI Prism machines can be done automatically.

2. As a safeguard, the internal-lane standard in each sample should be checked for correct size calling: This can efficiently be done by laying up panels of data for the standard (usually red) channels alone on the display. If any discrepancies in sizing are found, it is usually because the data for the run concerned have in some way failed to fit the existing size-standard template (e.g., because of run speed or late onset of data collection). The creation of a new size-standard file, specific to the run in question, should allow analysis to proceed normally.

3. The availability of the "gel image" on the 373 and 377 slab-gel machines allows the tracking of each lane to be verified. Any errors should be rectified by retracking and reanalyzing the lanes concerned. This does not apply to the single-channel 310.

4. The third most common data problem, besides sizing and tracking errors, is the appearance of signal crossover between color channels, sometimes described as "pull up peaks." This occurs when the signal in one channel is so intense that it saturates the normal baseline compensation in the software (the multicomponent matrix) and allows a ghost peak to appear in a neighboring channel (*see* **Fig. 4**). This problem can be solved by loading less material if the signal is too intense. If the signal is on scale, the preprocess/analysis parameters (depending on software version) should be checked to ensure that the correct matrix file was selected and used. *See* **Note 9** for more details.

3.4. Downstream Analysis

1. The Genescan analysis software presents each analyzed sample as an electropherogram trace of the colored bands' relative mobility, together with a table of calculated band sizes (**Fig. 5**). A genotype for an individual could be drawn manually from this data, by noting the principal band sizes and making allele assignments accordingly. To increase

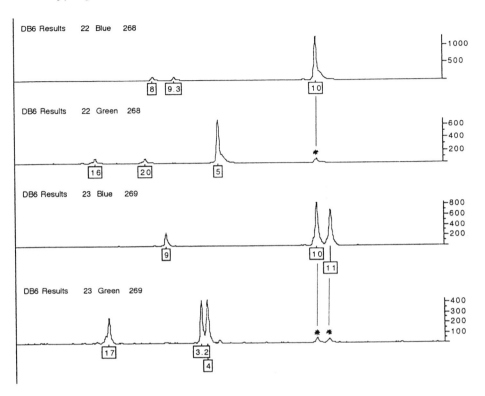

Fig. 4. A forensic Genescan data display, showing matrix correction errors: bleed through of signal from the blue channel to the green (marked *).

the level of automation and to reduce criticism of human intervention in assigning genotypes, a downstream analysis program is available. This is the Genotyper software, which is capable of taking the original Genescan results files, searching for size-labeled peaks in the defined ranges, and filtering the data to remove stutter bands, and so forth.

2. The valid peaks, once identified, can be marked with allele numbers rather than simple size calls, depending on the allele "bin size" they fall into (*see* **Note 10** and **Fig. 6**). The allele numbers can then be collated into a definitive genotype for the sample. A series of results can thus be reduced to a spreadsheet format (*see* **Fig. 7**).

4. Notes

1. It is very important to keep a batch log of all reagents, including the water supply, if any variations are suspected. All relevant details should be noted: supplier or source, batch number, date of supply, and date of introduction to the laboratory. Any anomalies in the performance of the system can be quickly checked against this, revealing potential causes of trouble at the outset, rather than after a lengthy process of testing all components one by one, by elimination from the system.

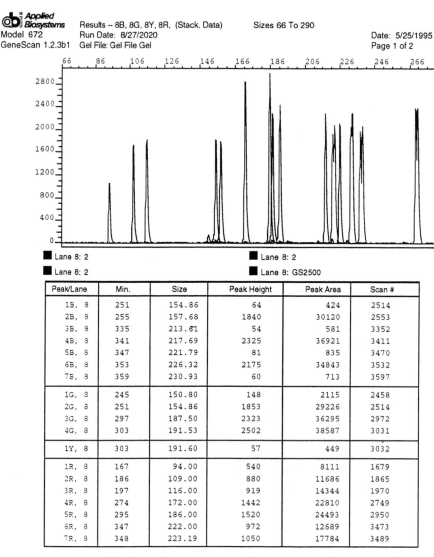

Fig. 5. A forensic Genescan display, showing the electropherogram of peaks with tabulated size data below.

2. Fluorescein-dye amidites for labeling primers during synthesis:

Dye amidite	Absorb max. (nm)	Em4iss max. (nm)	Chemical name
6-FAM	495	521	6-carboxy-fluorescein
TET	519	539	4,7,2',7'-tetrachloro-6-carboxy fluorescein

- F13A1
  - 10   All peaks from 206.30 to 208.40 bp in green with height >= 50
  - 11   All peaks from 210.00 to 212.50 bp in green with height >= 50
  - 12   All peaks from 214.50 to 216.50 bp in green with height >= 50
  - 13   All peaks from 218.60 to 220.60 bp in green with height >= 50
  - 14   All peaks from 222.60 to 225.60 bp in green with height >= 50
  - 15   All peaks from 226.60 to 230.00 bp in green with height >= 50
  - 16   All peaks from 230.70 to 234.50 bp in green with height >= 50
  - 17   All peaks from 234.90 to 237.40 bp in green with height >= 50
  - 3.2   All peaks from 181.10 to 182.50 bp in green with height >= 50
  - 4   All peaks from 183.00 to 184.60 bp in green with height >= 50
  - 5   All peaks from 186.90 to 188.50 bp in green with height >= 50
  - 6   All peaks from 190.80 to 192.40 bp in green with height >= 50
  - 7   All peaks from 194.50 to 196.30 bp in green with height >= 50
  - 8   All peaks from 198.40 to 200.40 bp in green with height >= 50
  - 9   All peaks from 202.30 to 204.40 bp in green with height >= 50
- FES
  - 10   All peaks from 219.70 to 221.10 bp in blue with height >= 50
  - 11   All peaks from 223.80 to 226.00 bp in blue with height >= 50
  - 12   All peaks from 227.90 to 230.90 bp in blue with height >= 50
  - 13   All peaks from 231.90 to 235.05 bp in blue with height >= 50
  - 14   All peaks from 235.10 to 238.50 bp in blue with height >= 50
  - 8   All peaks from 211.50 to 213.10 bp in blue with height >= 50
  - 9   All peaks from 215.60 to 217.00 bp in blue with height >= 50
- TH01
  - 11   All peaks from 178.20 to 178.90 bp in blue with height >= 50
  - 5   All peaks from 152.20 to 154.60 bp in blue with height >= 50
  - 6   All peaks from 156.60 to 158.80 bp in blue with height >= 50
  - 7   All peaks from 160.90 to 162.70 bp in blue with height >= 50
  - 8   All peaks from 165.30 to 166.70 bp in blue with height >= 50
  - 9   All peaks from 169.70 to 170.70 bp in blue with height >= 50
  - 9.3   All peaks from 173.00 to 174.80 bp in blue with height >= 50
- VWA
  - 13   All peaks from 134.40 to 135.60 bp in green with height >= 50
  - 14   All peaks from 138.30 to 139.70 bp in green with height >= 50
  - 15   All peaks from 142.20 to 143.60 bp in green with height >= 50
  - 16   All peaks from 146.20 to 147.60 bp in green with height >= 50
  - 17   All peaks from 150.50 to 151.70 bp in green with height >= 50
  - 18   All peaks from 154.40 to 155.80 bp in green with height >= 50
  - 19   All peaks from 158.50 to 159.90 bp in green with height >= 50
  - 20   All peaks from 162.60 to 164.20 bp in green with height >= 50
  - 21   All peaks from 166.60 to 168.60 bp in green with height >= 50

Fig. 6. A forensic Genotyper quadruplex category list.

HEX	537	556	4,7,2',4',5',7'-hexachloro-6-carboxyfluorescein

Fam and Hex can be used with filter set A (373, 310, or 377) with either Tamra or Rox size standards or filter set B (373) or C(310, 377) with a Tamra size standard. If Fam, Tet, and Hex are used together, filter set B (373) or C (310, 377) must be used, with a Tamra size standard. *See* **Note 9** below 3 for an explanatory dye/filter-set chart.

3. Genescan size standards are available labeled with two different rhodamine dyes, Tamra (emission max. 580 nm) and Rox (emission max. 605 nm). There are three size ranges:

File name	lane no.	sample no.	category	allele 1	allele 2
DB6 Results	22	268	FES	10	10
DB6 Results	22	268	TH01	8	9.3
DB6 Results	22	268	F13A1	5	5
DB6 Results	22	268	VWA	16	20
DB6 Results	23	269	FES	10	11
DB6 Results	23	269	TH01	9	9
DB6 Results	23	269	F13A1	3.2	4
DB6 Results	23	269	VWA	17	17

Fig. 7. A Genotyper results table for quadruplex results.

Standard	Size range	Type
GS 350	(35–350 bases)	d.s. DNA, labeled on one strand
GS 500	(35–500 bases)	d.s. DNA, labeled on one strand
GS 2500	(37–2860 bases)	d.s. DNA, labeled on both strands

For historical reasons, the double-labeled GS 2500 has been widely used in forensic work up to the present day. The newer single-labeled standards give better performance for microsatellite-based typing systems and should be preferred if starting a new project.

4. Formamide for the loading buffer should be top-quality and should be deionized before use (*see* **Subheading 3.**). Poor-quality formamide will adversely affect results. To aid the loading of 373 and 377 slab-gel sequencers, 5 mg/mL of dextran blue can be added to the formamide to give a strong color to the sample.

5. The standard procedure for preparing a gel for the 373 or 377 is quite involved and time-consuming. It is necessary to specify this routine in order to guarantee the performance of the machines. Some laboratories have decided to use bulk supplies of premixed acrylamide-gel reagents for reasons of time efficiency and batch control. This procedure is verified by the performance validation of every batch purchased before its introduction to routine use.

6. The recommended disk maintenance program is Norton Utilities for Macintosh, (Symantec Corp.). A similar program suitable for older Macintoshes is SilverLining (LaCie Corp.). It is also good practice to follow the Macintosh "rebuild desktop" routine at least once a month.

7. Gel-loading tips are available from Sorenson Bioscience Inc., Salt Lake City, UT, in both 0.4 and 0.2 mm thickness for 373 and 377 gels, respectively.

8. Filter set vs dye-usage chart.

	Bandpass filter (nanometers)				
Filter set A	531	—	560	580	610
Filter set B	531	545	560	580	—
Filter set C	532	543	557	584	—

Primer label
Dyes (amidites)   6-FAM   TET   HEX
Size std.   TAMRA   ROX

9. To ensure that the measured intensity of a band falls within the linear range of the detector, it should show a peak height between 40 and 4000 U on the Y-axis of the electropherogram. The multicomponent matrix file that is used to baseline data for each channel must be appropriate for the dyes and filter set used and for the type of gel used for separation. It should be noted that individual matrix files are specific to the machine that they were created on, so that care should be taken to keep results and matrix files together when analyzing data on a remote computer.

10. For useful discussions on the selection of bin sizes, and the statistical work on the precision and accuracy of band sizing with these systems, *see* **refs. 5,10.**

Acknowledgments

I wish to acknowledge the use of Genescan and Genotyper data for **Figs. 4, 6,** and **7** that was provided by Dr. Julia Anderson of the Metropolitan Laboratory of the Forensic Science Service, London, UK.

References

1. Smith, L. M., Sanders, J. Z., Kaiser, R. J., Hughes, P., Dodd, C., Connell, C. R., Heiner, C., Kent, S. B. H., and Hood, L. (1986) Fluorescence detection in automated DNA sequence analysis. *Nature* **321,** 674–679.

2. Mayrand, P. E., Robertson, J., Ziegle, J. S., Hoff, L. B., McBride, L. J., Chamberlain, J. S., and Kronick, M. N. (1991) Automated genetic analysis. *Extrait de la Revue: Annales de Biologie Clinique* **4,** 224–230.

3. Ziegle, J. S., Su, Y., Corcoran, K. P., Li, N., Mayrand, P. E., Hoff, L. B., McBride, L. J., Kronick, M. N., and Diehl, S. R. (1992) Application of automated DNA sizing technology for genotyping microsatellite loci. *Genomics* **14,** 1026–1031.

4. Fregau, C. J. and Fourney, R. M. (1992) DNA typing with fluorescently tagged short tandem repeats: a sensitive and accurate approach to human identification. *BioTechniques* **15,** 100–119.

5. Kimpton, C. P., Gill, P., Walton, A., Urquhart, A., Millican, E. S., and Adams, M. (1993) Automated DNA profiling employing multiplex amplification of short tandem repeat loci. *PCR Meth. Applicat.* **3,** 13–22.

6. Clayton, T. M., Whitaker, J. P., and Maguire, C. N. (1995) Identification of bodies from the scene of a mass disaster using DNA amplification of short tandem repeat (STR) loci. *Forensic Sci. Int.* **76,** 7–15.

7. Clayton, T. M., Whitaker, J. P., Fisher, D. L., Lee, D. A., Holland, M. M., Weedn, V. W., Maguire, C. N., DiZinno, J. A., Kimpton, C. P., and Gill, P. (1995) Further validation of a quadruplex STR DNA typing system: a collaborative effort to identify victims of a mass disaster. *Forensic Sci. Int.* **76,** 17–25.

8. Kimpton, C. P., Fisher, D., Watson, S., Adams, M., Urquhart, A., Lygo, J., and Gill, P. (1994) Evaluation of an automated DNA profiling system employing multiplex amplification for tetrameric STR loci. *Int. J. Legal Med.* **106,** 302–311.

9. Urquhart, A., Oldroyd, N. J., Kimpton, C. P., and Gill, P. (1995) Highly discriminating heptaplex short tandem repeat PCR system for forensic identification. *Biotechniques* **18,** 116–121.

10. Smith, R. N. (1995) Accurate size comparison of short tandem repeat alleles amplified by PCR. *Biotechniques* **18,** 122–128.

18

Separation of PCR Fragments
by Means of Direct Blotting Electrophoresis

Lotte Henke and Jürgen Henke

1. Introduction

For separation of alleles of PCR-dependent short tandem repeat (STR) poly-morphisms, high resolution with the power to discriminate fragments differing by one base pair is often desired. Furthermore, sequencing reactions also require unambiguous one base-pair separation. For both techniques, high sen-sitivity for detecting small amounts of DNA and the use of nonradioactive detection methods are welcomed. Because of the high costs of automatic sequenc-ing devices, the direct blotting electrophoresis system can be a useful alternative approach.

The technique described below is applicable to, for example, PCR-dependent VNTR-systems like ACTBP2, FXIIIA1, D21S11, TH01, and D18S21, with separation under denaturing conditions. One of the primers has to be labeled at the 5' end, which can be with either a digoxygenin or a biotin molecule. The efficiency of the PCR reaction should be monitored on a minigel.

The direct blotting system (**Fig. 1**) basically consists of a vertical gel and a moving belt at the bottom of the gel. A membrane is taped onto the belt. When the DNA fragments pass off the gel, the fragments are blotted directly onto the membrane *(1)*. The length of the gel (glass plates of 7, 9, 32, and 50 cm in length are available) and the speed of the belt can be adapted to a wide range of applications *(2,3)*.

2. Materials

1. Electrophoresis buffer, 1X TBE: 0.9 M Tris-(Hydroxymethyl)-aminomethane, 0.9 M boric acid, 0.025 M EDTA. Make up in 1000 mL of distilled water. The solution is adjusted to pH 8.3 by means of HCl. The buffer can be stored up to

From: *Methods in Molecular Biology, Vol. 98: Forensic DNA Profiling Protocols*
Edited by: P. J. Lincoln and J. Thomson © Humana Press Inc., Totowa, NJ

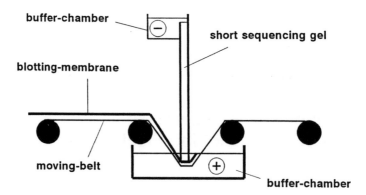

Fig. 1. Schematic illustration of a DBE device.

two months at room temperature. (The electrophoresis buffer has to be degassed prior to electrophoresis for approx 30 min in order to avoid gas bubbles in the lower buffer chamber.)

2. Loading dye: 0.025% bromphenol blue, 0.025% xylene cyanol dissolved in formamide.
3. Urea dilution buffer: 84 g urea, 18.7 g 10X TBE, 73.4 mL H_2O.

3. Methods

3.1. Gel Preparation

1. The gel solution consists of 7 mL 30% acrylamide, 0.8% bisacrylamide, and 43 mL urea dilution buffer. For 32 cm glass plates, 20 g degassed gel solution is mixed with 75 μL 10% ammonium persulfate and 17 μL TEMED (*N,N,N',N'*-Tetra-methylethylendiamine).

3.2. Sample Preparation

1. Because of the efficiency of the labeling reaction for the primer and the PCR itself, the amplification product can be diluted with distilled water up to 1:40 (DNA:water) before mixing with the loading dye.
2. Combine diluted product with loading dye (1:5, product:loading dye).
3. The samples are denatured by boiling for 5 min and stored on ice until loading. Loading should be performed within 10 min of denaturation; 1 μL of amplificate-loading buffer mix is loaded per slot.

3.3. Running Conditions

A 30 min prerun (500 V) before loading the samples provides improved resolution. The gels can be used twice if the electrophoresis buffer is changed. The electrophoresis is set at 17 mA (~1950 V, 33 W). The membrane (Hybond N, Amersham, Arlington Heights, IL) moves with a speed of 19 cm/h.

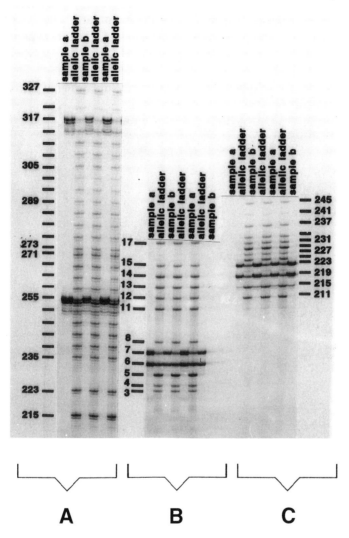

Fig. 2. Separation of amplification products by means of DB electrophoresis. **(A)** ACTBP2, **(B)** FXIIIA1, **(C)** D21S11.

3.4. Detection

The DNA fragments are crosslinked to the dried membrane by UV light. The visualization of the fragments is carried out by means of the DIG Nucleic Acids Detection Kit (Boehringer Mannheim, Mannheim, Germany).

An example of the resolution of the system is shown in **Fig. 2**. For this experiment, the 32 cm glass plates with 0.19 mm spacer were used. For loading FXIIIA1 and D21S11 samples, the electrophoresis was stopped 20 min after loading the SE33 samples. The moving of the membrane was started 125 min after loading the SE33 samples (size of the smallest fragment is approx 117 bp).

4. Note

1. As general advice, in order to obtain optimal resolution, highly purified reagents should be used!

Acknowledgments

The authors are indebted to Dr. Peter Berschick (Düsseldorf), Inge Kops, and Sabine Cleef (Cologne) for esteemed support and skillful technical assistance.

References

1. Beck, S. and Pohl, F. M. (1984) DNA sequencing with direct blotting electrophoresis. *EMBO J* **3,** 2905.
2. Berschick, P., Henke, L., and Henke, J. (1993) Analysis of short tandem repeat polymorphism SE33: a new high resolution separation of SE33 alleles by means of direct blotting electrophoresis. *Proc. 4th Int. Symp. Human Ident.*, Promega Corp., Madison, WI.
3. Schmitter, H., Berschick, P., Henke, L., and Henke, J. (1995) Molekularbiologische Untersuchungsverfahren in der forensischen Spurenanalyse (DNA-Analysen, DNA-Profiling), in *Humanbiologische Spuren* (Schleyer, F., Oepen, I., and Henke, J., eds.), Kriminalistik Verlag, Heidelberg, Germany, pp. 181–229.

19

Amplification and Sequencing of Mitochondrial DNA in Forensic Casework*

Robert J. Steighner and Mitchell Holland

1. Introduction

1.1. PCR Amplification of the Mitochondrial DNA Hypervariable Regions

Mitochondrial DNA (mtDNA) typing is increasingly used for the forensic identification of human remains *(1–9)*. This is especially true when only limited quantities of sample are present, such as when the sample has undergone extensive degradation and nuclear-typing methods are ineffectual. One characteristic of mtDNA responsible for the increasing reliance is the high copy number of mtDNA per cell, with mtDNA existing in hundreds if not thousands of copies per cell *(10–12)*. The discriminatory power of mtDNA testing arises from the polymorphic nature (between unrelated individuals) of the two hypervariable regions (HV1 and HV2) located within the D-loop of the mtDNA genome *(13–16)*. The haploid, maternal inheritance patterns of mtDNA transmission between generations, allow an inclusion or exclusion to be made when the sample sequence is compared to that of a maternal reference. Related individuals will share similar polymorphisms relative to a consensus standard *(17)*.

The initial DNA extraction and PCR amplification steps are crucial to obtaining mtDNA-sequence information from biological samples. A number of procedures for extraction of DNA from bone have been published, including the standard organic extraction (SDS/proteinase K and phenol-chloroform) pro-

*The opinions and assertions expressed herein are solely those of the author and are not to be construed as official or as the views of the United States Department of Defense or the United States Department of the Army.

From: *Methods in Molecular Biology, Vol. 98: Forensic DNA Profiling Protocols*
Edited by: P. J. Lincoln and J. Thomson © Humana Press Inc., Totowa, NJ

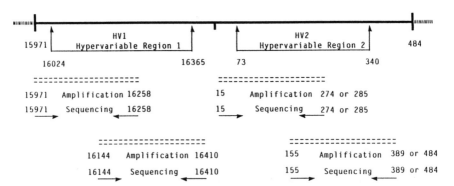

Fig. 1. Schematic diagram of the mtDNA D-loop region containing the two hypervariable regions amplified in mtDNA testing. HV1 extends from nucleotide 16024 to 16365, and HV2 from 73 to 340. Dotted lines represent the relative location of each primer set amplified and sequenced in casework. Primer sequences are listed in **Table 1**.

cedure, a silica-based extraction method, and chelation *(2,18–20)*. PCR amplification of degraded samples is generally more successful if smaller sequences are amplified. In our laboratory, each hypervariable region is amplified completely when amplified from fresh, or well-preserved templates, or in four separate reactions when amplified from degraded samples. Primer sets one (PS1) and two (PS2) are designed to amplify the HV1 region, while primer sets three (PS3) and four (PS4) amplify the HV2 region (**Fig. 1**). Each primer-set pair contains overlapping regions, providing an additional quality-control check of the data. Both strands of the PCR product are then cycle-sequenced, using both forward- and reverse-sequencing primers.

1.2. DNA Sequence Analysis of PCR Products

Cycle sequencing combined with automated sequence analysis is a high-throughput strategy for DNA typing of the mtDNA. In our laboratory, PCR products of the mtDNA control region are cycle-sequenced using dye-terminator chemistry and then analyzed on a 373A DNA Sequencer (Applied Biosystems, Foster City, CA). Dye-primer chemistry results in more normalized peak heights than that obtained with dye terminators; however, the technique also requires four separate sequencing reactions as well as increased quantities of template. The newest generation of cycle-sequencing polymerases, AmpliTaq, FS™ (Perkin–Elmer, Norwalk, CT) and Thermo Sequenase™ (Amersham, Arlington Heights, IL), greatly improve the disparity in peak heights found with the dye-terminator chemistry *(21)*. Since each technique has additional advantages and disadvantages, an evaluation regarding which strategy to pursue should be made based on the application *(22–24)*.

Analysis of the sequence electrophoretograms generated in cycle sequencing can require considerable manual editing. Typical problems include the incomplete removal of dye-terminators post-cycle sequencing, multiple PCR products in the template preparations, and mtDNA heteroplasmy (either length- or sequence-based). Additionally, some inherent limitations in the methodology include the 5' sequence-dependent pattern for insertion of dye-terminators resulting in uneven peak heights and the differential migration of the distinct dye-terminators potentially disrupting the spacing of the ladder (although software corrections usually correct for this latter effect) *(21,25)*.

2. Materials

2.1. PCR Amplification
of the Mitochondrial DNA Hypervariable Regions

1. The optimum amount of DNA template is between 10 and 1000 pg; however, amplification can be performed on <10 pg.
2. *Taq* DNA polymerase.
3. 2.5 mM deoxynucleoside triphosphates (dNTP) mixture (working concentration of 0.2 mM).
4. DNA-grade bovine serum albumin (BSA) at a working concentration of 4 μg/μL.
5. Amplification primers at a working concentration of 10 μM.
6. 10X PCR buffer (1X): 10 mM Tris-HCl, pH 8.3, 50 mL KCl, and 1.5 mM MgCl$_2$.
7. Sterile deionized water.
8. Agarose (electrophoresis grade).
9. Agarose gel-loading buffer (6X): 50% glycerol, 1.5 mM bromophenol blue, 100 mM EDTA.
10. 10X TBE (1X): 89 mM Tris-HCl, pH 8.3, 89 mM boric acid, and 2 mM Na$_2$EDTA.
11. 5 mg/mL ethidium bromide (working concentration 0.5 μg/mL).
12. DNA sizing ladder (123 base pair or other suitably sized control ladder).
13. 10% commercial bleach (7 mM sodium hypochlorite solution).
14. 667 Polaroid® film.
15. 95% ethanol.
16. Centricon® 100 Spin Dialysis Columns.
17. 9600 Perkin–Elmer Thermal Cycler or suitable equivalent from another manufacturer.

2.2. DNA Sequence Analysis of PCR Products

1. The optimal amount of PCR template is 10–200 ng.
2. Urea, ultrapure.
3. PRISM™ Ready Reaction DyeDeoxy Terminator Cycle Sequencing Kit.
4. Sequencing primers diluted to a working concentration of 10 μM.
5. Quick Spin™ G-50 Spin Dialysis Columns.

6. Nalgene® Disposable Filter Units, 0.2 μm.
7. 40% bis-acrylamide premix 19:1 5%C (electrophoresis purity).
8. Deionized formamide.
9. EDTA, 50 mM.
10. 10% ammonium persulfate (made fresh weekly and stored at 25°C).
11. TEMED ($N,N,N'N'$-tetramethylethylenediamine).
12. 373A DNA sequencer.
13. Hetovac (vacuum evacuator).
14. Heatblock.
15. DNA sequencing loading buffer (prepared fresh every day): 40 μL of 50 mM EDTA, 200 μL of deionized formamide (60 sequencing reactions).

3. Methods

3.1. PCR Amplification of the Mitochondrial Hypervariable Regions

1. Due to the high sensitivity of the PCR-amplification technique, rigorous care must be taken to eliminate all potential contaminating practices or procedures prior to amplification (*see* **Notes 1–3**). Wipe all pipeters, PCR tube racks, and the outside of gloves with 10% bleach. Allow bleach to dry before proceeding.
2. To prepare the PCR master-mix, add a volume of reagent equal to $N + 1$ times the volume added to a single reaction tube (where N is the total number of amplification reactions to be performed). Reaction components added to the master-mix include dH$_2$O, PCR buffer, dNTPs, the primers, and *Taq* polymerase (added last) (*see* **Note 4**).
3. Thirty-two PCR cycles are run when amplifying good-quality template and 38 cycles for poor-quality templates. Each PCR cycle is composed of a 94°C denaturation step (20 s), followed by a 56°C annealing step (10 s), and a 72°C extension step (30 s). To increase the specificity of the primer set I reaction, replace the 56°C annealing temperature with 62°C.
4. Analyze PCR products on 1–2% agarose gels (**Notes 5** and **6**). Analysis should focus on the specificity and robustness of the reaction (*see* **Table 1** for the size of expected products).
5. Unincorporated nucleotides and primers are removed from the PCR reactions by centrifugation through Centricon 100 spin dialysis columns (*see* **Note 7**). PCR products are added to 2 mL of sterile deionized water in a Centricon 100 spin dialysis column and centrifuged at a maximum of 1000g for 15–30 min. An additional 2 mL of sterile deionized water is added and the columns are recentrifuged. The samples are collected into the supplied retentate cups by inverting the columns and centrifuging for 1–5 min. Transfer the recovered PCR product to a microtube, estimate the volume, and store at 4°C.
6. The recommended quantity of DNA template to use for sequencing depends on the size of the template, but 10–100 ng (25 ng optimal) for the AmpliTaq polymerase sequencing kits and 1–20 ng for the AmpliTaq FS polymerase kit work well with the PCR products described in this procedure.

Table 1
Primer Sets Used for the PCR Amplification
and Sequencing of the Mitochondrial Hypervariable Regions[a]

Region	Primer set	Product size (base pairs)
HV1	F15971 / R16410	440
HV2	F15 / R381, R389, R484	367, 375, 470
	F29 / R381, R389, R484	353, 361, 456
PSI	F15971 / R16255, R16258	285, 288
PSII	F16144 / R16410	267
PSIII	F15 / R270, R274, R285	256, 260, 271
	F29 / R270, R274, R285	242, 246, 257
PSIV	F155 / R381,R389,484	227, 235, 330

Primer sequences	
F15971	5' TTA ACT CCA CCA TTA GCA CC 3'
R16255	5' CTT TGG AGT TGC AGT TGA TG 3'
R16258	5' TGG CTT TGG AGT TGC AGT TG 3'
F16144	5' TGA CCA CCT GTA GTA CAT AA 3'
R16410	5' GAG GAT GGT GGT CAA GGG AC 3'
F15	5' CAC CCT ATT AAC CAC TCA CG 3'
F29	5' CTC ACG GGA GCT CTC CAT GC 3'
R270	5' TGG AAA GTG GCT GTG CAG AC 3'
R274	5' TGT GTG GAA AGT GGC TGT GC 3'
R285	5' GTT ATG ATG TCT GTG TGG AA 3'
F155	5' TAT TTA TCG CAC CTA CGT TC 3'
R381	5' GCT GGT GTT AGG GTT CTT TG 3'
R389	5' CTG GTT AGG CTG GTG TTA GG 3'
R484	5' TGA GAT TAG TAG TAT GGG AG 3'

[a]Primer sequences are based on the numbering system of Anderson *(17)*. Numbering begins at the 5' end of each primer. F and R designations stand for forward (mitochondrial light) and reverse (mitochondrial heavy) strands, respectively.

3.2. DNA Sequence Analysis of PCR Products

1. At least two sequencing reactions are run per PCR product with forward and reverse primers. The sequence obtained from the reverse primer will complement that obtained from the forward primer, providing additional confirmation of the data. The sequencing primers may be the same primers used in the PCR amplification reaction; however, certain advantages exist in using sequencing primers distinct from those used in the PCR reactions (*see* **Note 9**). Before sequencing, analyze product gels and determine which samples to sequence based on success of the amplification. Prepare each sequencing reaction as follows in a 0.2 mL thin-walled PCR microamp tube. A pGEM-3Zf(+) DNA control reaction (provided with both terminator kits) is set up (according to the

manufacturer's instructions provided with the kit) as are the appropriate quality controls (*see* **Note 10**).

   a. 1 µL of 10 µ*M* sequencing primer (either a forward or reverse primer).
   b. 10–100 ng purified PCR product (1–20 ng if using the AmpliTaq FS kit).
   c. 9.5 µL of DNA Prism Ready Reaction mix (8.0 µL if using the AmpliTaq FS kit).
   d. Q.S. to 20 µL with sterile deionized water.
2. The cycle-sequencing conditions are 96°C for 15 s (DNA denaturation), 50°C for 5 s (annealing), and 60°C for 2 min (extension). Twenty-five cycles are run followed by a rapid ramp to a 4°C holding temperature.
3. Quick spin columns (Boehringer-Mannheim, Mannheim, Germany) are used to remove the unincorporated dye-terminators from sequenced products according to manufacturer's instructions (*see* **Note 11**).
4. Speed-desiccate the samples to dryness at room temperature. Desiccated samples may be stored for at least 1 mo at –20°C.
5. Prepare a 4.75% polyacrylamide-gel solution in 1X TBE. Allow the gel to polymerize for at least 1 h at room temperature. Ensure that plates are on a level surface during polymerization.
6. After polymerization, prerun the gel for at least 10 min at constant power of 28 W for ABI 373A stretch model or 30 W for regular 373A model. Ranges should be 980–1600 V, 16–21 mA, and temperature at 40°C (although 40°C may not be reached during the prerun).
7. Resuspend the samples in 4 µL of loading buffer.
8. Heat the products at 96°C for 2 min in a heat block and immediately place on ice before loading.
9. Gel run times are 8 h for the primer sets or 12 h for the entire HV1 or HV2 regions. Use the same run parameters as those used during the prerun.

4. Notes

4.1. PCR Amplification of the Mitochondrial DNA Hypervariable Regions

1. In order to prevent contamination, all steps in this procedure should be performed in a laminar flow or dead-space hood, in a PCR-product-free environment. This is especially critical when poor-quality template is being amplified. Individuals performing the reactions should avoid exposure to previously amplified DNA on the day of the experiment. No specimens or materials used for the extraction of DNA from biological specimens should be allowed in a postamplification PCR product room. Gloves, disposable lab coats and/or disposable sleeves must be changed after working with each individual specimen. Use specially designated PCR set-up pipetters and aerosol-resistant pipet tips to minimize contamination. Change pipet tips between each transfer or addition.
2. An extraction reagent blank control (prepared during sample extraction) should also be processed. This control will regulate for the presence of contaminants

during the sample extraction. Provided enough evidentiary sample remains, a second extraction of the sample, completely independent from the first extraction, should be performed. The sequence obtained from the second extraction will provide additional confirmation that the sequence obtained is authentic.

3. Include two negative controls with every set of amplification reactions. The first negative control (set up first) should test the reagents used in the PCR master mix. The second negative control should be set up last, in order to test the cleanliness of the master mix following additions to all other reaction tubes. A known human positive-control sequence should also be included with each amplification reaction. The positive control should be set up as the second-to-last tube. All other tubes should be set up and closed prior to adding the positive control (with the exception of the last negative control).

4. Preparing a PCR master mix containing all common reagents saves considerable time and helps normalize the reactions. The standard 50 µL PCR reaction contains 10 mM Tris-HCl, pH 8.3, 50 mM KCl, 1.5 mM MgCl$_2$, BSA at 4 µg/µL, a primer concentration of 0.4 µM (recommended primer sets and sequences may be found in **Table 1**), 0.2 mM dNTPs, and 2.5 U of *Taq* polymerase.

5. Amplification may not be successful due to PCR inhibition, which in some cases may be seen by the absence of primer-dimer on the product gel. If primer-dimer is expected, but not observed, inhibition of the reaction is suspected. *Taq* DNA polymerase can be elevated (not greater than a total of 12.5 U) or the volume of extract added to the reaction decreased.

6. In cases where low concentrations of template or badly degraded samples result in no product, it may be beneficial to run a second round of PCR, using an aliquot of the first reaction for the template. A seminested or nested primer strategy, where one or both of the second set of primers are internal to the first set, will help control for the contamination problems inherent with running a large number of PCR cycles. If nested reactions are run, the products from two completely independent series of nested reactions with the same sample should be compared to assure that the sequences are identical.

7. Centricon 100s are more effective in removing nucleotides than primers in some cases. The presence of PCR primers in the sequencing reactions may result in secondary-sequencing initiation sites at positions autonomous to the desired start position. The resulting data may contain higher background or display multiple overlapping peaks. This is especially critical when using dye-terminator chemistry, since all sequenced products are labeled. If this appears to be a problem, the use of lower concentrations of primers in the PCR reaction, multiple washings of the Centricon 100s, or gel purification of the PCR products may alleviate the problem.

8. In addition to the strategy presented here, a number of different amplification and sequencing schemes for the mitochondrial D-loop region have been published, each of which focuses on obtaining the sequence of the two hypervariable regions *(23,26–29)*. The technical working group for DNA analysis methods (TWGDAM) has established the recommended region for HV-1 sequence from base pairs

16024–16365 and 73–340 for the HV-2 region. The most straightforward procedures involve amplification of both HV-1 and HV-2 in two separate PCR reactions using primers "tailed" with a universal sequencing primer site *(26–28)*. The
PCR products are then analyzed with dye-labeled primers. Another procedure
employs a nested PCR strategy after first amplifying the entire D-loop region in
a single reaction *(29)*. The hypervariable regions are then amplified in the nested
PCR reactions. Although nested PCR may generate a cleaner product, it may also
be more susceptible to contamination problems due to the large number of total
amplification cycles performed. Furthermore, in highly degraded samples, it may
be difficult to amplify the entire D-loop region in a single amplification reaction.

4.2. DNA Sequence Analysis of PCR Product

9. Primer-dimer formation during PCR reactions may affect the resulting data generated during cycle sequencing. Primer dimers formed during PCR may not be
   eliminated by centrifugation through the Centricon 100s. Subsequent annealing
   of the sequencing primer to primer dimer instead of the sequence of interest will
   result in sequencing of the dimer. This may result in obscured data toward the
   beginning of a sequence (within the first 25–30 base pairs). Designing sequencing primers to avoid complementarity with PCR primers will reduce the problem.
10. In addition to the pGem control, a positive sequencing control should be
    sequenced to provide known sequence information for the region being analyzed.
    If the sequence obtained from the extraction reagent blank is different from the
    sample sequence(s), and the sample sequence(s) are consistent from one speci-
    men to the next over multiple extractions and/or amplifications, the sequence
    information from the sample can be considered authentic.
11. If using Boehringer-Mannheim G-50 gel-filtration columns, a single dH$_2$O wash
    prior to purification has been found to be superior to multiple washings. Multiple
    washings with dH$_2$O diminishes the columns' retentive capacity. Failure of the
    columns may be seen by yellow "smearing" on the gel image (red on the electro-
    phoretograms) caused by labeled thymine terminators eluting through the col-
    umn. Manifestations of incomplete terminator-dye removal include increased
    background throughout the entire lane, obscured sequence data at the beginning
    of the sequence, and/or formation of a particularly broad "thymine-smear" that
    migrates at a position equivalent to approx 200–240 base pairs on the gel.
12. Clean sequencing plates are essential. Do not air dry the plates after washing. If
    ethanol is not completely removed, a blue or green fluorescence may result when
    plates are scanned on the 373A sequencer. Some brands of urea and old
    formamide also result in a higher fluorescent background.
13. Sequencing with kits containing the Amplitaq, FS polymerase results in more
    normalized peak heights and easier sequence-editing than data generated with
    kits containing the Amplitaq polymerase. The concentration of dye-terminators
    is diminished due to the increased efficiency of the AmpliTaq, FS polymerase to
    incorporate dye-terminators in the cycle-sequencing reaction, ultimately simpli-
    fying sample cleanup (ethanol precipitation may replace gel filtration for remov-

ing dye-terminators postcycle sequencing). Our laboratory is in the process of converting to the Amplitaq, FS kit for all of our cycle-sequencing reactions.

14. The concentration of terminator mix used in the mtDNA sequencing reactions may be halved with the volume difference made up with dH$_2$O. Peak heights obtained with 0.5X terminator mix are approx 95% of the heights obtained with 1X terminator mix. No difference in the length of read of the small templates mentioned in this procedure was seen. This observation was obtained with the AmpliTaq cycle-sequencing kit; it is not known if similar results would be obtained with the AmpliTaq FS cycle-sequencing kit. Independent verification should be performed if longer cycles are sequenced.

15. Sequence information at each base position should be confirmed by data from both DNA strands when possible. A specific exception is when a polycytosine stretch prohibits confirmation in both directions. In these instances, confirming sequence information from a single DNA strand will be acceptable. In general, data should be obtained from at least two independent amplifications when sequence information from only one strand is being reported (*see* **Note 16**).

16. Patterns for the sequence-dependent insertion of terminators have been published and are reflected by uneven peak heights *(21)*. The patterns for the AmpliTaq FS kit have not been characterized as extensively as those for the AmpliTaq kits, therefore only a few rudimentary observations are presented. For AmpliTaq FS, the rules include Gs after Cs are small; Gs after As are small; Gs, As, and Ts after Gs are large; and Cs after Ts are "often" larger *(22)*. The information may be used to justify editing a base call, however, this should not be the sole criterion for base editing.

17. Sequence data at the poly-cytosine stretches are often difficult to confirm. In particular, the range between 16182 and 16193 in HV1 may be hard to interpret if there is a 16189 T-C polymorphism or length-based heteroplasmy present in the sample. Length-based heteroplasmy through this region is common and results in poor sequencing data due to the multiple overlapping peaks in the electrophoretogram at all positions distal to the insertion. Sequence data in positions leading up to the poly-cytosine stretch will be normal. If there is no confirming data from reverse strand, then length-based heteroplasmy is suspected.

Acknowledgment

Many thanks go out to our colleagues at the Armed Forces DNA Identification Laboratory AFDIL). The hard-working staff of AFDIL deserves much of the credit for the information presented in this chapter. In addition, thanks go to the Armed Forces Institute of Pathology, the Office of the Armed Forces Medical Examiner, and the American Registry of Pathology for their support.

References

1. Holland, M. M., Fisher, D. L., Mitchell, L. G., Rodriquez, W. C., Canik, J. J., Merril, C. R., and Weedn, V. W. (1993) Mitochondrial DNA sequence analysis of human skeletal remains: identification of remains from the Vietnam War. *J. Forensic Sci.* **38**, 542–553.

2. Holland, M. M., Fisher, D. L., Roby, R. K., Ruderman, J., Bryson, C., and Weedn, W. (1995) Mitochondrial DNA sequence analysis of human remains. *Crime Lab. Dig.* **22,** 109–115.

3. Stoneking, M., Melton, T., Nott, J., Barritt, S., Roby, R., Holland, M., Weedn, V., Gill, P., Kimpton, C., Aliston-Greiner, R., and Sullivan, K. (1995) Establishing the identity of Anna Anderson Manahan. *Nature Genet.* **9,** 9,10.

4. Sullivan, K. M., Hopgood, B., and Gill, P. (1992) Identification of human remains by amplification and automated sequencing of mitochondrial DNA. *Int. J. Legal Med.* **105,** 83–86.

5. Ivanov, P., Wadhams, M., Roby, R., Holland, M., Weedn, V., and Parsons, T. (1996) Mitochondrial DNA sequence heteroplasmy in the Grand Duke of Russia Georgij Romanov establishes the authenticity of the remains of Czar Nicholas II. *Nature Genet.* **12,** 417–422.

6. Ginther, C., Issel-Tarver, L., and King, M. C. (1992) Identifying individuals by sequencing mitochondrial DNA from teeth. *Nature Genet.* **2,** 135–138.

7. Gill, P., Ivanov, P. L., Kimpton, C., Piercy, R., Benson, N., Tully, G., Evett, I., Hagelberg, E., and Sullivan, K. (1994) Identification of the remains of the Romanov family by DNA analysis. *Nature Genet.* **6,** 130–135.

8. Boles, T. C., Snow, C. C., and Stover, E. (1995) Forensic DNA testing on skeletal remains from mass graves: a pilot project in Guatemala. *J. Forensic Sci.* **40,** 349–355.

9. Stoneking, M., Hedgecock, D., Higuchi, R. G., Vigilant, L., and Erlich, H. A. (1991) Population variation of human mtDNA control region sequences detected by enzymatic amplification and sequence-specific oligonucleotide probes. *Am. J. Hum. Genet.* **48,** 370–382.

10. Michaels, G. S., Hauswirth, W. W., and Lapis, P. J. (1982) Mitochondrial DNA copy number in bovine ocytes and somatic cells. *Developmental Biol.* **94,** 246–251.

11. Robin, E. D. and Wong, R. (1988) Mitochondrial DNA molecules and virtual number of mitochondria per cell in mammalian cells. *J. Cellular Physiol.* **136,** 507–513.

12. Bogenhagen, D. and Clayton, A. (1974) The number of mitochondrial deoxyribonucleic acid genomes in mouse L and human Hela cells. *J. Biol. Chem.* **249,** 7991–7995.

13. Lee, H. C., Paglairo, E. M., Berka, K. M., Folk, N. L., Anderson, D. T., Ruano, G., Keith, T. P., Phipps, P., Herrin, G. L., Garner, D. D., and Gaensslen, R. E. (1991) Genetic markers in human bone: I. Deoxyribonucleic acid (DNA) analysis. *J. Forensic Sci.* **36,** 320–330.

14. Cann, R. L., Brown, W. M., and Wilson, A. C. (1984) Polymorphic sites and the mechanism of evolution in human mitochondrioal DNA. *Genetics* **106,** 479–499.

15. Cann, R. L., Stoneking, M., and Wilson, A. C. (1987) Mitochondrial DNA and human evolution. *Nature* **325,** 31–36.

16. Horai, S. and Hayasaka, K. (1990) Intraspecific nucleotide sequence differences in the major noncoding region of human mitochondrial DNA. *J. Hum. Genet.* **46,** 828–842.

17. Anderson, S., Bankier, A. T., Barrell, B. G., deBruijn, M. H. L., Coulson, A. R., Drouin, I. C., Eperon, I. C., Nierlick, D. P., Roe, B. A., Sanger, F., Schreier, P. H., Smith, A. J. H., Staden, R., and Young, I. G. (1981) Sequence and organization of the human mitochondrial genome. *Nature* **290,** 457–465.

18. Hagelberg, E. and Clegg, J. (1991) Isolation and characterization of DNA from archaeological bone. *Proc. Roy. Soc. London* **224,** 45.

19. Hoss, M. and Paabo, S. (1991) DNA extraction from pleistocene bones by a silica based purification method. *Nucleic Acids Res.* **21,** 3913,3914.

20. Walsh, P. S. et al. (1991) Chelex 100 as a medium for simple extraction of DNA for PCR-based typing from forensic material. *Biotechniques* **10,** 506–513.

21. Parker, L. T., Deng, Q., Zakeri, C., Carlson, C., Nickerson, D. A., and Kwok, P. Y. (1995) Peak height variations in automated sequencing of PCR products using *Taq* dye-terminator chemistry. *Biotechniques* **19,** 116–121.

22. *ABI Prism™ Comparative PCR Sequencing: A Guide to Sequencing Based Mutation Detection.* (1995) No. 770901–001.

23. *ABI PRISM™ Mitochondrial DNA Sequencing.* Tech. booklet no. 237820–001.

24. *DNA Sequencing Chemistry Guide.* (1995) Perkin-Elmer/A. B. I. no. 903563 version A.

25. Metzker, M. L., Lu, J., and Gibbs, R. A. (1995) Electrophoretically uniform fluorescent dyes for automated DNA sequencing. *Science* **271,** 1420–1422.

26. Sullivan, K. M., Hopgood, R., Lang, B., and Gill, P. (1991) Automated amplification and sequencing of human mitochondrial DNA. *Electrophoresis* **12,** 17–21.

27. Hopgood, R., Sullivan, K. M., and Gill, P. (1992) Strategies for automated sequencing of human mitochondrial DNA directly from PCR products. *Biotechniques* **13,** 60–68.

28. Bertranpetit, P., Sala, J., Calafell, F., Underhill, P. A., Moral, P., and Comas D. (1994) Human mitochondrial variation and the origin of the Basques. *Ann. Hum. Genet.* **59,** 63–81.

29. Orrego, C. and King, M. C. (1990) Determination of familial relationships, in *PCR Protocols: A Guide to Methods and Applications* (Innis, M. A., Gelfand, D. H., Sninsky, J. J., and White, J. J., eds.), Academic, London, pp. 416–426.

20

Minisatellite Variant Repeat Unit Mapping Using PCR (MVR-PCR)

David L. Neil

1. Introduction

Since the discovery of DNA fingerprinting ten years ago *(1)*, the direct analysis of DNA variation has been successfully used in a wide range of applications and has become an important tool in forensic science *(2,3)*. Most DNA typing systems assay DNA fragment length at repeated loci showing tandem repeats, for example, minisatellites and microsatellites, which exhibit allelic variation in tandem-repeat copy number *(4–10)*. However, despite the great discriminatory power and sensitivity of this approach, it has inherent limitations that have prevented its full potential from being realized. The most informative loci, hypervariable minisatellites, have quasicontinous allele-length distributions, with many more alleles than can be resolved by agarose-gel electrophoresis, making unequivocal identification of individual alleles impossible *(3,6,11–16)*. This has led to protracted arguments over the definition of match criteria and the extent to which allele-size estimates are prone to error *(17–19)*. This problem can, in principle, be overcome by using microsatellites, which have fewer and smaller alleles that can be precisely sized on sequencing gels *(9,10)*. Although microsatellites have now been widely adopted in the construction of investigative forensic databases, these loci also have attendant drawbacks, the most important being restricted allelic variability, as compared to hypervariable minisatellites.

We have developed a technique that can, in principle, overcome many of these limitations by sampling sequence variation, rather than length differences, between hypervariable minisatellite alleles. In addition to variation in the number of tandem repeats in an allele, all minisatellites characterized to date also

From: *Methods in Molecular Biology, Vol. 98: Forensic DNA Profiling Protocols*
Edited by: P. J. Lincoln and J. Thomson © Humana Press Inc., Totowa, NJ

show variation in the sequence of the repeated unit. As a consequence, alleles at such loci can exhibit internal variation in the interspersion of the different repeat-unit types *(4–6,20–26)*. The positions of minisatellite variant repeats (MVRs) along a tandemly repeated array define an MVR map (**Fig. 1A**) that represents the internal structure of an allele with respect to the distribution of those repeat-unit types. We have developed a simple PCR-based technique (MVR-PCR) that can be used to obtain MVR map information from alleles at any suitable minisatellite locus *(27)*.

MVR-PCR uses different MVR-specific primers to recognize and prime from different repeat-unit types along a minisatellite allele (**Fig. 1B**). Separate amplifications between the different MVR-specific primers and a primer at a fixed site in the DNA flanking the minisatellite generate complementary sets of PCR products extending from the flanking primer site into the minisatellite repeat array and terminating at each type of variant repeat unit (**Fig. 1C**, *see*

Fig. 1. *(opposite page)* The principles of minisatellite repeat coding. (**A**) Schematic representation of a minisatellite locus suitable for diploid MVR coding. The flanking DNA is indicated by a plain line and the variant repeats by shaded and white boxes. The internal structure of minisatellite alleles with at least two variant repeat types can be described by encoding these as a binary string extending from the first repeat unit. If all repeat units are the same size, the repeat unit interspersion pattern of both alleles from an individual can be described simultaneously, by a ternary code, as shown. At each repeat unit position alleles may have both **a**-type repeats (code 1), both **t**-type repeats (code 2), or be heterozygous with one **a**-type and one **t**-type repeat (code 3). Beyond the end the shorter allele the hemizygous code from the longer is described by codes 4 (**aO**) and 5 (**tO**). (**B**) The consensus repeat units of MS32 and MS31 and sequences of MVR-specific primers. For each locus the two major sites of polymorphic base substitution are shown above and below the consensus, with the position used in MVR mapping shown in bold. Variant repeat-unit-specific oligonucleotide primers terminating at this polymorphic position are shown below each minisatellite sequence. Each primer has a sequence complementary to the minisatellite repeat unit (uppercase) preceded by a 20 nt 5' non-minisatellite extension with no known homologue in the human genome (lowercase). The "TAG" primer identical to this sequence, is also shown. (**C**) The principle of MVR-PCR, illustrated for allele 1 using a **t**-type MVR-specific primer. **i.** During the first PCR cycle the MVR specific primer (black), at low concentration, anneals to approximately one **t**-type repeat unit per molecule and extends into the flanking DNA. **ii.** During the next cycle DNA synthesis from the flanking primer creates a sequence (diagonal stripes) complementary to TAG when the 20 nt extension of the repeat unit primer is copied. **iii.** Products of the second round of PCR, which terminate with the flanking primer and TAG complements are now amplified efficiently by a high concentration of the flanking primer and TAG, creating a stable set of PCR products extending from the flanking primer site to each **t**-type repeat unit. Occasional internal priming from PCR products by the MVR-specific

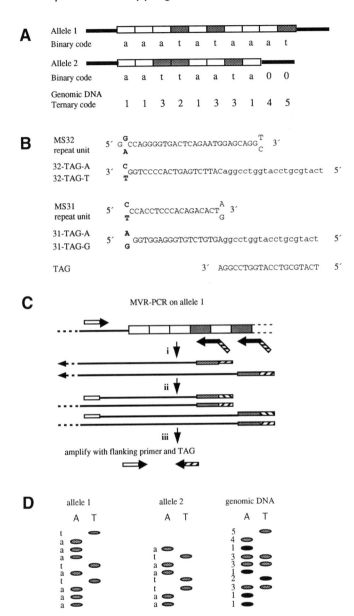

primer will generate authentic, but shorter, PCR products. The use of the other MVR-specific primer will create a complementary set of products terminating at each **a**-type repeat. **(D) i.** The profiles expected from Southern blot hybridization of MVR-PCR products generated from allele 1. **ii.** MVR-PCR mapping of allele 2. **iii.** MVR-PCR on genomic DNA from an individual with alleles 1 and 2. The binary single-allele codes and ternary code generated from genomic DNA are shown to the left of the respective maps.

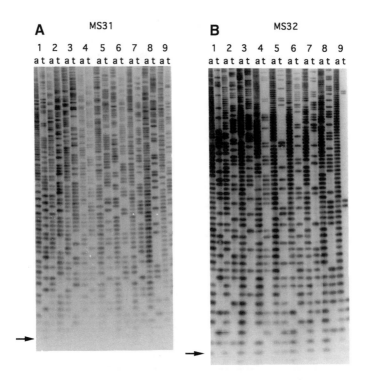

Fig. 2. Diploid MVR maps of MS31 and MS32. Duplex MVR-PCR was performed on 100 ng genomic DNA from 9 unrelated individuals (1–9). The MVR-PCR products were resolved by electrophoresis through an agarose gel transferred to Hybond-N (fp)™ membrane by Southern blotting. Hybridization with radiolabeled probe for MS31 **(A)** followed by stripping and reprobing with MS32 **(B)** revealed diploid MVR maps from each locus. Arrowed position indicates the first rung used for scoring MVR codes.

Subheading 3.1.). The sets of PCR products are then resolved side-by-side by electrophoresis through an agarose gel (*see* **Subheading 3.2.**). The resulting MVR maps can be revealed by ethidium-bromide staining *(28)*, by Southern-blot hybridization, and detection with radioactive or chemiluminescent probes, followed by autoradiography *(27,29)* (**Fig. 1D**, *see* **Subheadings 3.3. and 3.4.**), or by using nonisotopically labeled primers *(30)*. The rungs on the MVR ladder can then be read off as a simple digital code (**Figs. 1** and **2**, **Subheading 3.5.**). MVR-PCR of individual minisatellite alleles gives a binary code, with bands in one lane or the other (**Fig. 1A**). Polymorphic positions in the DNA flanking minisatellites can be exploited to design allele-specific flanking PCR primers that enable the mapping of single alleles directly from the genomic DNA of individuals heterozygous at such polymorphic positions *(31)*. At rare loci where most or all repeat units are the same length, MVR-PCR can be applied to genomic

DNA using a universal flanking primer, to give a diploid map of the interspersion patterns of repeats from two alleles superimposed. Diploid MVR maps can be described by a ternary code with bands in either or both lanes (**Fig. 1A,Diii**).

There are several criteria to which a highly informative locus for MVR-PCR must conform. It must be polymorphic, preferably with an allele-length heterozygosity >95%, to ensure that most or all alleles are rare. Repeat-unit heterogeneity must not be too extensive, and sites of variation must be suitably positioned to allow the design of repeat-unit-specific primers. Furthermore, there must be little or no repeat-unit-length variation if diploid MVR mapping is to be performed. Collectively, these criteria are rather stringent; only two out of 49 hypervariable minisatellites isolated in this laboratory, MS32 (D1S8) and MS31 (D7S21), have all of the aforementioned characteristics.

Single-allele MVR mapping of these two loci has shown levels of variability far in excess of those estimated from Southern-blot length analysis and has also revealed a gradient of variability along the alleles of these loci, with most MVR map variation confined to one ultravariable end of the array *(27,32)*. The high heterozygosity of these loci combined with polar allelic variability make diploid MVR maps generated from the ultravariable ends of MS31 and MS32 highly informative, almost to the point of individual specificity *(27,32)*.

Both MS31 and MS32 have two common sites of base substitutional polymorphism in their repeat units (**Fig. 1B**); MS32 has an A/G transition separated by 1 bp from a C/T transition while the variant positions at MS31 are adjacent, G/A followed by C/T *(6)*. However, only the more informative of the two sites (MS32 A/G and MS31 C/T) is assayed in each case. For historical reasons, MS32 repeat units with the "G" variant are called **a**-type repeat units and those with the "A" variant are called **t**-type repeat units. To ensure compatibility with computer software developed for analysis and manipulation of MS32 MVR-PCR allele codes, MS31 repeat units are also designated **a**-type ("T" variant) and **t**-type ("C" variant).

Both pairs of MS31 and MS32 repeat-unit-specific primers have a sequence complementary to the minisatellite repeat unit plus a noncomplementary 20 nt 5' extension ("TAG") and are identical, except for the 3'-most base, which corresponds to either form of the variant position being assayed (**Fig. 1B**). At the correct annealing temperature, these primers only allow extension from repeat units that they match perfectly. The TAG sequence allows MVR detection and subsequent amplification to be uncoupled by using different primers for each process. This prevents the progressive shortening of amplified products down to the first few repeat units, because of MVR-specific primers priming internally within extant PCR products at each PCR cycle *(7)*. PCR amplifications are carried out with a very low concentration of a repeat-unit-specific detector primer plus high concentrations of the two driver primers, one in the minisatellite

Table 1
Diploid MVR-PCR Primer Sequences

Primer use	Primer names	Primer sequence 5'–3'
MS31 5' flanking	31A	CCCTTTGCACGCTGGACGGTGGCG
MS31	31-TAG-A	tcatgcgtccatggtccggAGTGTCTGTGGGAGGTGGA[a]
MVR detection	31-TAG-G	tcatgcgtccatggtccggAGTGTCTGTGGGAGGTGGG
MS32 5' flanking	32OR	GAGTAGTTTGGTGGGAAGGGTGGT
MS32	32-TAG-A	tcatgcgtccatggtccggaCATTCTGAGTCACCCCTGGC
MVR detection	32-TAG-T	tcatgcgtccatggtccggaCATTCTGAGTCACCCCTGGT
MVR amplification	TAG	TCATGCGTCCATGGTCCGGA

[a]Uppercase denotes the region of MVR-specific primers complementary to the MS31A repeat unit. Lowercase denotes the TAG sequence.

flanking DNA plus the TAG sequence itself. (Primer sequences are shown in **Fig. 1B** and **Table 1**) 32-TAG-A terminates with a "C" and recognizes **a**-type repeat units; 32-TAG-T terminates with a "T" and primes from **t**-type repeats. 31-TAG-A ends with an "A" and detects **a**-type repeat units whereas MS31 **t**-type repeats are detected by 31-TAG-G, which ends with a "G". At each PCR cycle, an MVR-specific primer will prime from different cognate repeat units along the minisatellite input molecules and extend into the flanking DNA past the flanking priming site. At the next cycle, the flanking primer will prime synthesis of a strand complementary to these products back into the minisatellite, terminating by creating a sequence complementary to TAG, from which the TAG primer can now prime. The high concentration of the flanking primer and TAG now allows efficient amplification of this second PCR product during subsequent PCR cycles (**Fig. 1C**).

Since the same PCR parameters are used in diploid mapping of both MS31 and MS32, combinations of primers can be used to generate diploid codes from MS31 and MS32 alleles simultaneously *(32)*. "Duplex MVR-PCR" uses both MS31 and MS32 flanking primers, one of the MVR-specific primers from each locus and TAG. 31-TAG-A and 32-TAG-A are used in one PCR reaction with 31-TAG-G and 32-TAG-T in the other, to maintain the conventional order of **a**-type and **t**-type repeat-unit lanes on MVR-PCR gels. Examples of diploid codes generated from MS31 and MS32 by duplex MVR-PCR are shown (**Fig. 2**). Southern-blot analysis by sequential hybridization with MS31 (**Fig. 2A**), followed by MS32 (**Fig. 2B**), reveals complete sets of PCR products from each locus with no evidence of interlocus interference or cross-hybridization, indicating that repeat units from both loci amplify, and are detected, independently.

Diploid profiles from both MS31 and MS32 are encoded in the same way. For individuals with alleles containing both **a**-type and **t**-type repeats, the diploid map is described by a ternary code in which each rung from the bottom of the ladder upward is coded as **1** (a band in the **a**-lane only), **2** (a band in the **t**-lane only), or **3** (a band in both lanes). Positions that cannot be scored reliably, for example, those corresponding to PCR products too small to hybridize efficiently with the probe, are designated as ambiguous by a "**?**". Coding commences from the second repeat unit (arrowed in **Fig. 2**), since the hybridization signal from the first repeat is often too faint to allow reliable scoring; this start position is confirmed by reference to a standard DNA sample of known code run on all gels. Diploid MVR profiles from both loci can generally be reliably scored at least 50 repeat units into the minisatellite array. Diploid MVR-PCR typing of MS31 and MS32 in several hundred unrelated Caucasians, followed by ternary-code comparison, showed that there were on average 30 code mismatches per pair of individuals over the first 50 repeat units. No two individuals shared the same MVR code at either locus and all individuals could be distinguished using only the first 17 repeat positions (data not shown).

Both MS31 and MS32 alleles also contain a small proportion (around 1%) of repeat units that are not amplified by either MVR-specific primer. These "null" or **0**-type repeats can be scored by dosage at MS32 creating three additional coding states, namely **4** (a faint band in the **a**-lane), **5** (a faint band in the **t**-lane), and **6** (no bands at that position). Bands corresponding to states **4** and **5** are half the intensity of those for states **1** and **2**, while code **6** positions, which are rare, appear as a gap on the ladder *(29)*. Coding states **4**, **5**, and **6** will also be generated beyond the end of the shorter of a pair of alleles, since the code above this position will be derived from only one allele (**Fig. 1A**). No PCR products will appear beyond the end of the longer allele, generating a **66666...** code. Because of variation in band intensity at MS31, dosage cannot be used to identify reliably the positions of null repeats, except for code **6** positions *(32)*. However, the potential problem of misscoring weak bands as nulls does not compromise the use of diploid coding for individual identification, since individual specificity at both loci remains when band-intensity information is removed by converting all code **4** (**a0**) and code **5** (**t0**) positions to codes **1** (**aa**) and **2** (**tt**), respectively, to generate quaternary codes (**1**, **2**, **3**, and **6**) corresponding to bands present only in the **a**-track, only in the **t**-track, in both tracks, and in neither track, respectively.

Diploid MVR-PCR provides a novel and simple method for generating highly discriminatory digital information directly from human DNA and therefore potentially provides a powerful new DNA profiling system for use in forensic DNA typing *(27,32)*. The technique overcomes many of the limitations associated with DNA typing systems that involve allele-length measurements and also offers additional advantages. Most importantly, MVR-PCR obviates the problem of DNA-profile matching and allele-size measurement based on error-prone DNA fragment length estimation. MVR maps are resolved on agarose gels but do not require band-size measurements, meaning that code generation does not require standardization of electrophoretic systems, is immune to gel distortions and band shifts, and does not require side-by-side comparisons of DNA samples on the same gel. Furthermore, MVR maps are simple to encode in a standardized digital format, making them highly amenable to computer databasing and analysis, simplifying sample comparison and facilitating rapid database searches and the dissemination of MVR information between laboratories. These codes are ideal for objectively determining a match between a forensic sample and a criminal suspect, since match criteria are unambiguous. Diploid MVR-map databases will provide a simple method for determining the statistical significance of a match between a forensic sample and a suspect, by counting the frequency (probably zero) of the particular MVR code in the appropriate database.

Besides providing enormous, although as yet unquantified, exclusionary power, preliminary investigations suggest that MVR-PCR is particularly well-suited to many types of forensic analysis. It is applicable to very small quantities of DNA *(33)* as well as to degraded DNA and, importantly, to mixed DNA samples of the type often encountered in forensic casework *(34)*. In the latter case, single-allele MVR-PCR can often be used to recover MVR information from each individual contributing to the mixture *(31,34)*. The technique can also be used in parentage testing, with the caveat that these loci have relatively high rates (1%) of *de novo* mutation to new MVR structures *(27)*. In addition, the design of the MVR system means that it is also suitable for automation. For example, the use of ethidium-bromide staining *(28)* or nonisotopically labeled primers, together with in-gel or microcapillary tube *(30,35–37)* detection by laser scanning, may allow the scoring of MVR codes directly into a computer database. As such, diploid MVR-PCR provides a DNA-typing system as widely applicable and sensitive as many currently used in forensic casework.

2. Materials
2.1. Hardware and Consumables

All chemicals, reagents, glass, and plasticware used for MVR-PCR are standard and can be purchased from recognized suppliers of molecular-biology reagents (Applied Biotechnologies Limited, Boehringer Mannheim, Fisons, FMC bioproducts, Gibco-BRL, New England Biolabs, Serva, Sigma, Pharmacia, and Perkin-Elmer/Cetus).

2.2. Apparatus
2.2.1. PCR Amplification from Genomic DNA

1. Thermal Cycler. We use GeneAmp PCR system 9600 or DNA thermal cycler 480, Perkin–Elmer/Cetus (*see* **Note 2**).

2. Spectrophotometer to determine the PCR primer and template DNA concentration (*see* **Note 7**).

2.2.2. Agarose Gel Electrophoresis

1. Horizontal gel electrophoresis tank (*see* **Note 3**).
2. Electrophoresis power pack (e.g., BioRad Laboratories, Hercules, CA, Model 200/2.0).
3. Gel molds and combs (*see* **Note 17**).

2.2.3. Southern Blot Hybridization (see **Note 4**)

1. An orbital shaker for gel pretreatments (depurination, denaturation, and neutralization).
2. Capillary-action Southern-blot-hybridization platform.
3. Hybridization membrane (e.g., Hybond-N (fp) (Amersham, Arlington Heights, IL).
4. Chromatography paper. Whatman 3MM Chr.
5. Blotting paper (e.g., Quickdraw [Sigma, St. Louis, MO]).
6. 80°C oven to dry hybridization membranes.
7. UV source for DNA crosslinking, e.g., UV Crosslinker (Stratagene, La Jolla, CA), UV transilluminator (Chomato-vue C-63, UV Products Inc.)
8. Siliconized Pasteur pipets for probe recovery.
9. Containers for Southern-blot hybridization. For safety and convenience we use hybridization bottles mounted in a heated rotisserie oven (Hybaid, Teddington, UK); dedicated perspex boxes or sealed plastic bags may also be used. If boxes or bags are to be used, a 65°C shaking waterbath or hybridization oven will be needed.
10. A boiling waterbath or 100°C heated block are required for probe denaturation.

2.2.4. Autoradiography

1. A hand-held β-monitor to estimate required exposure times/conditions.
2. X-ray film, cassettes, and intensifying screens for autoradiography. Standard films (e.g., Fuji RX) are suitable, and X-ray cassettes are available from numerous manufacturers (e.g., Genetic Research Instrumentation Limited.)

2.3. Reagents and Solutions

2.3.1. PCR Amplification from Genomic DNA

1. Human genomic DNA (100–400 ng) (*see* **Note 7**).
2. Oligonucleotides. Primers for MVR-PCR (**Table 1**) were synthesized using an Applied Biosystems (Foster City, CA) Model 380B DNA synthesizer and were ethanol-precipitated and dissolved in PCR clean water prior to use. Kits containing MVR-PCR primers and probes are available from Cellmark Diagnostics (Abingdon Oxon, OX14 1YX, UK).
3. PCR buffer. The buffer used for all PCR in our laboratory is prepared as an 11.1X stock containing: 499.5 mM Tris-HCl, pH 8.8, 121 mM $(NH_4)_2SO_4$, 49.95 mM $MgCl_2$, 74.37 mM β-mercaptoethanol, 48.84 μM EDTA, pH 8.0, 11.1 mM dATP, 11.1 mM dCTP, 11.1 mM dGTP, 11.1 mM dTTP (purchased as 100 mM solutions from Pharmacia (Uppsala, Sweden) and stored at –20°C), 1.254 mg/mL bovine

serum albumin (BSA) (enzyme-grade BSA, 10 mg/mL from Pharmacia, stored at –20°C). Buffer is stored as 200 µL aliquots at –20°C and is stable for at least 3 mo with repeated freezing and thawing.

4. Thermostable DNA polymerase (e.g., Amplitaq, Perkin–Elmer (Cetus), Thermostable DNA polymerase, Applied Biotechnologies Limited. (Stored at –20°C) (*see* **Note 8**).

2.3.2. Agarose-Gel Electrophoresis

1. Low-EEO, high-gelling-temperature agarose (e.g., Seakem LE [Fisons]or Sigma Type 1).
2. Electrophoresis buffer: in 45 mM Tris-borate, 45 mM boric acid, 1 mM EDTA, 0.5 µg/mL ethidium bromide (0.5X TBE, pH 8.3). This is prepared as a 10X stock and stored at room temperature. Ethidium bromide (stored as a 5 mg/mL stock at 4°C) is added to a concentration of 0.5 µg/mL before electrophoresis.
3. 5X gel loading buffer: 12.5% Ficoll 400, 0.1% bromophenol blue in 0.2 M Tris-acetate, pH 8.3, 0.1 M sodium acetate, 1 mM EDTA (5X TAE).
4. DNA size markers: 100 ng/µL ϕX DNA digested with *Hae*III stored at –20°C.

2.3.3. Southern Blot Hybridization

1. Depurinating solution: 0.25 M HCl.
2. Denaturing solution: 0.5 M NaOH, 1.5 M NaCl.
3. Neutralizing solution: 0.5 M Tris-HCl, pH 7.5, 1.5 M NaCl.
4. High-salt transfer buffer (20X SSC): 3 M NaCl, 0.3 M trisodium citrate, titrated to pH 7.0 with HCl.
5. Phosphate/SDS hybridization solution: 0.5 M sodium phosphate, pH 7.2, 7% sodium dodecyl sulfate (SDS), 1 mM EDTA. Sodium phosphate is made as a 1 M stock by dissolving 97.1 g anhydrous disodium hydrogen orthophosphate and 49.3 g sodium dihydrogen orthophosphate in 1 L water and titrating to pH 7.2 with orthophosphoric acid if necessary. This solution is stored at 4°C and prewarmed to prevent precipitation before adding an equal volume of 14% SDS and EDTA to make the working solution (*see* **Note 9**).

2.3.4. Radioactive Probe Labeling

1. Probe DNA: MS31A is detected by the 5.7 kb *Sau*3AI minisatellite insert isolated from the plasmid pMS31; MS32 is detected by the 5 kb *Dra*I fragment from plasmid pMS32 *(6)*. Kits containing MVR-PCR primers and probes are available from Cellmark Diagnostics. Probes are diluted to a concentration of 2–10 ng/µL with water or TE and stored at –20°C (*see* **Note 10**).
2. Probe-labeling buffer: Probes are labeled using the random hexamer oligonucleotide priming method to incorporate α-^{32}P-dCTP. There are many commercially available random hexamer probe-labeling kits (e.g., Pharmacia).
3. Oligonucleotide-labeling buffer (OLB) is made by mixing the following solutions A, B, and C in the ratio 2A:5B:3C. A: 1.25 M Tris-HCl, pH 8.0, 125 mM MgCl$_2$, 0.18% (v/v) β-mercaptoethanol, 0.5 mM dNTPs, store at –20°C. B: 2 M

HEPES titrated to pH 6.6 with NaOH, store at 4°C. C: Hexadeoxyribonucleotides (Pharmacia), evenly suspended in 3 mM Tris-HCl, 0.2 mM EDTA, pH 7.0, at 90 OD U/mL, store at –20°C. OLB is divided into 60 µL aliquots and stored at –20°C; it is stable for at least three months with repeated freezing and thawing.
4. Klenow fragment of DNA polymerase I (Pharmacia).
5. α-^{23}P-dCTP (110 TBq/mmol, Amersham).
6. BSA 10 mg/mL, enzyme grade (Pharmacia).

2.3.5. Radiolabeled Probe Recovery

1. Oligolabeling stop solution (OSS): 20 mM NaCl, 20 mM Tris-HCl, pH 7.5, 2 mM EDTA, 0.25% SDS, 1 µM dCTP.
2. 2 M sodium acetate, pH 7.0.
3. 3 mg/mL high-molecular-weight herring-sperm DNA (Sigma). Supplied as a powder, dissolved in water, and stored at –20°C
4. 100% ethanol.
5. 80% ethanol.

2.3.6. Post-Hybridization Washing Solution

1. Hybond-N (fp) membranes are washed to high stringency following hybridization in 0.1X SSC, 0.01% SDS at 65°C. This is made up as required from stocks of 20X SSC and 10% SDS.

2.3.7. Autoradiography

1. Proprietary X-ray developer, stop and fixing solutions; e.g., Kodak LX24 developer, FX40 fixer and HX40 hardener (Kodak).

3. Methods
3.1. MVR-PCR Amplification from Genomic DNA

1. Aliquot 1 µL of 100–400 ng/µL genomic DNA into two tubes per individual. DNA from an individual of known MVR code is included on all gels to standardize code registration.
2. Two MVR-PCR reaction mixes are required per individual, containing either one or the other repeat unit-specific primer together with the other reaction components. These are made up as master mixes that are aliquoted into the reaction tubes containing the template DNA and also into empty tubes for the appropriate zero-DNA controls. Add 6 µL of the following MVR-PCR reaction mixes per individual to the template DNA, using the appropriate flanking and MVR-specific primers for MS31 or MS32 (*see* **Notes 11–16**).

a-type repeat units	t-type repeat units
a. 0.63 µL 11.1X PCR buffer	0.63 µL 11.1X PCR buffer
b. 0.7 µL 2.5 µM MS32OR/MS31A	0.7 µL 2.5 µM MS32OR/MS31A
c. 0.7 µL 2.0 µM TAG primer	0.7 µL 2.0 mM TAG primer

    d. 0.7 μL 100 n*M* MS32-Tag-A/   0.7 μL 200 n*M* MS32-Tag-T/
        0.7 μL 500 n*M* MS31-Tag-A   0.7 μL 250 n*M* MS31-Tag-G
    e. 0.25 U *Taq* polymerase   0.25 U *Taq* polymerase
    f. H$_2$O to 6 μL   H$_2$O to 6 μL

3. a. Transfer the tubes to a thermal cycler.
    b. Program 20 cycles of: 94°C for 30 s, 68°C for 30 s, and 70°C for 3 min, 1 cycle of 68°C for 1 min and 70°C for 10 min.

3.2. Agarose Gel Electrophoresis

1. Prepare the gel plate and comb (*see* **Notes 11–16**).
2. Make up a 1% agarose gel. Suspend 1 g agarose powder per 100 mL 0.5X TBE + ethidium bromide (200 mL for a 20 cm gel, 400 mL for a 40 cm gel), heat to boiling in a microwave until the agarose is dissolved, cool to 50–60°C, and pour into the gel former.
3. Once set, transfer the gel to the electrophoresis tank and add 0.5X TBE buffer + ethidium bromide to just cover the surface of the gel. Cover the top surface of the gel with a glass/perspex plate, so that the buffer covers the gel but not the upper plate, and the wells are not covered by the upper plate.
4. Add 2 μL 5X TAE loading dye to each PCR reaction and transfer to a well in the agarose gel. **a**-type and **t**-type reactions from each individual are run in adjacent lanes.
5. Run the gel until the lowest rung of the MVR-PCR ladder is estimated to be close to the bottom of the gel by comparison with 1 μg ϕX/*Hae*III size markers. For diploid mapping, this is when the 118 bp marker is at the bottom of the gel (*see* **Note 20**). For a 40 cm gel this takes approx 16 h at 130 V.

3.3. Southern Blot Transfer and Hybridization

3.3.1. Southern Blot Transfer (see **Notes 22–24**)

1. After electrophoresis, turn the gel, still between the glass plates, upside down and transfer to a plastic tray.
2. Add depurinating solution to just cover the gel, shake gently using an orbital shaker for 5 min twice, with a change of solution after 5 min.
3. Carefully remove the depurination solution and cover the gel with denaturing solution. Denature for 10 min twice, with a change of solution after 10 min.
4. Remove the denaturing solution and cover the gel with neutralizing solution. Neutralize for 10 min twice, with a change of solution after 10 min.
5. Prepare a Southern-blot hybridization platform. This is made by draping a wick of Whatmann 3MM paper, saturated with 20X SSC, over a glass sheet longer and wider than the gel to be blotted, into a reservoir of 20X SSC in a plastic tray. Put an additional piece of 3MM cut larger than the gel to be blotted onto the 3MM and carefully remove all bubbles by rolling with a glass rod or pipet.
6. Transfer the gel to the hybridization platform, remove any bubbles by rolling, and surround with clingfilm to prevent direct transfer of buffer between the wick and the dry blotting material above.

7. Wet a piece of hybridization membrane, cut to the same size as the gel, with 10X SSC. Float the membrane onto the surface of a tray of 10X SSC to ensure the membrane is wetted evenly and completely, then submerge.
8. Place the wet hybridization membrane on top of the gel and remove any bubbles by rolling.
9. Wet two pieces of 3MM cut to the same size as the gel with 20X SSC and place on top of the hybridization membrane, remove bubbles by rolling.
10. Place a stack of absorbent paper, e.g., paper towels or Quickdraw™ (Sigma) on top of the 3MM paper.
11. Cover with a glass plate and place an evenly distributed weight of approx 1 kg/20 cm^2 on top.
12. Leave the blot to transfer for 2.5 h, with a change of blotting towels after 0.5 h.

3.3.2. Fixation of DNA to Hybridization Membrane

1. Remove the hybridization membrane from the blotting apparatus, rinse in 3X SSC and dry in an 80°C oven (approx 30 min).
2. Place the membrane in a DNA crosslinker or on a transilluminator, with the DNA side facing the UV source. The UV dose required for most efficient crosslinking depends on the type of membrane used and the strength of the UV source. Optimal exposure time can be determined empirically for a UV transilluminator or set with a UV crosslinker to deliver a measured UV dose.

3.3.3. Probe Labeling

1. Aliquot probe-DNA solution containing 10 ng probe (10 ng probe DNA is sufficient to hybridize up to 5 hybridization membranes) into a 1.5 mL microcentrifuge tube, make up to 20 μL with H$_2$O.
2. Denature probe DNA by placing in a boiling waterbath or 100°C heated block for 3 min.
3. Immediately add 6 μL OLB, 1.2 μL 10 mg/mL BSA.
4. Add 2.5 μL α-^{32}P-dCTP (110 TBq/mmol, Amersham).
5. Add 2.5 units of the Klenow fragment of DNA polymerase.
6. Incubate for 1 h at 37°C or overnight at room temperature. We generally find better label incorporation with the latter method and routinely obtain specific activities of labeled probe >10^9 dpm/μg.

3.3.4. Probe Recovery

1. Add 70 μL OSS to the labeling reaction.
2. Add 40 μL H$_2$O.
3. Add 30 μL 2 *M* sodium acetate, pH 7.0.
4. Add 30 μL 3 mg/mL carrier herring-sperm DNA.
5. Add 500 μL 100% ethanol, shake to coagulate precipitated DNA, carefully remove ethanol using a siliconized Pasteur pipet.
6. Rinse DNA in 80% ethanol to remove unincorporated α-^{32}P-dCTP, remove ethanol and dissolve in 500 μL H$_2$O.

3.3.5. Hybridization

1. Prewarm hybridization solution to 65°C.
2. Transfer crosslinked hybridization membranes to a hybridization bottle or box, add sufficient hybridization solution to cover the membranes, prehybridize by incubating at 65°C with rotation/shaking for a minimum of 10 min.
3. Denature the radiolabeled probe by placing in a boiling waterbath or 100°C block for 3 min.
4. Remove the prehybridization solution and add the minimum volume of hybridization solution that will just cover the membranes.
5. Add the probe DNA to the hybridization solution immediately after denaturation.
6. Incubate overnight at 65°C in a rotisserie or shaking waterbath.

3.3.6. Post-Hybridization Washing

Excess and nonspecifically hybridized probe is removed by washing the hybridization membranes at high stringency.

1. Heat 1 L high-stringency washing solution to 65°C.
2. Add 100 mL wash solution directly to the membranes in their hybridization bottles, or transfer filters to a plastic box for washing, repeat every 10 min until washing solution is finished.
3. Remove membranes from the last wash into a tray of 3X SSC to rinse off remaining SDS, transfer to an 80°C oven to dry.

3.4. Autoradiography

1. Once the filters are dry, attach them to a sheet of 3MM cut to the same size as the X-ray cassette. A 35 × 43 cm cassette will take two 20 × 30 cm hybridization membranes.
2. Use a hand-held β-monitor to estimate the exposure time that will be required. The hybridized filters from MVR gels generally give 100–500 cps, in which case expose overnight at room temperature without an intensifying screen. If the membranes count <50 cps expose overnight at –80°C with an intensifying screen.
3. After developing the first exposure, adjust exposure times/conditions accordingly to give required band intensities for subsequent exposures.

3.5. MVR-Code Analysis

1. Diploid MVR-maps are converted into a digital code and entered into a database using software written in Microsoft QuickBasic™ by A. J. Jeffreys, available from the author or from Cellmark Diagnostics on request. For individuals with alleles containing both **a**-type and **t**-type repeats, the diploid map is described by a ternary code in which each rung on the ladder can be coded as, **1** (both alleles **a**-type at that position, a band only in the **a**-track), **2** (both **t**-type, a band only in the **t**-track) or **3** (heterozygous, a band in both tracks).

4. Notes

1. The techniques of PCR amplification, agarose-gel electrophoresis, Southern-blot transfer and hybridization, with either radioactive or nonisotopic probes, are almost universally used in molecular-biology laboratories. The precise apparatus and techniques used to carry out these routine procedures may vary from lab to lab; therefore, the general items of equipment suggested for use here may be substituted for by a similar alternative without affecting the overall results of the MVR analysis.

2. MVR-PCR amplification from genomic DNA. Because MVR-PCR operates optimally over an annealing temperature window of around 2°C, a thermal-cycling machine capable of accurate temperature control is required.

3. We use custom-built gel tanks constructed in-house to take an agarose gel 40 cm long and 20 cm wide; however, MVR maps can also be resolved on standard 20 cm agarose gels in tanks from most molecular biology-equipment suppliers, e.g., BioRad.

4. We have extensively tested different hybridization membranes and found Hybond-N (fp) (Amersham) to be the best for MVR-PCR.

5. If a transilluminator is used for crosslinking, it should be regularly calibrated, using test blots to determine the length of UV exposure required for optimal crosslinking.

6. MVR-PCR amplification from genomic DNA. As with any PCR technique, precautions should be taken to ensure that all tools and reagents used for PCR are free of contaminating DNA, contaminants that may inhibit PCR, and from other DNA molecules (e.g., recombinant molecules, other human DNA or PCR products). This is facilitated by the temporal and spatial separation of DNA preparation and subsequent PCR amplification. To minimize sources of PCR contamination, all solutions used for PCR are made using reagents transferred directly from containers as supplied by the manufacturer using sterile 25 mL plastic sample bottles (Sterilin) or plastic tips taken directly from the bags in which they are supplied. Dedicated pipets and plasticware contained in a laminar-flow hood are used in the preparation of all PCR reagents and in setting up PCR reactions. All reactions should be performed with the appropriate zero-DNA controls.

7. We have found that organically extracted DNA provides the best template for MVR-PCR. DNA is diluted to a concentration of approx 500 ng/μL with PCR clean water and stored at 20°C. MVR-PCR can be performed on 10 ng or less input DNA and is also possible using subnanogram quantities of DNA; however, in the latter case, some MVR code information may be lost. Control template DNA of known MVR code is available in an MVR-PCR kit from Cellmark Diagnostics. MVR-PCR can tolerate large variations in the quantity of input DNA; therefore, precise quantification is not needed. We generally estimate DNA concentration by comparison with size markers or genomic DNAs of known concentration on agarose gels.

8. We have tested several kinds of *Taq* polymerase and found Amplitaq™ (Perkin–Elmer Cetus) and the thermostable DNA polymerase from Advanced Biotechnologies Limited to be the best for amplification of minisatellite repeats in our particular buffer.

9. A high-stringency phosphate/SDS hybridization buffer is recommended for use with all highly repetitive minisatellite probes, since they have the potential to cross-hybridize to other loci at lower stringency.

10. Both radioactive and nonisotopic probes are suitable for MVR-PCR analysis. Since the precise methods used for probe-labeling are not critical, the labeling protocols and reagents already routinely used in your laboratory are probably best for using in MVR-PCR.

11. To reduce the overall consumption of reaction components to a minimum, 7 µL reactions are performed. Larger reactions can be performed and may be required when dealing with very low quantities of input DNA. In this case, the quantities given can be scaled up accordingly.

12. The quality of MVR mapping data was found to be improved by reducing *Taq* polymerase concentration. With less *Taq* polymerase, the signal from misprimed PCR products is reduced; however, the overall yield of genuine products also decreases; 0.25 U/ 7–10 µL reaction was determined by titration to be optimal for MVR-PCR.

13. Duplex MVR-PCR of both MS31 and MS31 can be performed simultaneously by using both flanking primers together with Tag and repeat unit-specific primers for one of the repeat types at each of these loci in the same tube.

14. All temperatures and times quoted are for the Geneamp 9600 Perkin–Elmer/Cetus, using thin-walled 0.2 mL tubes. PCR parameters are the same for MVR-PCR amplification of both MS31 and MS32. Denaturing and annealing times may have to be altered (probably increased) for optimal MVR-PCR efficiency in other thermal cycling machines, especially if thicker walled 0.5 mL microfuge tubes are used for PCR reactions.

15. By increasing the number of cycles (to 30), increasing the concentration of repeat unit-specific primers (0.7 µL 2.5 µ*M* primer per reaction) and omitting the TAG primer, it is possible to visualize MVR maps on ethidium-bromide-stained gels *(28)*. However, it is only possible to obtain information from the first 15–20 repeat units using this procedure.

16. The use of allele-specific flanking primers in MVR-PCR allows the amplification of single alleles directly from the genomic DNA of an individual previously identified as being heterozygous at one of several polymorphic positions in the DNA flanking MS31 and MS32 *(27,32)*.

17. Precautions should be taken to maximize MVR map resolution and minimize aberrant migration, e.g., "smiling," caused by local variations in voltage and/or gel thickness. Using a 40 cm-long gel rather than the conventional 20 cm format increases the resolution of minisatellite repeat units, particularly toward the top of the gel. However, adequate resolution for forensic analysis can be obtained using a 20 cm gel. We routinely use a comb to form 26 wells, approx 0.15 × 0.4 × 0.4 cm per gel. MVR-PCR gels are poured on 40 × 20 cm glass plates, using a piece of masking tape securely round the edge to contain the gel. Using glass plates reduces the possibility that variation in thickness across the gel will occur as a result of bowing of plastic gel plates when they come into contact with hot agarose.

18. A spirit level is used to ensure that both gels and gel tanks are horizontal before electrophoresis.

19. Gels are poured in a cold room and covered in a thin layer of 0.5X TBE buffer, once set, to prevent the upper surface from drying out.

20. A glass plate is put on the top surface of the gel to maximize the current passing through the gel. This not only improves resolution and reduces the loss of small fragments due to diffusion, but also reduces the extent to which the lanes spread out toward the bottom of the gel. The upper plate can be removed to observe the extent of migration of DNA size markers.

21. In practice, the use of size markers is not necessary once the distance moved by the bromophenol blue in the loading buffer is known relative to the position of the first repeat unit. However, ethidium bromide is still included in the running buffer to enhance resolution.

22. Gels are Southern-blotted from below, since the loaded DNA is closer to this surface of the gel and will therefore transfer from this side more efficiently. Inverting the gel while it is still between the glass plates is much easier than inverting it later.

23. The precise method used for Southern blotting is not critical for MVR-PCR. If duplex MVR-PCR has been performed, it is possible to do a bidirectional sandwich blot to a hybridization membrane placed on each side of the gel, and subsequently hybridize each of these with a different probe.

24. Both radioactive and chemiluminescent probes have been used successfully in MVR mapping. The protocols described here are for radioactive probe labeling. Protocols and probes for nonisotopic detection of MVR maps are available from Cellmark Diagnostics.

References

1. Jeffreys, A. J., Wilson, V., and Thein, S. L. (1985) Individual-specific "fingerprints" of human DNA. *Nature* **316,** 76–79.

2. Jeffreys, A. J. and Pena, S. J. D. (1993) Brief introduction to human DNA fingerprinting, in *DNA Fingerprinting, State of the Science* (Pena, S. D. J., Chakraborty, R., Epplen, J. T., and Jeffreys, A. J., eds.), Birkhaüser-Verlag, Basel, pp. 1–20.

3. Jeffreys, A. J., Turner, M., and Debenham, P. (1991) The efficiency of multilocus DNA fingerprint probes for individualization and establishment of family relationships, determined from extensive casework. *Am. J. Hum. Genet.* **48,** 824–840.

4. Jeffreys, A. J., Wilson, V., and Thein, S. L. (1985) Hypervariable "minisatellite" regions in human DNA. *Nature* **314,** 67–73.

5. Nakamura, Y., Leppert, M., O'Connell, P., Wolff, R., Holm, T., Culver, M., Martin, C., Fujimoto, E., Hoff, M., Kumlin, E., and White, R. (1987) Variable number of tandem repeat (VNTR) markers for human gene mapping. *Science* **235,** 1616–1622.

6. Wong, Z., Wilson, V., Patel, I., Povey, S., and Jeffreys, A. J. (1987) Characterization of a panel of highly variable minisatellites cloned from human DNA. *Ann. Hum. Genet.* **51,** 269–288.

7. Jeffreys, A. J., Wilson, V., Neumann, R., and Keyte, J. (1988) Amplification of human minisatellites by the polymerase chain reaction, towards DNA fingerprinting of single cells. *Nucleic Acids Res.* **16,** 10,953–10,971.

8. Boerwinkle, E., Xiong, W., Fourest, E., and Chan, L. (1989) Rapid typing of tandemly repeated hypervariable loci by the polymerase chain reaction, application to the apolipoprotein B 3' hypervariable region. *Proc. Natl. Acad. Sci. USA* **86,** 212–216.

9. Litt, M. and Luty, J. A. (1989) A hypervariable microsatellite revealed by in vitro amplification of a nucleotide repeat within the cardiac muscle actin gene. *Am. J. Hum. Genet.* **44,** 397–401.

10. Weber, J. L. and May, P. E. (1989) Abundant class of human DNA polymorphisms which can be typed using the polymerase chain reaction. *Am. J. Hum. Genet.* **44,** 388–396.

11. Jeffreys, A. J., Brookfield, J. F. Y., and Semeonoff, R. (1985) Positive identification of an immigration test case using human DNA fingerprints. *Nature* **317,** 818,819.

12. Jeffreys, A. J., Royle, N., Patel, I., Armour, J. A. L., MacLeod, A., Collick, A., Gray, I., Neumann, R., Gibbs, M., Crosier, M., Hill, M., and Signer, E. (1991) Principles and recent advances in DNA fingerprinting, in *DNA Fingerprinting, Approaches and Applications* (Burke, T., Dolf, G., Jeffreys, A. J., and Wolff, R., eds.), Birkhaüser-Verlag, Basel, pp. 1–19.

13. Baird, M., Balazs, I., Giusti, A., Miyazaki, L., Nicholas, L., Wexler, K., Kanter, E., Glassberg, J., Allen, F., Rubinstein, P., and Sussman, L. (1986) Allele frequency distribution of two highly polymorphic DNA sequences in three ethnic groups and its application to the determination of paternity. *Am. J. Hum. Genet.* **39,** 489–501.

14. Balazs, I., Baird, M., Clyne, M., and Meade, E. (1989) Human population genetic studies of five hypervariable DNA loci. *Am. J. Hum. Genet.* **44,** 182–190.

15. Odelberg, S. J., Plaetke, R., Eldridge, J. R., Ballard, L., O'Connell, P., Nakamura, Y., Leppert, M., Lalouel, J.-M., and White, R. (1989) Characterization of eight VNTR loci by agarose gel electrophoresis. *Genomics* **5,** 915–924.

16. Smith, J. C., Anwar, R., Riley, J., Jenner, D., Markham, A. F., and Jeffreys, A. J. (1990) Highly polymorphic minisatellite sequences, allele frequencies and mutation rates for five locus specific probes in a Caucasian population. *J. Forensic Sci. Soc.* **30,** 19–32.

17. Lander, E. S. (1989) DNA fingerprinting on trial. *Nature* **339,** 501–505.

18. Budowle, B., Giusti, A. M., Waye, J. S., Baechtel, F. S., Fourney, R. M., Adams, D. E., Presley, L. A., Deadman, H. A., and Monson, K. L. (1991) Fixed-bin analysis for statistical evaluation of continuous distributions of allelic data from VNTR loci, for use in forensic comparisons. *Am. J. Hum. Genet.* **48,** 841–855.

19. Monckton, D. G. and Jeffreys, A. J. (1993) DNA profiling. *Curr. Opin. Biotech.* **4,** 660–664.

20. Bell, G. I., Selby, M. J., and Rutter, W. I. (1982) The highly polymorphic region near the human insulin gene is composed of simple tandemly repeated sequences. *Nature* **295,** 31–35.

21. Capon, D. J., Chen, E. Y., Levinson, A. D., Seeberg, P. H., and Goeddel, D. V. (1983) Complete nucleotide sequence of the T24 human bladder carcinoma oncogene and its normal homologue. *Nature* **302**, 33–37.
22. Owerbach, D. and Aagaard, L. (1984) Analysis of a 1963bp polymorphic region flanking the human insulin gene. *Gene* **32**, 475–479.
23. Wong, Z., Wilson, V., Jeffreys, A. J., and Thein, S. L. (1986) Cloning a selected fragment from a human DNA "fingerprint," isolation of an extremely polymorphic minisatellite. *Nucleic Acids Res.* **14**, 4605–4616.
24. Jarman, A. P., Nicholls, R. D., Weatherall, D. J., Clegg, J. B., and Higgs, D. R. (1986) Molecular characterization of a hypervariable region downstream of the human α-globin gene cluster. *EMBO J.* **5**, 1857–1863.
25. Page, D. C., Bieker, K., Brown, L. G., Hinton, S., Leppert, M., Lalouel, J.-M., Lathrop, M., Nystrom-Lahti, M., De La Chappelle, A., and White, R. (1987) Linkage, physical mapping and DNA sequence analysis of pseudoautosomal loci on the human X and Y chromosomes. *Genomics* **1**, 243–256.
26. Gray, I. C. and Jeffreys, A. J. (1991) Evolutionary transience of hypervariable minisatellites in man and the primates. *Proc. Roy. Soc. London B* **243**, 241–253.
27. Jeffreys, A. J., MacLeod, A., Tamaki, K., Neil, D. L., and Monckton, D. G. (1991) Minisatellite repeat coding as a digital approach to DNA typing. *Nature* **354**, 204–209.
28. Yamamoto, T., Tamaki, K., Kojima, T., Uchihi, R., Katsuyama, Y., and Jeffreys, A. J. (1994) DNA typing of the D1S8 (MS32) locus by rapid detection minisatellite variant repeat (MVR) mapping using polymerase chain reaction (PCR) assay. *Forensic Sci. Int.* **66**, 69–75.
29. Giles, A. F., Booth, K. J., Parker, J. R., Garman, A. J., Carrick, D. T., Akhavan, H., and Schapp, A. P. (1990) Rapid simple non-isotopic probing of Southern blots for DNA fingerprinting, in *Advances in Forensic Haemogenetics, 3* (Polensky, H. F. and Mayer, N. R., eds.), Springer Verlag, Berlin.
30. Brumbaugh, J. A., Jang, G. Y., and Schmidt, M. A. (1993) Rapid non-radioactive automated digital DNA fingerprinting. *Am. J. Hum. Genet.* **51**, supplement, A210.
31. Monckton, D. G., Tamaki, K., MacLeod, A., Neil, D. L., and Jeffreys, A. J. (1993) Allele-specific MVR-PCR analysis at minisatellite D1S8. *Hum. Mol. Genet.* **2**, 513–519.
32. Neil, D. L. and Jeffreys, A. J. (1993) Digital DNA typing at a second hypervariable locus by minisatellite variant repeat mapping. *Hum. Mol. Genet.* **2**, 1129–1135.
33. Hopkins, B., Williams, N. J., Webb, M. B. T., Debenham, P. G., and Jeffreys, A. J. (1994) The use of minisatellite variant repeat polymerase chain reaction (MVR-PCR) to determine the source of saliva on a used postage stamp. *J. Forensic Sci.* **39**, 526–531.
34. Tamaki, K., Huang, X.-L., Yamamoto, T., Uchihi, R., Nozawa, H., and Katsumata, Y. (1995) Applications of minisatellite variant repeat (MVR) mapping for maternal identification from remains of an infant and placenta. *J. Forensic Sci.* **40**, 695–700.
35. Robertson, J., Ziegle, J., Kronick, M., Madden, D., and Budowle, B. (1991) Genetic typing using automated electrophoresis and fluorescence detection, in *DNA Fingerprinting, Approaches and Applications* (Burke, T., Dolf, G., Jeffreys, A. J., and Wolff, R., eds.), Birkhäuser-Verlag, Basel, pp. 391–398.

36. Sullivan, K. M., Hopgood, R., Lang, B., and Gill, P. (1991) Automated amplification and sequencing of human mitochondrial DNA. *Electrophoresis* **12,** 17–21.
37. Sullivan, K. M., Walton, A., Kimpton, C., Tully, G., and Gill, P. (1993) Fluorescence-based DNA segment analysis in forensic science. *Biochem. Soc. Trans.* **21,** 116–120.

21

Sex Determination by PCR Analysis of the X-Y Amelogenin Gene

Atsushi Akane

1. Introduction

PCR-based sex determination was first accomplished by amplifying multiple-copy sequences in the Y-chromosomal DYZ1 locus (1,2). These methods are quite sensitive (amplifiable from trace samples) due to the multiplicity, but it is impossible to tell whether the template DNA is from a female or whether the analysis has ended in failure when no fragment is amplified by PCR. Simultaneous amplification of X- and Y-chromosomal genes is thus necessary for a reliable sex test, in which the X-specific product may act as an internal control for PCR. From this point of view, centromeric alphoid repeats on the X and Y chromosomes have been amplified separately (3) or in the same reaction mixture (4) using two pairs of primers. These methods are also sensitive. However, a difference in the copy numbers of the repeated sequences (i.e., 5000 and 100 copies in X and Y centromeres, respectively) is likely to cause a difference in the sensitivity of PCR amplification so that the X product does not act strictly as an internal control. Sex determination by PCR analysis of X-Y homologous genes is thus most trustworthy, because X and Y sequences are of equal (single) copy number and because they can be amplified using one set of primers.

Among known X-Y homologous genes, the amelogenin gene is the most suitable for the sex test by PCR: Following a single PCR with one pair of primers, X- and Y-specific products with different sizes are simultaneously detected (2,5) because of difference in the lengths of corresponding introns (6). However, the sensitivity of amplification of this single-copy gene is relatively low (2). Then downsizing of PCR products with the use of a different set of primers (7,8) and nested PCR technique (9,10) has been applied to improve the

From: *Methods in Molecular Biology, Vol. 98: Forensic DNA Profiling Protocols*
Edited by: P. J. Lincoln and J. Thomson © Humana Press Inc., Totowa, NJ

sensitivity. By the latter method, the target gene is amplified with a set of outer primers, and then reamplified using a set of inner, nested primers. This technique also improves PCR specificity, since the nested primers anneal to the sequence internal to the two outer primers so as to identify the subset of PCR products that corresponds to the target fragment *(11)*. Thus, the protocol of hemi- (semi-)nested PCR using a single inner-nested primer *(12)* is also described herein. Hemi-nesting is simpler than, and as powerful as, double-nesting using three nested primers *(9)*. The lengths of PCR products are 719–988 base pairs (bp), which can be detected by submerged agarose-gel electrophoresis.

2. Materials

1. *Taq* DNA polymerase. Store at –20°C.
2. 10X conc. PCR buffer: 100 m*M* Tris-HCl, 15 m*M* MgCl$_2$, 500 m*M* KCl, pH 8.3. Prepare with Milli-Q quality water, autoclave, and store at –20°C until needed.
3. dNTP mix: 10 m*M* each of dATP, dCTP, dGTP, and dTTP in sterile Milli-Q water. Store at –20°C.
4. 0.5 µ*M* primer AMXY-1F: 5'-CTG ATG GTT GGC CTC AAG CCT GTG-3'. Store at –20°C.
5. 0.5 µ*M* primer AMXY-2R: 5'-TAA AGA GAT TCA TTA ACT TGA CTG-3'. Store at –20°C.
6. 0.5 µ*M* nested primer AMXY-4R: 5'-TTC ATT GTA AGA GCA AAG CAA ACA-3'. Store at –20°C.
7. 1.6 mg/mL bovine serum albumin (BSA): Sigma (St. Louis, MO) A-3350 or equivalent. Prepare with sterile Milli-Q water and store at –20°C.
8. Mineral oil: Sigma M-5904 or equivalent. Store ambient.
9. Gel-loading buffer (10X): 0.03% bromophenol blue, 0.03% xylene cyanol FF, 40% (w/v) sucrose in Milli-Q water. Dissolve at 65°C and store at 4°C.
10. Electrophoresis buffer (1X TBE): 89 m*M* Tris-borate, 10 m*M* EDTA, pH 8.0. Prepare with deionized water and store ambient.
11. Agarose gel: Heat 1.5% SeaKem™ LE agarose (FMC Corp. BioProducts, Rockland, ME) or equivalent in the electrophoresis buffer by a microwave oven, mix well, and cool the solution to 65°C. Add appropriate volume of 10 mg/mL ethidium bromide to a final concentration of 0.5 µg/mL. Pour the agarose solution into the gel tray of electrophoresis apparatus, and position the sample comb before the gelation of agarose to make slots of 25–30 µL vol.
12. Molecular-weight standard marker, e.g., 100 bp ladder (Pharmacia Biotech Inc., Piscataway, NJ). Dilute appropriately with Milli-Q water and add 1/10 vol of the gel-loading buffer. Store at 4°C.

3. Methods

3.1. Single (First) PCR

1. Put sample DNA into each reaction tube placed on ice (*see* **Note 1**). Put a male genomic DNA into another tube as a positive control (*see* **Note 2**). Adjust the

volume to 20 μL by adding sterile Milli-Q water. Also put 20 μL of the water into another tube as a negative control (*see* **Note 3**). The number of samples containing both positive and negative controls is represented as *n* hereafter.

2. Add 10X (*n*+1) μL of the 10X conc. PCR buffer into 53.5X (*n*+1) μL of sterile Milli-Q water (master mixture, *see* **Note 4**) in a microtube placed on ice.
3. Add 10X (*n*+1) μL of 1.6 mg/mL BSA into the master mixture (*see* **Note 5**).
4. Add 2X (*n*+1) μL of dNTP into the master mixture.
5. Add 2X (*n*+1) μL each of primers AMXY-1F and AMXY-2R into the master mixture.
6. Add 0.5X (*n*+1) μL of *Taq* DNA polymerase into the master mixture.
7. Add 80 μL of the mixture into each sample.
8. Overlay with 100 μL of mineral oil to reduce evaporation of the mixture during the amplification (*see* **Note 6**) and centrifuge briefly.
9. PCR: After preheating at 95°C for 2 min, perform consecutive incubations at 94°C for 1 min, 55°C for 1 min (*see* **Note 7**), and 72°C for 2 min 30 times, followed by postincubation at 72°C for 7 min.
10. Mix 20 μL each of the samples with 2 μL of the gel-loading buffer and centrifuge briefly.
11. Apply the samples to the slots of agarose gel submerged in the electrophoresis buffer and electrophorese at 90 V for 1 h. Detect 977 bp X- and 788 bp Y-specific fragments in the gel visualized by ultraviolet (UV) radiation (*see* **Notes 2** and **8**).

3.2. Nested (Second) PCR

1. Put 1 μL each of the first reaction samples into each reaction tube placed on ice (*see* **Note 6**).
2. Add 10X (*n*+1) μL of the 10X conc. PCR buffer into 82.5X (*n*+1) μL of sterile Milli-Q water (master mixture) in a microtube placed on ice.
3. Add 2X (*n*+1) μL of dNTP into the master mixture.
4. Add 2X (*n*+1) μL each of primers AMXY-1F and AMXY-4R into the master mixture.
5. Add 0.5X (*n*+1) μL of *Taq* DNA polymerase into the master mixture.
6. Add 98 μL of the master mixture into each sample.
7. Overlay with 100 μL of mineral oil (*see* **Note 6**) and centrifuge briefly.
8. PCR: After preheating at 95°C for 2 min, perform consecutive incubations at 94°C for 1 min, 55°C for 1 min, and 72°C for 2 min 25 times, followed by postincubation at 72°C for 7 min.
9. Mix 20 μL each of the reaction mixture with 2 μL of the gel-loading buffer and centrifuge briefly.
10. Apply the samples to the slots of agarose gel submerged in the electrophoresis buffer and electrophorese at 90 V for 1 h. Detect 908 bp X- and 719 bp Y-specific fragments in the gel visualized by UV radiation (*see* **Notes 2** and **8**).

4. Notes

1. Because the sensitivity of amplification is relatively low *(2)*, 250–500 ng/reaction of sample DNA is recommended to be used as a template, if available. Smaller samples should be analyzed by nested PCR.

2. Nonspecific PCR products may be rarely detected between X and Y fragments *(2)*. This product is possibly confused with Y fragment in a female sample. Thus the X- and Y-specific fragments should be identified by comparing electrophoretic mobilities with the fragments amplified from a male template (positive control).

3. To check the contamination of PCR reagents, a mock reaction is needed as a negative control in which template DNA is replaced with sterile Milli-Q water.

4. When several samples are tested, a master mixture containing the PCR buffer, dNTP, primers, and *Taq* polymerase should be made to simplify repeated procedures and to avoid mis-addition of the reagents. The master mixture must be prepared immediately before use. The mixture should be made for (sample number + 1) reactions, because a slight loss of the mixture will occur during the division into tubes.

5. Amplification of the amelogenin gene may be inhibited by some contaminants such as the heme compounds and by degradation of template DNA *(2,9,12–14)*. Addition of BSA is quite effective to encounter inhibition of PCR by the contamination *(13,14)*. When contamination is unlikely, BSA can be replaced with sterile Milli-Q water. Filtration using a microconcentrator such as Centricon® 100 (Amicon, Inc., Beverly, MA) is also effective to remove the contaminants *(13)*. DNA degraded to 150–180 bp fragments cannot be eliminated by the microconcentrator, but can be removed by the gel-filtration technique *(12)*.

6. The mineral oil may bother the transfer of PCR products. The oil can be removed by the addition of 150 µL of chloroform, but the solvent may be carried over to the second reaction when nested PCR is performed. It is convenient to use AmpliWax™ PCR Gem 100 (Perkin-Elmer) instead of mineral oil, because the wax solidifies at room temperature. Otherwise, the reaction mixture is not overlaid with mineral oil and PCR is carried out using a thermal cycler with a heated cover, such as GeneAmp® PCR System (Model 9600 or 2400, Perkin-Elmer) and PTC-200 DNA Engine™ (MJ Res., Inc., Watertown, MA). The heated cover prevents evaporation of the mixture.

7. Low annealing temperature (55°C) is suitable to amplify the amelogenin gene. When the primers are annealed at 65°C, nonspecific shorter fragments may be detected significantly.

8. It is very important to carefully observe the intensities of PCR products in the gel. Usually both X and Y sequences are amplified equally from the male template DNA. A shorter Y fragment may be amplified more effectively than an X fragment under some conditions. If the X fragment is much more intensive than the Y fragment, the template DNA might be from a female and the sample or reagents might have been contaminated by exogenous male DNA.

References

1. Kogan, S. C., Doherty, M., and Gitschier, J. (1987) An improved method for prenatal diagnosis of genetic diseases by analysis of amplified DNA sequences. *New Engl. J. Med.* **317,** 985–990.

2. Akane, A., Shiono, H., Matsubara, K., Nakahori, Y., Seki, S., Nagafuchi, S., Yamada, M., and Nakagome, Y. (1991) Sex identification of forensic specimens by polymerase chain reaction (PCR): two alternative methods. *Forensic Sci. Int.* **49,** 81–88.

3. Witt, M. and Erikson, R. P. (1989) A rapid method for the determination of Y-chromosomal DNA from dried blood specimens by the polymerase chain reaction. *Hum. Genet.* **82,** 271–274.

4. Gaensslen, R. E., Berka, K. M., Grosso, D. A., Ruano, G., Pagliaro, E. M., Messina, D., and Lee, H. C. (1992) A polymerase chain reaction (PCR) method for sex and species determination with novel controls for deoxyribonucleic acid (DNA) template length. *J. Forensic Sci.* **37,** 6–20.

5. Nakahori, Y., Hamano, K., Iwaya, M., and Nakagome, Y. (1991) Sex identification by polymerase chain reaction using X-Y homologous primers. *Am. J. Med. Genet.* **39,** 472–473.

6. Nakahori, Y., Takenaka, O., and Nakagome, Y. (1991) A human X-Y homologous region encodes "amelogenin." *Genomics* **9,** 264–269.

7. Sullivan, K. M., Mannucci, A., Kimpton, C. P., and Gill, P. (1993) A rapid and quantitative DNA sex rest: fluorescence-based PCR analysis of X-Y homologous gene amelogenin. *Biotechniques* **15,** 636–641.

8. Mannucci, A., Sullivan, K. M., Ivanov, P. L., and Gill, P. (1994) Forensic application of a rapid and quantitative DNA sex test by amplification of the X-Y homologous gene amelogenin. *Int. J. Legal Med.* **106,** 190–193.

9. Akane, A., Seki, S., Shiono, H., Nakamura, H., Hasegawa, M, Kagawa, M., Matsubara, K., Nakahori, Y., Nagafuchi, S., and Nakagome, Y. (1992) Sex determination of forensic samples by dual PCR amplification of an X-Y homologous gene. *Forensic Sci. Int.* **52,** 143–148.

10. Nakagome, Y., Seki, S., Nagafuchi, S., Nakahori, Y., and Sato, K. (1991) Absence of fetal cells in maternal circulation at a level of 1 in 25 000. *Am. J. Med. Genet.* **40,** 506–508.

11. Erlich, H. A., Gelfand, D., and Sninsky, J. J. (1991) Recent advances in the polymerase chain reaction. *Science* **252,** 1643–1651.

12. Akane, A., Matsubara, K., Takahashi, S., and Kimura, K. (1994) Purification of highly degraded DNA by gel filtration. *Biotechniques* **16,** 235–237.

13. Akane, A., Shiono, H., Matsubara, K., Nakamura, H., Hasegawa, M., and Kagawa, M. (1993) Purification of forensic specimens for the polymerase chain reaction (PCR) analysis. *J. Forensic Sci.* **38,** 691–701.

14. Akane, A., Matsubara, K., Nakamura, H., Takahashi, S., and Kimura, K. (1994) Identification of the heme compound copurified with deoxyribonucleic acid (DNA) from bloodstains, a major inhibitor of polymerase chain reaction (PCR) amplification. *J. Forensic Sci.* **39,** 362–372.

22

Species Determination
by Analysis of the Cytochrome b Gene

Rita Barallon

1. Introduction

Samples likely to be used in forensic analysis are often available in minute quantities; for example, one strand of hair. In order to be able to assign a species of origin to this material using molecular-biology techniques, it is necessary to extract the DNA. When the material is limiting, the number of cells present and thus the number of copies of genomic DNA present will be very small. The DNA, depending on the state of the original material, may be degraded, and so the amount and quality of the extracted DNA may be very poor.

Mitochondrial DNA is not contained within the nucleus of the DNA but is found in the cytoplasm. Unlike the nuclear DNA, mitochondrial DNA is found in high copy number, on average between 1000 and 10,000 copies of the mitochondrial genome per cell. Thus, this highly repetitive DNA is considerably more abundant in extracted DNA than is nuclear DNA. This is advantageous when dealing with highly processed or degraded samples.

These organelles are of particular interest because they are enveloped in a protective double membrane that is resistant to considerable environmental stress. Each mitochondrion possesses a copy of circular DNA (mtDNA), which is matrilinearly inherited and undergoes no recombination during reproduction *(1–3)*. The mtDNA polymerase has poor fidelity, thus mutations may arise during replication. In addition, since there is no DNA repair system, mutations are permanent.

In regions of mitochondrial DNA, which are highly conserved in nature, the few differences present may be used to characterize the DNA in terms of species of origin, using techniques such as the polymerase chain reaction (PCR) *(4)*, where a specific region of the DNA is amplified almost a million-fold, and cleaved amplified polymorphic sequences (CAPS) *(5,6)*, where the amplified

From: *Methods in Molecular Biology, Vol. 98: Forensic DNA Profiling Protocols*
Edited by: P. J. Lincoln and J. Thomson © Humana Press Inc., Totowa, NJ

product is then subjected to restriction endonuclease activity in order to generate species-specific profiles.

A particularly diverse region is the cytochrome b coding region. Cytochrome b is one of nine to ten proteins that compose the complex II of the mitochondrial oxidative phosphorylation system and is the only one encoded by the mitochondrial genome *(7)*. Consequently, cytochrome b sequences have been obtained from many taxa *(8)* and used extensively for phylogenetic analysis to, e.g., differentiate five types of mammalian DNA *(9)* derived from blood samples.

2. Materials
2.1. DNA Extraction from Blood

1. 5 g Chelex® 100 to 100 mL sterile distilled water.
2. Lysis buffer: 0.32 M sucrose, 10 mM Tris-HCl, pH 7.5, 5 mM MgCl$_2$, 1% Triton® X-100.
3. Digestion buffer: 10 mM Tris-HCl, pH 8.3, 50 mM KCl, 25 mM MgCl$_2$, 0.1 mg/mL gelatin, 0.45% Nonidet P40 (Sigma), 0.45% Tween®-20.
4. Proteinase K (10 mg/mL): Add 50 mg of proteinase K to 5 mL of sterile deionized water (*see* **Note 3**).

2.2. DNA Extraction from Hair

1. Extraction buffer: 10 mM Tris-HCl, pH 8.0, 5 mM Na$_2$EDTA, pH 8.0, 0.1 M NaCl, 2% (w/v) SDS, 39 mM DTT, 200 µg/mL Proteinase K.
2. TE$_8$ buffer: 10 mM Tris-HCl, 1 mM Na$_2$ EDTA, pH 8.0.
3. Sepharose CL6B spun columns *(13)*: Typically 100 g of the Sepharose CL-6B (Pharmacia) is equilibrated with TE$_8$ buffer. Using a hypodermic needle, pierce a small hole at the bottom of a 0.7 mL microtube. Place the pierced tube in a 2 mL microtube. Add a drop of 0.4 mm glass beads (autoclaved and stored in TE$_8$ buffer at room temperature). Add 500 µL of equilibrated Sepharose CL-6B. Spin, with the caps removed, in a bench-top microfuge at 2500g for 2 min. Replace the 2 mL tube with a fresh, sterile 1.5 mL microtube (cap removed). Gently pipet the wash (maximum vol/column is 50 µL). Spin at 2500g for 2 min. Store eluate at 4°C or –20°C until required.

2.3. PCR

1. 10X PCR buffer: 500 mM KCl, 100 mM Tris-HCl, pH 8.3, 15 mM MgCl$_2$, 0.01% w/v gelatin. Make up to 100 mL with distilled water. Aliquot 1 mL portions in screw-cap microtubes and sterilize.

2.4. Species Identification by Caps

1. 10X TBE: 108 g Tris base, 55 g boric acid, 9.3 g Na$_2$EDTA, water to 1 L.
2. Bromophenol blue buffer: 0.25% bromophenol blue, 30% glycerol in TE, pH 8.0; 0.25% xylene cyanol(ff) (optional).

3. Ethidium bromide: 10 mg tablets may be purchased from Sigma and made up in water. Wrap the container in foil. Care must be taken when handling ethidium bromide as it is a mutagen.

3. Methods

3.1. DNA Extraction from Blood

There are several methods available for extraction of DNA from blood. The two protocols described have given satisfactory results in our laboratory. The first method, based on the method of Walsh et al. *(10)*, uses Chelex 100.

3.1.1. DNA Extraction from Blood Using Chelex 100

1. Pipet 1 mL sterile distilled water into a sterile 1.5 mL microtube. Then add 3–300 μL whole blood.
2. Gently mix the sample and incubate at room temperature for 30 min with occasional inversion.
3. The samples are centrifuged at 15,000g for 2 min.
4. Remove all but 20–30 μL of the supernatant from each sample and discard.
5. Add 5% Chelex 100 to a final volume of 200 μL (*see* **Note 1**).
6. Incubate at 56°C for 20 min.
7. Vortex at high speed for 5–10 s.
8. Boil samples in a water bath for 8 min (*see* **Note 2**).
9. Vortex samples at high speed for 5–10 s.
10. Centrifuge samples at 15,000g for 2 min.

Use 2.5 μL of the supernatant for PCR. The rest of the sample may be stored at 4°C or frozen. Before using stored samples repeat the centrifugation step.

3.1.2. DNA Extraction from Blood Using Proteinase K

1. Add 500 μL of the lysis buffer into a sterile 1.5 mL microtube. Then add 3–300 mL of whole blood.
2. Centrifuge at 15,000g for 20 s.
3. Carefully remove and discard the supernatant.
4. Add 1 mL of the lysis buffer and resuspend the pellet by vortexing for 30 s.
5. Repeat steps 2–4 twice.
6. Centrifuge samples at 15,000g for 20 s.
7. Carefully remove and discard the supernatant.
8. Add 0.5 mL of digestion buffer and 3 μL of Proteinase K (10 mg/mL).
9. Incubate at 60°C for 1 h.
10. Incubate the sample at 95°C for 10 min to inactivate the Proteinase K.
11. Briefly centrifuge to sediment the contents of the tube and store at –20°C.
12. Use 2.5 μL of the supernatant in a 25 μL PCR.

3.2. DNA Extraction from Hair

Several methods have been described for the extraction of DNA from hair. If the sample contains a root bulb, good-quality chromosomal DNA can be obtained. However, if only the hair shaft is available, then the yield of DNA will be limited and will be mitochondrial DNA.

For all the extraction methods described, the hair samples must be prepared as follows: Using a clean glass plate (washed in absolute alcohol). The hair shaft is cut into small portions (0.5 cm in length from the root) and transferred into sterile 1.5 mL microtubes. The hair samples are washed in sterile distilled water then absolute alcohol. The microtubes are then left to dry at 65°C. DNA may then be extracted by the following three methods.

3.2.1. DNA Extraction from Hair Using Phenol/Chloroform (12)

1. To the cleaned and dried hair samples, 1.0 mL of the extraction buffer is added and the sample incubated overnight at 65°C in a rotating oven (e.g., Hybaid hybridization oven).
2. Add 500 µL of a phenol/chloroform mixture. Mix by inversion and centrifuge at 15,000*g* for 10 min.
3. Transfer the top aqueous layer to a fresh, sterile microtube and repeat **step 2**.
4. Add an equal volume of chloroform to the aqueous layer obtained in **step 3**. Mix by inversion and centrifuge at 15,000*g* for 10 min. Transfer the aqueous top layer to a sterile microtube.
5. Precipitate the DNA by adding Propan-2-ol (6/10 vol) and 3M sodium acetate, pH 5.2 (1/10 vol). Mix by inversion and leave overnight at –20°C.
6. Spin at 15,000*g* for 10 min. Carefully discard the supernatant.
7. Wash the DNA pellet by adding –20°C 70% ethanol (in distilled water). Spin at 15,000*g* for 10 min. Decant supernatant and dry the pellet briefly at 65°C or by leaving on the bench for 10–15 min.
8. Resuspend pellet in 20 µL TE$_8$ buffer.
9. Further purify DNA using Sepharose CL6B spin columns.

3.2.2. DNA Extraction from Hair Using Chelex 100

This method is obtained from the CETUS "Rapid DNA extraction" protocol.

1. The hair is prepared as outlined and transferred into a 1.5 mL microtube.
2. Add 5% (w/v) Chelex 100 and incubate at 56°C overnight.
3. Boil the tubes in a water bath for 8 min.
4. Vortex tubes to mix contents then centrifuge at 15,000*g* for 2 min.
5. Use the supernatant for PCR.

3.2.3. DNA Extraction from Hair Using a Microwave Oven (14)

1. Prepare the hair portions as outlined previously. Transfer to 0.7 mL microtube.
2. Irradiate at full power in a domestic microwave for 2 min.

3. Add PCR reaction mixture directly to the microtube.
4. Centrifuge at 15,000*g* for 2 min to sediment the hair. Use directly in a PCR.

3.3. DNA Extraction from Urine (15)

1. Collect 50 mL of urine in a sterile tube (e.g., Falcon® or any tube that can withstand centrifugation).
2. Spin for 10 min at 3000*g*.
3. Carefully discard the supernatant (*see* **Note 4**).
4. Resuspend the pellet in 1 mL of 0.9% saline and transfer to a sterile 1.5 mL microtube.
5. Centrifuge at 15,000*g* for 10 min, then discard the supernatant.
6. Resuspend pellet in 1 mL of 0.9% saline. Centrifuge at 15,000*g* for 10 min.
7. Repeat **step 6**.
8. Resuspend the pellet in 500 µL sterile distilled water.
9. Boil for 10 min. Use the supernatant for PCR.

3.4. DNA Extraction from Tissue

The protocol is in the QIAamp tissue kit from QIAGEN.

1. Cut 25 mg of tissue (fresh or frozen) into small pieces. Place in a 1.5 mL microtube. Add 180 µL of ATL buffer.
2. Add 20 µL of Proteinase K.
3. Mix by vortexing and incubate at 55°C for 1–3 h in a reciprocal (shaking) water bath until all the tissue has lysed. Remove occasionally during this incubation step and mix sample by vortexing.
4. Add 200 µL AL buffer mix immediately by vortexing. Incubate at 70°C for 10 min (**Note 5**).
5. Add 210 µL of ethanol. Mix immediately by vortexing.
6. Apply lysate to a QIAamp column. Centrifuge for 1 min at 6000*g*.
7. Place QIAamp column in a fresh microtube, add 500 µL AW buffer. Centrifuge 6000*g* for 1 min.
8. Repeat **step 7**.
9. Place spin column in a 1.5 mL microtube and elute the DNA by adding 200 µL sterile distilled water preheated to 70°C (alternatively for long-term storage, use 200 µL Tris-HCl, pH 9.0). Spin at 6000*g* for 1 min.
10. Transfer eluate to a labeled microtube and repeat **step 9**.
11. Pool the eluate and use for PCR.

3.5. PCR

The reaction conditions for PCR will vary depending on the size of primers, the polymerase, and the type of thermal cycler being used. For further details on thermostable polymerases and thermal cyclers, *see* **Note 9**. Also, for more information about PCR in general, refer to McPherson et al. *(16)* and Innis et al. *(17)*.

1. The conditions described below are those used to generate an approximately 357 bp fragment of the cytochrome b gene. The primers *(18)* are: Cytochrome bH: 5' CCCTCAGAATGATATTTGTCCTCA 3' and Cytochrome bL: 5' CCATCCAACATCTCAGCATGATGAAA 3'.

2. Usually 100 ng template DNA is sufficient for a PCR.
   a. Add 2.5 µL of each primer (20 µ*M*).
   b. Add water to make the volume to 25 µL.
   c. Add 50 µL of mineral oil (Sigma).
   d. (NB: This step is not required if the thermal cycler has a heated lid.)

3. Heat the solution in a thermal cycler at 95°C for 4 min. The tubes are then maintained at 80°C for 30 min to allow the addition of the enzyme dNTP mixture (this hot-start step is not always necessary, *see* **Note 6**).

4. While the solution is at 80°C, 25 µL of the following (as a mixture) is added (total reaction volume 50 µL).
   a. 8 µL dNTP (200 µ*M* final concentration).
   b. 5 µL 10X PCR buffer.
   c. 0.25 µL *Taq* polymerase (Perkin-Elmer).
   d. 11.75 µL sterile distilled water.
   Positive controls (e.g., DNA extracted from authenticated species) and a negative control usually require that sterile water be included in any PCR assay to confirm the specificity of the reaction.

5. The PCR cycle used for these set of primers are (*see* **Note 8**):
   a. 1 cycle: 95°C for 4 min, 50°C for 10 s, 72°C for 1 min.
   b. 28 cycles: 95°C for 1 min, 50°C for 10 s, 72°C for 1 min.
   c. 1 cycle: 95°C for 1 min, 50°C for 10 s, 72°C for 8 min.

6. Samples can then be analyzed on a 1% agarose gel in 1X TBE buffer containing 1 µg/mL ethidium bromide. The solution of TBE containing the agarose is heated in a microwave until the agarose has dissolved. The solution is then allowed to cool to approx 50°C. To 100 mL of agarose solution add ethidium bromide to final concentration of 1 µg/mL; a gel is cast in a plastic tray with the ends sealed with tape. A comb with the appropriate numbers of wells is inserted before pouring the molten agarose.

7. Once the gel has set, it is placed in a tank and covered in 1X TBE containing 1 µg/mL ethidium bromide (*see* **Note 7**). The buffer level should be approx 1–2 mm above the gel surface.

8. The DNA samples, containing 3–5 µL bromophenol blue (if mineral oil was used in the PCR, it is not necessary to first, remove the oil), are loaded onto the agarose gel. A suitable DNA size marker is also loaded. (These are commercially available from a number of sources.) A current is applied, ensuring that the DNA runs from negative (black lead) to positive (red lead).

9. The gel is visualized and photographed on a transilluminator under ultraviolet light (short wavelength 245 nm).

3.6. Species Identification by Caps

The amplified DNA is then subjected to restriction endonuclease analysis. The choice of restriction enzyme may be determined from sequence information available from a number of databases, such as DNAStar, GenBank, EMBL, and so forth.

It is not necessary to remove excess nucleotides from the amplified product. Typically approx 20 μL of PCR product is used.

1. In an microtube, 1 μL of restriction enzyme (usually supplied as 10 U/μL) is used per digest. These enzymes are thermo-labile and need to be stored at –20°C.
2. 2 μL of 10X buffer, supplied with the enzyme, is added, followed by the amplified DNA.
3. A control reaction is also prepared in which the restriction enzyme is omitted.
4. The microtubes are then incubated in a water bath at the appropriate temperature (as indicated by the manufacturers of the restriction enzymes—this is usually 37°C for the majority of enzymes).
5. After 1–2 h incubation, the tubes are removed and 5 μL bromophenol is added to each tube.
6. The samples are then subjected to gel electrophoresis using 2% agarose made up in 1X TBE. A typical result of CAPS analysis is illustrated in **Fig. 1**.

4. Notes

1. Chelex tends to settle out unless kept stirred constantly.
2. If not using screw cap tubes, pierce the microtube caps with a hypodermic before placing the tubes in a boiling water bath to relieve pressure in the microtube.
3. Proteinase K may be kept at 4°C for 2–3 mo. The shelflife is extended by storage at –20°C but repeated freeze and thawing should be avoided. It is therefore best to store aliquots of the proteinase K at –20°C.
4. Sensible precautions must be taken when working with biological fluids, i.e., wearing of gloves and safety glasses at all times and using a biological safety-cabinet, particularly during the initial extraction process. The supernatant is discarded into a 10% solution of hypochlorite. Any spillages are mopped up and the area swabbed down with a 10% hypochlorite solution. Tissues, gloves, and any plasticware should be placed along with any other waste in double autoclave bags, immediately sealed and autoclaved. No glassware should be used during routine DNA extraction from urine.
5. The use of RNase A is optional. A 20 mg/mL solution is prepared in sterile distilled water, then boiled at 100°C for 15 min to destroy any DNases. The solution may then be stored frozen at –20°C.
6. The hot-start step is not necessary but has been found to improve specificity and as a consequence, sensitivity *(19)*. Incorporating all of the components of a PCR increases the potential of mispriming before the thermal cycling has been initiated *(20)*.
7. The best results are obtained by excluding ethidium bromide in the electrophoresis tank and poststaining the gel in 1X TBE containing 0.5 μg/mL ethidium bromide for 30 min. The gel is then photographed in the normal way.
8. The range of thermal cyclers currently available today is beyond the scope of this chapter. The type of thermal cycler purchased will depend on available funds and the number of samples expected to be processed. For example, avail-

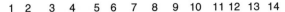
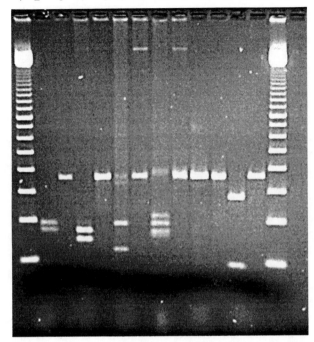

Fig. 1. Gel electrophoresis of CAPS profiles using the restriction enzymes *Alu*I and *Hae*II. In the examples illustrated, it was possible to distinguish between fish, mammal, and bird. In this case, a profile for woodcock was not obtained. However, by careful choice of restriction enzymes it is possible to generate a range of species-specific profiles. 1. Molecular-weight marker. 2. Skipjack-digested with *Hae*II. 3. Skipjack-undigested. 4. Human-digested with *Hae*II. 5. Human-undigested. 6. Cat-digested with *Alu*I. 7. Cat-undigested. 8. Dog-digested with *Alu*I. 9. Dog-undigested. 10. Woodcock-digested with *Alu*I. 11. Woodcock-undigested. 12. Quail-digested with *Alu*I. 13. Quail-undigested. 14. Molecular-weight marker.

able from Perkin–Elmer (Applied Biosystems Division) is the 2400 machine that offers a cooling facility allowing samples to be left overnight and has the capacity for 24 samples. Alternatively, also from Perkin-Elmer, is the 9600 machine, which has the same facilities as the 2400 but has the capacity for 96 samples.

References

1. Wilson, A. C., Cann, R. L., Carr, S. M., George, M., Gyllensten, U., Helm-Bychowski, K., Higuchi, R. G., Palumbi, S. R., Prager, E. M., Sage, R. D., and Stoneking, M. (1985) Mitochondrial DNA and two perspectives on evolutionary genetics. *Biol. J. Linn. Soc.* **26,** 375–400.

2. Avise, J. C., Arnold, J., Ball, R. M., Bermingham, E., Lamb, T., Neigel, J. E., Reeb, C. A., and Saunder, N. C. (1987) Intraspecific phylogeography—The mitochondrial-DNA bridge between population-genetics and systematics. *Annu. Rev. Ecol. Sys.* **18,** 489–522.
3. Moritz, C., Dowley, T. E., and Brown, W. M. (1987) Evolution of mitochondrial-DNA—Relevance for population biology and systematics. *Annu. Rev. Ecol. Sys.* **18,** 269–292.
4. Mullis, K. B. and Faloona, F. A. (1987) Specific synthesis of DNA *in vitro* via a polymerase-catalysed chain reaction. *Meth. Enzymol.* **155,** 335–350.
5. Akopyanz, N., Bukanov, N. O., Westblom, T. U., and Berg, D. E. (1992) PCR-based RFLP analysis of DNA sequence diversity in the gastric pathogen *helicobacter pylori. Nucleic Acids Res.* **20,** 6221–6225.
6. Konieczny, A. and Ausubel, F. M. (1993) A procedure for mapping Arabidopsis mutations using codominant ecotype-Specific PCR-based markers. *Plant J.* **4,** 403–410.
7. Hatefi, Y. (1985) The mitochondrial electron transport and oxidative phosphorylation system. *Annu. Rev. Biochem.* **54,** 1015–1069.
8. Irwin, D. M., Thomas, D. K., and Wilson, A. C. (1991) Evolution of the cytochrome b gene of mammals. *J. Molec. Evol.* **32,** 128–144.
9. Tsuchida, S., Umenishi, F., and Ikemoto, S. (1992) Species identification by polymerase chain reaction and direct sequencing, in *Advances in Forensic Haemogenetics* (Rittner, C. and Schneider, P. M., eds.), Springer-Verlag, Berlin, 118–120.
10. Walsh, P. S. Metzger, D. A., and Higuchi, R. (1991) Chelex® 100 as a medium for simple extraction of DNA for PCR-based typing from forensic material. *Biotechniques* **10,** 506–513.
11. Higuchi, R., von Beroldingen, C. H., Sensabaugh, F., and Erlich, H. (1988) DNA typing from single hairs. *Nature* **332,** 543–545.
12. Uchihi, R., Tamaki, K., Kojina, T., Yamamoto, T., and Katsumata, Y. (1992) Deoxyribonucleic acid (DNA) typing of human leukocytes antigen (HLA)-DQA1 from single hairs (in Japanese). *J. Forensic Sci.* **37,** 853–859.
13. Weising, K., Beyermann, B., Ramser, J., and Kahl, G. (1991) Plant DNA fingerprinting with radioactive and digoxigenated oligonucleotide probes complementary to simple repetitive DNA sequences. *Electrophoresis* **12,** 159–169.
14. O'Hara, M., Kurosu, Y., and Esumi, M. (1994) Direct PCR of whole blood and hair shafts by microwave treatment. *Biotechniques* **17,** 726–728.
15. Holland, M., Roy, R., Fraser, M., and Lui, R. (1993) Application of serological and DNA methods for the identification of urine specimen donors. *Forensic Sci. Rev.* **5,** 1–14.
16. McPherson, M. J., Quirke, P., and Taylor, G. R. (1991) *PCR: A Practical Approach.* IRL, New York.
17. Innis, M. A., Gelfand, D. H., Sninsky, J. J., and White. T. J. (1990) *PCR Protocols: A Guide to Methods and Applications.* Academic, New York.
18. Kocher, T. D., Thomas, W. K., Meyer, A., Edwards, S. V., Pääbo, S., Villablanca, F. X., and Wilson, A. C. (1989) Dynamics of mitochondrial DNA evolution in animals: amplification and sequencing with conserved primers. *Proc. Natl. Acad. Sci. USA* **86,** 6196–6200.

19. Mullis, K. B. (1991) The PCR in an anemic mode: how to avoid cold oligodeoxy-ribonuclear fision. *PCR Meth. Applicat.* **1,** 1–4.
20. Powell, S. J. (1995) Protocol optimization and reaction specificity, in *PCR Essential Data* (Newton, C. R., ed.), Wiley, New York, pp. 75,76.

23

Use of the AmpliType PM + HLA DQA1 PCR Amplification and Typing Kits for Identity Testing

Michael L. Baird

1. Introduction

Direct analysis of the composition of DNA has been used for forensic and paternity analysis since 1985 *(1)*. Since each person, except for identical twins, has a unique DNA composition, methods that allow the detection of differences in the DNA are useful to resolve identification issues involving the origin of forensic biological samples or to resolve paternity disputes. Methods that detect insertions, deletions, or sequence changes are used for identity testing. In forensics, because all cells from a specific person have the same DNA, comparisons between a known reference sample (i.e., from a victim or suspect) and an evidentiary sample of unknown origin can provide evidence to help to identify the origin of the evidentiary sample. If the DNA profiles of the evidentiary sample and the known sample differ, then the evidentiary sample did not originate from the known individual. If the DNA profiles are the same, the known individual is not excluded as the source of the evidentiary sample and a likelihood of finding the DNA profile can be calculated based on population-genetics principles and an appropriate database. In cases of questioned parentage, since we obtain half of our DNA from each of our biological parents, comparisons between the DNA of a child and an alleged parent can help to resolve questions of paternity or maternity. For example, in a paternity test, if the child's paternal allele is not present in the alleged father, he is excluded as the biological father (barring mutation or recombination). If the child's paternal allele is present in the alleged father, he is not excluded as the biological father and a paternity index (PI) can be calculated as the genetic odds in favor of paternity.

From: *Methods in Molecular Biology, Vol. 98: Forensic DNA Profiling Protocols*
Edited by: P. J. Lincoln and J. Thomson © Humana Press Inc., Totowa, NJ

Table 1
AmpliType Genetic Loci Characteristics

	HLA DQA1	LDLR	GYPA	HBGG	D7S8	GC
Chromosomal location	6p21.3	19p13.1 –13.3	4q28 –31	11p15.5	7q22 –31.1	4q11 –13
PCR product size in bp	239/242	214	190	172	151	138
Number of alleles	7[a]	2	2[b]	3	2	3

[a]The HLA DQA1 4.2 and 4.3 alleles are detected but not distinguished from each other by this assay.
[b]The GYPA A and A prime alleles are detected but not distinguished from each other by this assay.

The first methodology used to answer questions about DNA-based identification was RFLP analysis, a technique first published in 1975 *(2)*. This approach relies on the separation and identification of DNA fragments generated by restriction endonuclease digestion of the DNA. Analysis of the human genome led to the discovery of a number of highly polymorphic loci that were composed of insertions and deletions of DNA known as variable number tandem repeats (VNTRs) *(3)*. Such loci were very powerful in resolving identification issues. However, RFLP analysis requires relatively large amounts (0.5 µg) of high-mol-wt DNA in order to obtain results. The amount and quality of DNA isolated from forensic evidentiary samples often did not contain a sufficient quantity of DNA for analysis. A new testing method was required to obtain results from such samples.

The second method used to analyze DNA for identification was polymerase chain reaction (PCR), first published in 1985 *(4)*. This method allows the amplification of small quantities of DNA, amounts insufficient for RFLP analysis. The first locus examined extensively for identification was the HLA DQA1 locus, originally called HLA DQalpha. This locus was made commercially available as a kit in a reverse-dot-format by Perkin-Elmer. This kit was followed by a second reverse dot blot format kit known as Polymarker, which allowed the simultaneous amplification and analysis of five loci. The most recent kit, AmpliType PM+DQA1, allows the simultaneous amplification of all six loci. The remainder of this chapter deals with the use of this most recent kit. Using the AmpliType PM + DQA1 PCR Amplification and Typing Kit (Perkin-Elmer Corp., Norwalk, CT), it is possible to amplify six loci simultaneously: HLA DQA1 *(5)*, low-density lipoprotein receptor (LDLR) *(6)*, glycophorin A (GYPA) *(7)*, hemoglobin G gammaglobin (HBGG) *(8)*, D7S8 *(9)*, and group-specific component (Gc) *(10)*. **Table 1** lists the chromosomal positions of these loci.

All six of these loci are typed using the reverse-dot-blot approach where an allele-specific oligonucleotide probe is immobilized to a nylon strip. The reverse dot blot allows the determination of alleles without separation by electrophoresis. During the amplification process, a biotin molecule is incorporated into the synthesized DNA. Strips are prepared that contain the specific alleles to be detected as oligonucleotide sequences attached to a poly dT tail used to bind the oligonucleotide sequence to the membrane. During the hybridization reaction, sequences complementary from the sample bind to the immobilized oligonucleotide on the membrane. The biotin molecule is subsequently recognized by streptavidin, which is in turn attached to an enzyme (alkaline phosphatase) that can produce a color reaction in the presence of a substrate-like horseradish peroxidase. Thus, when an amplified DNA binds to the membrane, a sandwich is formed that ultimately produces a color change to form a blue dot. The intensity of the blue dot is determined by the amount of starting DNA and the time allowed to produce a color reaction. Controls are built into the strip to provide a threshold of intensity for interpretation.

2. Materials

2.1. Reagents Supplied the AmpliType PM+DQA1 PCR Amplification and Typing Kit

1. AmpliType PM+DQA1 PCR Reaction Mix (2.4 mL). This contains the enzyme AmpliTaq DNA polymerase, $MgCl_2$, dATP, dGTP, dCTP, dTTP, and 0.08% sodium azide in a buffer and salt. This should be stored at 2–8°C.
2. AmpliType PM+DQA1 Primer Set (1.2 mL). This contains 12 biotinylated primers and 0.05% sodium azide in buffer and salt. This should be stored at 2–8°C.
3. Control DNA (0.2 mL). This contains 100 ng/mL of human genomic DNA in 0.05% sodium azide and buffer. The genotype of this control DNA is as follows: LDLR, BB; GYPA, AB; HBGG, AA, D7S8, AB; GC, BB; DQA1, 1.1, 4.1. This should be stored at 2–8°C.
4. Mineral oil (5 mL). This is supplied in a dropper bottle and should be stored at 2–30°C. Do not expose to strong ultraviolet light.
5. AmpliType PM and HLA DQA1 DNA probe strips (50 strips each). These strips are provided in a screw-top tube with a packet of desiccant. Store the strips at 2–8°C in the screw-top tube with the desiccant and protect from light.
6. Enzyme conjugate: HRP-SA (2.0 mL). This contains horseradish peroxidase and streptavidin (HRP-SA) enzyme conjugate in a buffer with preservative. Store at 2–8°C.
7. Chromogen: TMB (60 mg). This contains powdered 3,3',5,5'-tetramethlybenzidine (TMB). Dissolve before use as indicated in **Subheading 3.2.** Store at 2–8°C.

2.2. Reagents Not Supplied with Kit

The following reagents are not included in the kit but are required for PCR amplification; PCR product-gel analysis, hybridization, and color development. This list does not include reagents required for DNA isolation.

1. Agarose. For example, NuSieve GTG and SeaKem GTG from FMC.
2. Alcohol. 95% ethanol and 70% isopropanol.
3. 0.5X TBE running buffer: 44.5 mM Tris-borate, 1 mM EDTA, pH 8.0.
4. Gel loading buffer: 0.2% bromophenol blue, 50% glycerol, 20 mM Tris-HCl, pH 8.0, 2.5 mM EDTA.
5. Ethidium bromide, 10 mg/mL.
6. Citrate buffer: 0.1 M sodium citrate, pH 5.0.
7. Glycerol.
8. Tris base.

2.3. DNA Analysis

1. AmpliType DNA Typing Tray.
2. Aspirator apparatus.
3. Electrophoresis equipment including gel trays, boxes, and power supplies.
4. Forceps with nonpointed tips.
5. Wratten 22, orange filter for use with a Polaroid camera.
6. Microcentrifuge.
7. Pipetters to deliver 2–10, 10–100, and 50–200 µL. These should be positive displacement pipetters, and there should be one set each for use pre-PCR and post-PCR amplifications.
8. Shaker, variable speed with orbital platform.
9. Thermocycler and equipment including thin-walled reaction tubes and temperature-verification system.

In addition, standard laboratory supplies like gloves, lab coats, and protective eyewear are required during the analysis.

3. Method

3.1. Amplification

The following protocol describes the PCR amplification procedures required using a Perkin–Elmer 480 thermal Cycler and the typing procedures specific for the AmpliType PM+DQA1 Amplification and Typing Kit. An area should be dedicated for preparation of the PCR reactions. Ideally, this should be in a separate room inside a biological, laminar-flow hood. All equipment and supplies used to set up the amplification reactions should be kept in this dedicated facility. Pipet tips plugged with hydrophobic filters should be used to prevent contamination. Use of disposable gloves and dedicated lab coats is required.

1. Prepare the DNA samples for PCR amplification. The quantity of each sample should be determined so the final DNA concentration is in the range of 0.1–0.5 ng/µL to allow addition of 2–10 ng in 20 µL to the PCR reaction. If the DNA in a sample is degraded, which is especially true in forensic samples, it may be necessary to add more than 10 ng of DNA.

2. Determine the number of samples to be amplified including positive and negative controls. The positive control can be the Control DNA included in the kit or can be another DNA sample, well-characterized for the loci in question. One of the negative controls consists of 20 µL of DI H_2O in place of the DNA sample. Other negative controls can be included that check the integrity of the solutions used for DNA isolation.

3. The required number of reaction tubes containing 40 µL of aliquoted AmpliType PM+DQA1 PCR Reaction Mix (contains AmpliType *Taq* enzyme, $MgCl_2$, and dNTPs) are labeled on the sides and placed in a rack.

4. Add 40 µL of AmpliType PM+DQA1 Primer Set to each tube. It is important to begin the PCR amplification within 20 min after the addition of the primer set to minimize the formation of primer dimers and other nonspecific PCR products.

5. Carefully add two drops of mineral oil to each tube from the dropper supplied with the kit. Be careful not to touch the reaction tubes with the dropper. Cap tubes loosely, and do not vortex, mix, or spin.

6. Add 20 µL of sample DNA or DI H_2O for negative control to the appropriate tubes by carefully inserting the pipet tip through the mineral oil. Discard the pipet tip, and recap the tube before proceeding to the next sample so that no more than one tube is open at any time to prevent contamination. This results in a final reaction volume of 100 µL.

7. Place all tubes into the thermal cycler block and start the 32-cycle amplification. Using the DNA Thermal Cycler 480, the following parameters are required for each of the 32 cycles: denature at 94°C for 60 s, anneal at 60°C for 30 s, and extend at 72°C for 30 s.

8. Verify the cycling parameters by monitoring the first cycle, and check the tubes after the first cycle to ensure they are all seated tightly in the block.

9. After the PCR amplification is completed, remove the tubes from the thermal cycler. Add 5 µL of 200 m*M* disodium EDTA to each tube. The samples are now ready for analysis by gel electrophoresis, DNA hybridization, and color development. Amplified samples containing 9.5 m*M* EDTA may be stored at 2–8°C for 2 mo or –20°C for 6 mo. Store amplified DNA samples separate from all PCR amplification reagents, extracted DNA samples, and casework samples.

3.2. Product-Gel Loading and Electrophoresis

The presence and size of PCR products generated post-amplification can be determined by agarose-gel electrophoresis. This should be performed prior to denaturing the samples for DNA hybridization to ensure sharp product bands on the gel.

1. Prepare a 3% NuSieve/1% SeaKem agarose solution in 0.5X TBE gel running buffer. Melt the agarose and add 5 µL of a 10 mg/mL stock of ethidium bromide to each 100 mL of agarose. (Ethidium bromide is a mutagen. Avoid contact with skin.) Cast the agarose gel (i.e., 5.5 × 9.0 × 0.45 mm), and insert a gel comb at one end. After the gel has solidified at room temperature, remove the comb, place in a gel box, and add sufficient 0.5X TBE gel running buffer to cover the gel.

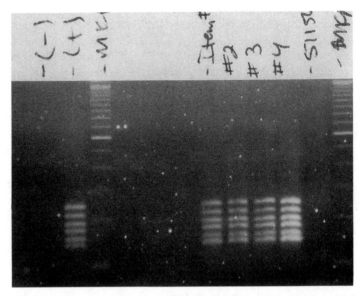

Fig. 1. Photograph of an ethidium bromide stained 4.0% agarose gel showing the six PCR product bands from the AmpliType PM+DQA1 Amplification and Typing Kit. Lanes labeled Item #1, #2, #3, and #4 contain bands that correspond to the following PCR products: 1) HLA DQA1 (242/239 bp); 2) LDLR (214 bp); 3) GYPA (190 bp); 4) HBGG (172 bp); 5) D7S8 (151 bp); 6) Gc (138 bp). Other lanes contain negative control (–), positive control (+), and 100 base-pair ladder (MRK).

2. Add 2 μL of gel loading buffer to an 0.5 mL microcentrifuge tubes. Add 5 μL of each amplified DNA sample to each of the tubes containing gel loading buffer. Mix and add the entire 7 μL to the appropriate well in the gel. Include a sizing standard like the 123 bp ladder (Life Technologies).

3. Connect the power supply so that the DNA travels toward the positive electrode. Run the gel at 115 V (7.5 V/cm) for about 1 h or until the bromophenol blue dye from the loading buffer has run 7.5 cm down the gel to allow adequate resolution of the six amplified product bands.

4. After the gel has run, disconnect the power supply and remove the gel. Wearing UV protective eyewear and gloves, photograph the gel by placing it on a UV transilluminator box under a fixed Polaroid camera with a Kodak 22 or 23A Wratten filter. Photograph the gel in the dark under UV illumination with type 55, 57, or 667 black and white Polaroid film.

5. The following six bands should be present in samples in which the DNA amplified: 242/239 bp (HLA DQA1), 214 bp (LDLR), 190 bp (GYPA), 172 bp (HBGG), 151 bp (D7S8), and 138 bp (Gc) (**Fig. 1**). Primer-dimer bands and unincorporated primers may appear as broad bands near the bottom of the gel in the region of lower molecular weight.

3.3. DNA Hybridization

The AmpliType DNA Hybridization process involves three steps performed sequentially: hybridization of amplified DNA probe strips; binding of HRP-SA enzyme conjugate to hybridized PCR products; and stringent wash to remove nonspecifically bound PCR products. Color development is performed after the stringent wash step.

1. Heat a shaking water bath to 55°C and maintain the temperature between 54°C and 56°C.

2. Warm the hybridization solution and the wash solution provided in the kit to 55°C to dissolve all solids. Allow the tube with the AmpliType DNA Probe Strips to equilibrate to ambient temperature to prevent condensation inside the tube, and remove the required number of strips from the tube. Label each strip on the right edge using the pen provided with the kit. (Some inks may effect the typing results.)

3. Place one strip each in the same orientation into each well of the AmpliType DNA Typing Tray. Tilt the typing tray towards the labeled end of the strips and add 3 mL of prewarmed hybridization solution to each well.

4. Denature the amplified DNAs by incubating at 95°C for 3–10 min.

5. Withdraw 20 μL of amplified DNA and immediately add it below the surface of the hybridization solution in the well of the appropriate DNA PM probe strip. Cap tube and return to the 95°C heat block 3–10 min.

6. Withdraw another 20 μL of amplified DNA from the tube and immediately add it below the surface of the hybridization solution in the well of the appropriate DNA HLA DQA1 probe strip.

7. The remaining amplified DNA can be stored at 2–8°C for 2 mo or at –20°C for 6 mo. Store amplified DNA samples separate from all PCR amplified reagents, extracted DNA samples, and casework samples.

8. Place a lid on the tray and mix by carefully rocking the tray. Ensure that each strip is completely wet. Once wet, strips should remain wet through the conclusion of the color development and photography steps.

9. Place the tray with the typing strips into a 55°C rotating water bath and place a weight on the cover to ensure that the tray does not float. Resume the rotation of the water bath at 50–70 rpm. Be sure that water does not splash into the wells of the tray. Incubate at 55°C for 15 min.

10. About 5 min before the end of the hybridization reaction, prepare the enzyme conjugate solution in a glass flask using the following formula to determine the volume of each component required: number of strips times 3.3 mL equals the volume of hybridization solution; number of strips times 27 μL equals the volume of enzyme conjugate: HRP-SA. Mix the solutions thoroughly to ensure that solids are in solution and leave at room temperature until ready to use.

11. After hybridization is completed, remove the tray from the water bath and aspirate the contents of each well from the labeled end of the strip while tilting the tray slightly. Remove condensation from the tray lid with a clean lab wipe.

12. Add 5 mL of prewarmed 55°C wash solution to each well. Rinse by gently rocking the tray for several seconds. Aspirate off this solution from each well.

13. Add 3 mL of the enzyme conjugate solution to each well and cover with the lid. Transfer to the rotating 55°C water bath. Place a weight on the tray and adjust the rotation of the water bath to 50–70 rpm. Incubate the enzyme conjugate solution with the DNA probe strips at 55°C for 5 min.

14. After incubation, remove the tray and aspirate the contents of each well from the labeled end of the strips while tilting the tray slightly. Remove condensation from the tray lid with a clean lab wipe. Add 5 mL of prewarmed wash solution into each well. Rinse by gently rocking the tray for several seconds. Aspirate the solution from the wells.

15. Add another 5 mL of prewarmed wash solution, cover the tray, and incubate at 55°C in shaking water bath as before. Incubate the strips for 12 min. The temperature and timing of this stringent wash step are critical.

16. After incubation, remove the tray from the water bath and aspirate the contents from the labeled end of the strips. Add 5 mL of wash solution to each well. Gently rock for several seconds then aspirate.

3.4. Color Development

1. Add 5 mL of citrate buffer to each well. Cover the tray with lid and place on an orbital shaker set at 50 rpm at room temperature for 5 min.

2. Prepare the color development solution during this wash step and use within 10 min. Add the following reagents in the order listed to a glass flask and mix thoroughly be swirling. Protect from light and do not vortex. Use the following formulas to determine the volumes required: number of strips times 5 for the volume of citrate buffer; number of strips times 5 μL of 3% hydrogen peroxide; number of strips times 0.25 mL of Chromogen TMB Solution.

3. Remove the tray from the orbital shaker. Remove the lid and aspirate off the contents from each well. Add 5 mL of freshly prepared color development solution to each well. Replace the lid and cover the lid with aluminum foil. Develop the strips at room temperature by rotating on an orbital shaker set at approx 50 rpm for 20–30 min. Develop until the S and C dots are visible.

4. Stop the color development by removing the solution from the well. Immediately dispense 5 mL of DI H_2O into each well. Place tray on an orbital shaker set at approx 50 rpm for 5–10 min. Remove the DI H_2O from the wells and repeat the wash steps three times.

3.5. Photography and Storage of Strips

1. Photographs should be taken for a permanent record of the typing results. Photographs must be taken while the DNA probe strips are still wet. Place wet strips on a flat nonabsorbent surface (i.e., a black surface like an exposed X-ray film). Use a Polaroid camera with type 55, 57, or 667 black and white film or Type 59 or 559 color film. An orange filter (Wratten 22 or 23A) will enhance contrast. Follow the film exposure and development instructions.

2. After photography, the strips may be air-dried and stored. The dot intensities fade upon drying.

3.6. Reuse of Typing Trays

The AmpliType DNA Typing Trays are designed to be disposable, but may be reused.

1. To reuse, immediately wash the trays and lids as follows. Add approx 5–10 mL of 95% ethanol or 70% isopropanol to each well. Do not use detergent or bleach. Cover the tray and gently agitate 15–30 s to dissolve any residual Chromogen:TMB. Remove the lid and visually inspect each well for the presence of a blue or yellow color that indicates the presence of Chromogen:TMB.
2. Repeat the ethanol wash if necessary until no color is present and then rinse each well and the tray lid with DI H_2O and dry before reuse.

3.7. Interpretation of PM Typing Results

Results are interpreted by observing the pattern and relative intensities of blue dots on the wet AmpliType PM and AmpliType HLA DQA1 DNA Probe Strips to determine which alleles are present in the DNA sample.

The AmpliType PM DNA Probe Strips have been spotted with a total of 14 sequence-specific oligonucleotide probes to distinguish the alleles of five genetic loci (a mixture of two probes is spotted at the GYPA A allele position).

1. To read the developed AmpliType PM DNA Probe Strip, the S dot is examined first and then each locus is examined separately. The S dot is designed to be the lightest typing dot on the PM DNA Probe Strip and acts as a minimum dot-intensity control for the remaining probes. It is recommended that a DNA probe strip with no visible S dot not be typed for any locus.
2. When an S dot is visible on the AmpliType PM DNA Probe Strip, the intensities of the dots at the remaining twelve positions are compared to the intensity of the S dot. Those dots that appear either darker than or equivalent to the S dot are considered positive. Each positive dot indicates the presence of the corresponding allele. Dots that are lighter than the S dot should be interpreted with care *(11,12)*.
3. The dots on the AmpliType PM DNA Probe Strip correspond to the following alleles:
   a. The A dot for each locus is positive in the presence of the A allele. (**Note:** The A dot for the GYPA locus is positive in the presence of both the A allele and the A prime allele. Both the GYPA AB and GYPA A prime B heterozygotes have balanced intensities, but additional GYPA A and B variant alleles, observed in <8% of African-American populations, may produce a slightly imbalanced heterozygous signal.)
   b. The B dot for each locus is positive for the B allele.
   c. The C dot for HBGG and Gc loci is positive for the C allele. An example of the results of an analysis of PM is shown in **Fig. 2**.

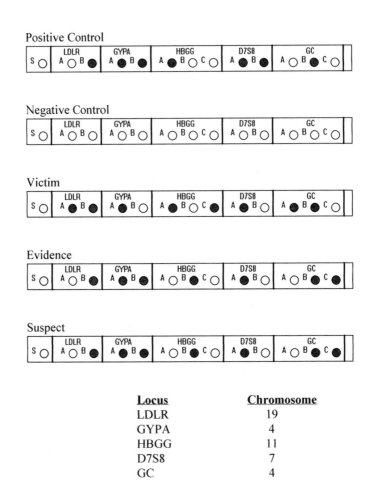

Fig. 2. AmpliType PM DNA Probe Strip typing results for Polymarker loci: low-density lipoprotein receptor (LDLR), glycophorin A (GYPA), hemoglobin G gammaglobin (HBGG), D7S8, and group-specific component (Gc). Genotypes of the typing strips follows: Positive control: LDLR, BB; GYPA, AB; HBGG, AA; D7S8, AB; Gc, BB. Negative control: no reaction. Victim: LDLR, AB; GYPA, AA; HGBB, AC; D7S8, AA; Gc, AB. Evidence and suspect: LDLR, BB; GYPA, AB; HBGG, BB; D7S8, AA; Gc, BC. These results are consistent with the evidence and the suspect having a common origin. The frequency of occurrence of the genotype of the suspect is 0.0011813 (1 in 847) in African-American, 0.0029653 (1 in 337) in Caucasian, 0.0039052 (1 in 256) in SE Hispanic, and 0.0032357 (1 in 309) in SW Hispanic populations.

3.8. Interpretation of HLA DQA1 Results

The AmpliType HLA DQA1 DNA Probe Strips have been spotted with a total of eleven sequence-specific oligonucleotide probes to detect eight alleles of the HLA DQA1 locus.

1. To read the developed AmpliType HLA DQA1 Probe Strip, the C dot is examined first. The control probe C on the AmpliType HLA DQA1 Probe Strip detects all of the HLA DQA1 alleles and is identical to the standard probe S on the AmpliType PM DNA Probe Strip. The C dot is designed to be the lightest typing dot on the strip, and it indicates that adequate amplification and typing of the HLA DQA1 in the sample has occurred. If the C dot is absent, an accurate determination of the type cannot be made.

   An accurate interpretation of the HLA DQA1 results depends on the presence and intensity of the C dot. The intensities of the dots at the remaining ten positions are compared to the intensity of the C dot. Those dots that appear either darker than or equivalent to the C dot are considered positive. Each positive dot indicates the presence of the corresponding HLA DQA1 allele. Dots with signals less than the C dot should be interpreted with care *(13–15)*.

2. Dots on the AmpliType HLA DQA1 Probe Strip correspond to the following alleles: the 1 dot is positive in the presence of the HLA DQA1 1.1, 1.2, and 1.3 alleles; the 2 dot is positive only in the presence of the HLA DQA1 2 allele; the 3 dot is positive only in the presence of the HLA DQA1 3 allele; the 4 dot is positive in the presence of the HLA DQA1 4.1, 4.2, and 4.3 alleles.

3. Four HLA DQA1 subtyping probes differentiate the HLA DQA1 1.1, 1.2, and 1.3 alleles as follows: The 1.1 dot is positive only in the presence of the HLA DQA1 1.1 allele (**Note:** A faint 1.1 dot will appear with some HLA DQA2 pseudogene alleles.) *(16)*; the 1.3 dot is positive only in the presence of the HLA DQA1 1.3 allele (**Note:** There is no probe that detects only the HLA DQA1 1.2 allele.); the 1.2, 1.3, 4 dot is positive in the presence of HLA DQA1 1.2, 1.3, 4.1, 4.2, and 4.3 alleles (**Note:** The 1.2, 1.3, 4 dot can be lighter than the C dot when the genotype has an HLA DQA1 4.2 or 4.3 allele because the HLA DQA1 4.2 and 4.3 alleles each have single partially destabilizing mismatch in the 1.2, 1.3, 4 probe *(17)*. The partially destabilizing mismatch allows these two alleles to bind to this probe weakly relative to the HLA DQA1 1.2, 1.3, and 4.1 alleles). The All but 1.3 dot is positive in the presence of all HLA DQA1 alleles except 1.3. This probe is necessary to differentiate the 1.2, 1.3 genotype from the 1.3, 1.3 genotype (**Note:** The All but 1.3 dot can be equal or lighter than the C dot when the genotype has an HLA DQA1 1.3 allele paired with an HLA DQA1 4.1,4.2, and 4.3 allele because the HLA DQA1 4.1, 4.2, and 4.3 alleles have a single partially destabilizing mismatch to the All but 1.3 probe *(17)*. The partially destabilizing mismatch allows these three alleles to weakly bind to this probe relative to the HLA DQA1 1.1, 1.2, 2, and 3 alleles.)

4. Two additional HLA DQA1 subtyping probes differentiate the HLA DQA1 4.1 allele from the HLA DQA1 4.2 and 4.3 alleles. The 4.1 dot is positive only in the presence of the HLA DQA1 4.1 allele; the 4.2/4.3 dot is positive in the presence of HLA DQA1 4.2 and 4.3 alleles. An example of AmpliType HLA DQA1 Probe Strips is shown in **Fig. 3**.

3.9. Frequency of Occurrence Calculations

1. The number of alleles detected by the PM portion of the kit is two of three at each locus. Two alleles designated A and B are detected at the LDLR, GYPA, and

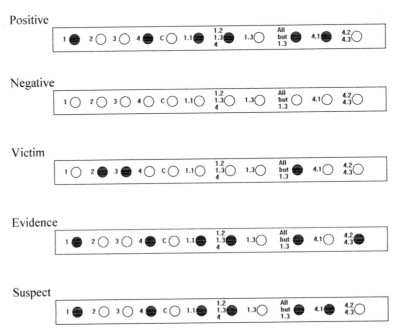

Fig. 3. AmpliType DQA1 DNA Probe Strip typing results for the HLA DQA1 locus. Genotypes of the typing strips follows: Positive control: 1.1, 4.1. Negative control: no reaction. Victim: 2, 3. Evidence: 1.1, 4.2/4.2, and Suspect: 1.1, 4.1. These results are consistent with the evidence and the suspect not having a common origin. The frequency of occurrence of the genotype of the evidence is 0.020750 (1 in 48) in African-American, 0.008848 (1 in 113) in Caucasian, and 0.023100 (1 in 43) in Hispanic populations.

D7S8 loci while three alleles designated A, B, and C are detected at the HBGG and GC loci. The number of genotypes at these loci is either three for LDLR, GYPA, and D7S8 (AA, AB, and BB) and six for HBGG and GC (AA, AB, AC, BB, BC, and CC). The frequencies of the alleles at these loci are listed in **Table 2**. The frequency of each genotype is calculated using the Hardy-Weinberg Equation. The number of genotypes is a function of the number of alleles detected and is calculated as $n(n+1)/2$ where n is the number of alleles. The total number of possible genotype combinations at the five loci detected by the Polymarker kit is 972.

2. The number of alleles detected by the HLA DQA1 portion of the kit is seven. The frequency of the alleles at this locus are listed in **Table 3**. The frequency of occurrence is calculated using the Hardy-Weinberg Equation. The number of genotypes at this locus is 28. The total number of genotypes for the PM+DQA1 kit is 27,216.

3. Generally, the allele dot intensities across the Polymarker typing strip are balanced. Thus, the dot intensities of each allele of a heterozygote at one locus (i.e., LDLR) are equivalent to the dot intensities of each allele of a heterozygote at

Table 2
The Allele Frequency Distribution Among United States Groups
of African-American, Caucasian, and Two Hispanic Populations
for the Polymarker Loci[a]

Allele	African-American[b]	Caucasian	SE Hispanic	SW Hispanic
LDLR A	0.224	0.453	0.415	0.563
LDLR B	0.776	0.547	0.585	0.438
GYPA A	0.479	0.584	0.532	0.656
GYPA B	0.521	0.416	0.468	0.344
HBGG A	0.507	0.470	0.426	0.344
HBGG B	0.197	0.524	0.548	0.609
HBGG C	0.297	0.007	0.027	0.047
D7S8 A	0.614	0.615	0.585	0.682
D7S8 B	0.386	0.385	0.415	0.318
Gc A	0.103	0.257	0.277	0.271
Gc B	0.707	0.172	0.223	0.208
Gc C	0.190	0.571	0.500	0.521

[a]Low density lipoprotein receptor (LDLR), glycophorin A (GYPA), hemoglobin G gamma-globbin (HBGG), D7S8, and group-specific component (Gc) *(12)*.
[b]Database size: African-American, 290; Caucasian, 296; Southeastern Hispanic, 188; Southwestern Hispanic, 192.

another locus (i.e., Gc). The dot intensity of the allele of a homozygote is more intense (generally twofold) than the alleles of a heterozygote. Unbalanced dots do occur in the analysis of forensic samples. This may be because of the presence of more than one DNA in different amounts. If more than two alleles are detected at the HBGG of Gc loci, this is consistent with the presence of more than one DNA. Mixtures are often difficult to resolve using the reverse-dot-blot approach and may require additional analysis. The intensity of dots should be used as a guide for resolving such mixtures.

4. The frequencies of the alleles detected at the Polymarker loci and HLA DQA1 locus have been determined (**Tables 2** and **3**). There is no detectable deviation from Hardy Weinberg Equilibrium for these loci in the four population groups based on the homozygosity test, likelihood ratio test, or the exact test *(12)*.

4. Notes

The following section deals with troubleshooting the AmpliType PM+DQA1 Amplification and Typing Kit.

1. No signal or faint signal: If no signal or a faint signal is detected from both the control DNA as well as the sample DNAs at all loci, there may be no or insufficient PCR amplification. Running a portion of the amplification reaction on a

Table 3
The Allele Frequency Distribution Among United States Groups
of African-American, Caucasian, and Hispanic Populations
for the HLA DQA1 Locus[a]

Allele	African-American[b]	Caucasian	Hispanic
1.1	0.125	0.158	0.105
1.2	0.329	0.190	0.130
1.3	0.058	0.073	0.053
2	0.130	0.145	0.115
3	0.090	0.192	0.218
4.1	0.185	0.214	0.269
4.2/4.3	0.083	0.028	0.110

[a](Perkin-Elmer AmpliType PM+DQA1 PCR Amplification and Typing Kit package insert)

[b]Database size: African-American, 200; Caucasian, 200; Hispanic, 200.

product gel should reveal whether there is any PCR product. If PCR product is observed, the hybridization conditions may be improper. Repeat the hybridization reaction with PCR product.

If no amplified product is observed after agarose-gel electrophoresis on the product gel, there may have been an insufficient amount of DNA added to the PCR reaction mix. The DNA sample should be quantitated and 2–10 ng of DNA used in a repeat test. No amplified product may also be a result of not adding the AmpliType PM+DQA1 primers to the reaction mix. Add the primers and repeat the reaction. A lack of amplified product might also be because of a failure of the thermocycler or the reaction tubes not seated tightly in the thermocycler.

If an amplification product is observed on the product gel, but no signal or faint signal is detected on the typing strips, the hybridization and/or wash conditions may be too stringent. This results from too high temperature (>55°C), too low salt concentration, or too long wash. Check temperatures and prepare new solutions and repeat the analysis. This could also be caused by inadequate agitation of the typing strips during the hybridization. No signal or a faint signal could result from the amplified DNA not being added or the amplified DNA not being denatured. Repeat of the analysis and checking that the heating block used to denature the DNA is at 95°C, and the samples remain 3 min is advised. Also, lack of signal or faint signal could result from inadequate enzyme conjugate, development solution, or Chromogen:TMB. Preparation of new solutions and a repeat of the analysis is advised.

If the positive control produces signal, but no signal is detected from the test samples, either an inhibitor may be present in the sample(s) or the test sample DNA is degraded. If an inhibitor is suspected, washing the sample in a Centricon® 100 column might remove it. Also, the addition of 16 μg BSA to the reaction might stabilize the enzyme. If the DNA is degraded, reanalysis with more DNA might yield results.

2. High DNA probe strip background color: High-DNA probe-strip background may be caused by low or lack of SDS in the hybridization and/or wash solution or inadequate agitation. Prepare new solutions, check equipment, and repeat analysis. This may also be caused by too much HRP-SA being added to the enzyme conjugate solution. Prepare new solution and repeat analysis. Also, exposure to light during color development can cause increased background. Be sure that water used for the water rinses is deionized or glass-distilled. Stored strips can have increased background if exposed to strong light and oxidizing agents.

3. Presence of additional dots in control: Additional or unexpected dots can appear because of cross-hybridization caused by the hybridization and/or wash temperatures being too low, the salt concentrations too high, or the time too short. Preparation of new solutions and monitoring of the equipment is advised. This could also be a result of contamination of the control DNA sample.

4. Signals weaker than the S or C dots: Signals weaker than the S or C dots may be a result of hybridization and/or stringent-wash temperatures being too high or too low, hybridization and/or wash solution salt concentrations too high or too low, stringent wash time too long or too short, or a mixed sample (i.e., contaminant). Repeat testing with new solutions and monitoring of equipment is advised. Weak signals may also result from failure to add EDTA to the reaction prior to the heat-denaturation step. If this is the case, add EDTA and repeat test. A faint 1.1 allele can be caused by the amplification of an HLA DQA2 pseudogene *(16)*.

5. More than two alleles present: More than two alleles may be detected at the HLA DQA1, HBGG, and GC loci. This may be a result of cross-hybridization caused by hybridization and/or stringent-wash temperatures being too low, hybridization and/or wash-solution salt concentration too high, or stringent-wash time too short. Repeat testing with new solutions and monitoring of equipment is advised. The presence of more than two alleles may be because of a mixed sample. Also the amplification of an HLA DQA2 pseudogene may result in a faint 1.1 dot *(16)*.

6. Some, not all, loci observed on product gel: The lack of detection of some PCR products by agarose gel electrophoresis may be a result of the test sample DNA being degraded. Evaluate the amplified product by agarose-gel electrophoresis to determine whether all six product bands are present. If degraded, repeat the amplification with more DNA. The lack of detection of some PCR products may be caused by the presence of an inhibitor (i.e., heme or dyes). Repeat testing with less DNA sample, Centricon 100 washed DNA, or with the addition of 16 µg BSA is advised. Alternatively, if the thermocycler did not sufficiently denature the input DNA and/or PCR product during amplification, some PCR products may not appear on the agarose gel. Repeat the testing and check the equipment.

7. Some, but not all, loci produce dots: The lack of some alleles on the typing strips may indicate that not all loci amplified. Check the PCR amplification product by agarose-gel electrophoresis for the presence of all six product bands. Lack of some alleles may be a result of the amplified DNA not being denatured. Monitor the equipment and repeat the analysis.

8. Imbalanced dot intensity: An imbalanced dot intensity may be because of the hybridization and/or stringent-wash temperature being too high or too low, the hybridization wash-solution salt concentration too high or too low, the stringent-wash time too long or too short, or EDTA was not added to the reaction prior to the heat-denaturation step of the DNA hybridization. Repeat analysis with new solutions and monitor the equipment. A mixed sample could also result with an imbalanced dot intensity proportional to the starting DNAs.
9. Weak or absent 4.1 dot on control DNA: A weak or absent 4.1 allele can result from EDTA not being added to the reaction prior to the heat-denaturation step of the DNA hybridization. Add EDTA and repeat the analysis.
10. 1.2, 1.3, and 4 dots weaker than C: A weak 1.2, 1.3, 4 dot on the HLA DQA1 typing strip can result from a sample that has an HLA DQA1 4.2 or 4.3 allele paired with an HLA DQA1 1.1, 2, 3, 4.2, or 4.3 allele.
11. 1.1 Dot weaker than C, but not signal for 1 dot: A weak 1.1 dot can result from the amplification of an HLA DQA2 pseudogene.
12. All but 1.3 signal weaker than C: A weak all but 1.3 dot can result from a sample that has an HLA DQA1 1.3 allele paired with an HLA DQA1 4.1, 4.2, or 4.3 allele.

References

1. Jeffreys, A., Wilson, V., and Thein, S. (1985) Individual specific fingerprints of human DNA. *Nature* **361,** 75–79.
2. Southern, E. (1975) Detection of specific sequences among DNA fragments separated by gel electrophoresis. *J. Mol. Biol.* **98,** 503–527.
3. Nakamura, Y., Leppert, M., O'Connell, P., Wolff, R., Holm, T., Culver, M., Martin, C., Fujimoto, E., Hoff, M., Kumlin, E., and White, R. (1987) Variable number of tandem repeat (VNTR) markers for human gene mapping. *Science* **235,** 1616–1622.
4. Saiki, R. K., Scharf, S., Faloona T., Mullis, K. B., Horn, G. T., Erlich, H. A., and Arnheim, N. (1985) Enzymatic amplification of beta globin genomic sequences and restriction analysis for diagnosis of sickle cell anemia. *Science* **230,** 1350–1354.
5. Gyllensten, U. B. and Erlish, H. A. (1988) Generation of single strand DNA by the polymerase chain reaction and its application to direct sequencing of the HLA-DQ alpha locus. *Proc. Natl. Acad. Sci. USA* **85,** 7652–7656.
6. Yamamoto, T., Davis, C. G., Brown, M. S., Schneider, W. J., Casey, M. L., Goldstein, J. L., and Russell, D. W. (1984) The human LDL receptor: a cysteine-rich protein with multiple Alu sequences in its mRNA. *Cell* **39,** 27–38.
7. Siebert, P. D. and Fukuda, M. (1987) Molecular cloning of guman glycophorin B cDNA: nucleotide sequence and genomic relationship to glycophorin A. *Proc. Natl. Acad. Sci. USA* **84,** 6735–6739.
8. Slighton, J. L., Blechl, A. E., and Smithies, O. (1980) Human fetal ^{G}gamma and ^{A}gamma globin genes: complete nucleotide sequences suggest that DNA can be exchanged between these duplicated genes. *Cell* **21,** 627–638.
9. Horn, G. T., Richards, B., Merrill, J. J., and Klinger, K. W. (1980) Characterization and rapid diagnostic analysis of DNA polymorphisms closely linked to the cystic fibrosis locus. *Clin. Chem.* **36,** 1614–1619.

10. Yang, F., Brune, J. L., Naylor, S. L., Cupples, R. L., Naberhaus, K. H., and Bowman, B. H. (1985) Human group-specific component (Gc) is a member of the albumin family. *Proc. Natl. Acad. Sci. USA* **82,** 7994–7998.

11. Herrin, G., Fildes, N., and Reynolds, R. (1994) Evaluation of the AmpliType PM DNA test system on forensic case samples. *J. Forensic Sci.* **39,** 1247–1253.

12. Budowle, B., Lindsey, J. A., DeCou, J. A., Koons, B. W., Guisti, A. M., and Comey, C. T. (1995) Validation and population studies of the loci LDLR, GYPA, HBGG. D7S8, and GC (PM loci), and HLA-DQalpha using a multiplex amplification and typing procedure. *J. Forensic Sci.* **40,** 45–54.

13. Blake, E., Mihalovich, J., Higuchi, R., Walsh, P. S., and Erlich, H. (1992) Polymerase chain reaction (PCR) amplification and human leukocyte antigen (HLA)-DQalpha oligonucleotide typing on biological evidence samples: casework experience. *J. Forensic Sci.* **37,** 700–726.

14. Comey, C. T., Budowle, B., Dams, D. E., Baumstark, A. L., Lindsey, J. A., and Presley, L. A. (1993) PCR amplification and typing of the HLA DQalpha gene in forensic samples. *J. Forensic Sci.* **38,** 239–249.

15. Higuchi, R. and Kwok, S. (1989) Avoiding false positives with PCR. *Nature* **339,** 237,238.

16. Crouse, C. A., Vincek, V., and Caraballo, B. K. (1994) Analysis and interpretation of the HLA DQalpha 1.1 weak-signal observed during the PCR-based typing method. *J. Forensic Sci.* **39,** 41–51.

17. Roy, R. and Reynolds, R. (1995) AmpliType PM and HLA DQalpha typing from pap smear, semen smear and post-coital slides. *J. Forensic Sci.* **40,** 266–269.

24

The Use of Capillary Electrophoresis in Genotyping STR Loci

John M. Butler

1. Introduction

Capillary electrophoresis (CE) is an emerging technology that has the potential to increase the speed and automation of DNA-typing procedures *(1,2)*. Although slab-gel electrophoresis is widely used, there are several advantages to analyzing DNA in a capillary format. First and foremost, the injection, separation, and detection can be fully automated. Rapid separations (from high voltages) are permitted as a result of efficient heat dissipation in the capillary. In addition, quantitative information is available in a single step in conjunction with computer collection of the data. Finally, only minute quantities of sample are consumed and very little waste is generated. These advantages make CE an efficient means of DNA analysis, particularly for DNA fragments generated by the polymerase chain reaction (PCR) (*see* **Note 1**).

PCR-based DNA-typing assays, particularly those involving short tandem repeats (STRs), are becoming increasingly popular in human-identity testing *(3,4)*. The use of PCR permits greater sensitivity and speed over traditional RFLP techniques. However, most methods for separating STR alleles involve slab-gel electrophoresis and thus are labor-intensive (e.g., gels must be poured and samples must be loaded manually).

Although CE has shown advantages compared to traditional gel techniques, the implementation of CE into DNA-typing procedures has developed slowly. Challenges involving injection, resolution, and sizing of DNA fragments have had to be addressed. Getting detectable levels of amplified DNA into the capillary has been a significant problem *(6,9)*. The high ionic strength of PCR samples can result in low amounts of DNA being injected, primarily because the KCl and $MgCl_2$ in PCR buffers (e.g., >50 mM Cl$^-$) interferes with the elec-

From: *Methods in Molecular Biology, Vol. 98: Forensic DNA Profiling Protocols*
Edited by: P. J. Lincoln and J. Thomson © Humana Press Inc., Totowa, NJ

trokinetic injection process. The rate at which a DNA sample can be loaded onto the CE column is inversely proportional to the ionic strength. Two approaches have been used: removal of the salt ions through float dialysis *(6)* and changing the sample/buffer ionic-strength ratio *(2)*. The second approach, which is described in this chapter, may be performed by diluting the PCR sample in deionized water and utilizing a sample stacking process *(2,13)*. Unfortunately, dilution reduces the amount of sample available and thus requires more sensitive detection, such as laser-induced fluorescence (LIF).

Originally, polyacrylamide gel-filled capillaries were used for high-resolution DNA separations *(6,15)*. More recently, open capillaries filled with a soluble, sieving polymer have been demonstrated to work effectively *(5,9)*. Gel-filled capillaries are not robust in nature and typically last only a few runs *(6)*, whereas column lifetimes of >2000 runs have been observed with refillable polymer systems *(13)*.

With traditional slab-gel methods, DNA-fragment sizing is performed by running a size standard in a neighboring lane. Similarly, when genotyping STRs, allelic ladders may be run for comparison purposes in adjacent lanes to an individual sample *(4)*. However, because CE instruments process samples in a sequential rather than a parallel fashion, the results between multiple separations need to be related. For example, a standard would be run followed by several samples. A high degree of precision is required between runs in order to relate the results of one run to the next.

Several groups have examined the possibility of using CE for genotyping STR loci, such as HUMTH0I *(2,5–8)* and SE33 *(9)*, or larger VNTRs, such as D1S80 *(9–11)* or apoB *(11,12)*. Most of the work in this area has focused on demonstrating the 4 bp resolution needed to resolve tetranucleotide STRs with various sieving matrices or gel-filled columns. Actual typing has been demonstrated by using dual internal standards to relate the sizes of an allelic ladder to an individual sample, which was run in a subsequent analysis *(2,7)*. Another approach involved the use of multiple fluorescent tags to distinguish between DNA standards and samples that are run at the same time *(8)*.

The method described in this chapter utilizes soluble polymer sieving buffers in an open, coated capillary. With a replaceable separation medium (e.g., *see* **Subheading 2.**), column lifetime improves and the possibility of cross-contamination between runs is reduced. A fluorescent intercalating dye is also used to improve resolution and sensitivity. Because of the linear relationship between migration time and DNA size in the STR region (i.e., 100–400 bp), DNA fragments are sized using 150 and 300 bp internal standards to form a calibrating line. An allelic ladder is usually run every 10 samples to calibrate the CE system. Using this method, STR alleles that differ in size by as little as 3 bp may be effectively separated and accurately typed *(2,7)* (*see* **Note 2**).

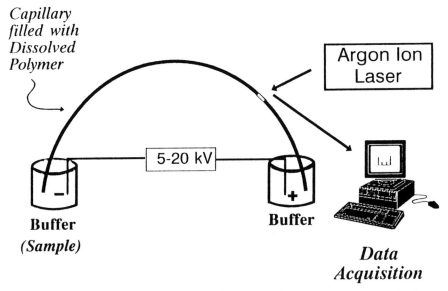

Fig. 1. Schematic of a CE instrument with laser-induced fluorescence detection.

2. Materials

1. The work described in this chapter was performed on a Beckman P/ACE™ 2050 with a Laser Module 488 argon ion laser (Beckman Instruments, Fullerton, CA). Beckman P/ACE Windows software (version 3.0) was used to control the CE instrument, and the data were collected on Waters Millennium 2010 software (version 2.0, Waters Chromatography, Bedford, MA). Beckman System Gold software (Beckman Instruments) may also be used to collect the peak migration time. The principal elements of this CE instrument consist of two buffer vials, a capillary filled with a sieving polymer, a high-voltage power supply, an argon-ion laser, and a computer for data collection (**Fig. 1**).
2. Column: 50 μm id × 27 cm, DB-17 coated (J&W Scientific, Folsom, CA).
3. Buffer: 100 m*M* Tris-borate, 2 m*M* EDTA, pH 8.2, with 1% hydroxyethyl cellulose (HEC), viscosity: 86–113 cP for a 2% solution at 25°C (Aldrich Chemical). The HEC (5 g) was typically stirred overnight at room temperature in 500 mL of the Tris-borate-EDTA solution.
4. Intercalating dye: 500 ng/mL YO-PRO-1 (Molecular Probes, Eugene, OR). The CE run buffer was usually prepared by adding 11.9 μL of the 1 m*M* YO-PRO-1 stock solution into 15 mL of the above HEC buffer. The dye is light-sensitive and a possible carcinogen and should be treated appropriately. Aluminum foil may be placed around the container to prevent degradation from light.
5. Internal DNA standards: 150 and 300 bp DNA fragments (BioVentures Inc., Murfreesboro, TN). These standards may be purchased at concentrations of 100 ng/μL and then diluted with water to the concentration needed.

6. Appropriate STR allelic ladder. For the method described here, it must be clean under nondenaturing (native) conditions.
7. Methanol, HPLC grade.
8. Vials: both amber and clear, 4 mL wide-mouth with threads.
9. Sample vials: 0.2 mL MicroAmp™ Reaction Tubes (Perkin-Elmer, Norwalk, CT).

3. Method

1. Prepare the capillary by cutting it to the desired length and removing approx 5 mm of the polyimide coating for the detection window (*see* **Note 3**). Place the capillary in an LIF capillary cartridge (Beckman Instruments). The cartridge will allow liquid to flow around the capillary and maintain a constant-temperature environment.
2. Using a transfer pipet, fill three 4-mL amber vials with the HEC buffer containing the intercalating dye. Be sure to remove all bubbles from the solution surface, because they may interfere with the flow of electrical current. Two buffer vials will be used as the inlet and outlet vials during the separation; the third will be used to fill the capillary with fresh separation media between each run (**Note 4**). Fill a fourth vial with methanol. When the vials are prepared, place them in the CE autosampler.
3. Wash the capillary with methanol for 10 min followed by a 10-min flush with the run buffer to equilibrate the column.
4. Remove 1 μL of the DNA sample generated by the PCR reaction and dilute it with 48 μL of deionized water. A modified PCR tube serves as an effective CE sample vial. If UV detection (rather than LIF) is used, the sample may need to undergo more extensive cleanup procedures, such as float dialysis, to remove salt ions *(6,9)*.
5. Add the 150 and 300 bp DNA fragments in 1 ng quantities (i.e., 1 μL of a solution containing 1 ng/μL of each fragment) to the above DNA sample. Adding this 1 μL should bring the total CE sample volume to 50 μL.
6. Place the prepared sample vial on a spring inside a 4 mL wide-mouth vial. Screw a silicon rubber cap on the 4 mL vial to prevent evaporation. Load the sample into the CE autosampler tray. Autosampler trays on CE instruments will usually hold more than 20 samples at a time.
7. Program the CE instrument method as follows:
   a. Set the detector to collect data at a rate of 5 points/s.
   b. Set the column temperature to 25°C.
   c. Rinse the capillary for 1 min with methanol.
   d. Fill the capillary with run buffer containing the entangled HEC polymer and YO-PRO-1 intercalating dye for 2 min.
   e. Dip the tip of the capillary inlet in a vial of deionized water for 5 s; *see* **Note 5**.
   f. Inject the sample for 5 s at 1 kV.
   g. Apply a separation voltage of 5 kV (185 V/cm). The current should rise to ~6.7 μA (*see* **Note 6**).
   h. Repeat **steps (c)–(g)** for each sample.
8. Label the samples in the sequence spreadsheet program and start the sample sequence. Watch to see that the current is steady when the voltage is applied to the first

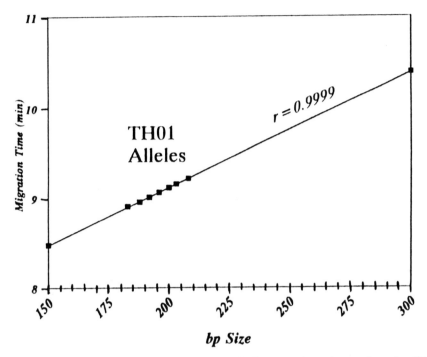

Fig. 2. The region between 150 and 300 bp is linear and can be used to size STR alleles, in this case an allelic ladder from the HUMTH01 locus (179–203 bp).

sample. If the current does not stabilize within the first minute of the run, the capillary is plugged and needs to be rinsed with methanol and buffer again.
9. Following the completion of the CE separations, data analysis may be performed:
   a. Use the migration times of the 150 and 300 bp fragments to make a calibrating line (**Fig. 2**).
   b. Calculate the allele sizes for both the allelic ladder and individual samples using the following equation:

$$\text{DNA fragment size} = 150\ \text{bp} + [(MT_{DNA} - MT_{150})^* q] \qquad (1)$$

   where MT is the peak migration time of a particular component and

$$q = (300 - 150)/(MT_{300} - MT_{150}) \qquad (2)$$

   c. Determine the genotype by comparing the calculated size of the sample's DNA fragment(s) to the allelic ladder. It is often useful to define the precision of the CE system by making 50 injections of the allelic ladder (**Fig. 3**). A match window can then be defined by 3.3 times a single standard deviation from the average of each allele size *(2)*.
   d. Running an allelic ladder prior to each set of ten samples has been shown to provide satisfactory calibration for accurate genotyping *(2)* (*see* **Table 1**).

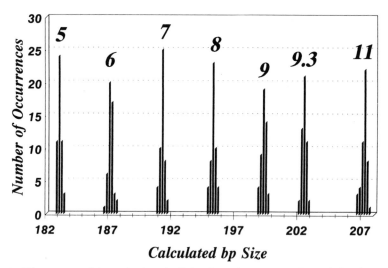

Fig. 3. Histogram of the calculated allele sizes from 50 replicate injections of the HUMTH01 allelic ladder *(2)*. Each bar represents a 0.2 bp bin. Conditions: *capillary:* 50 μm id × 27 cm DB-17; buffer: 1% HEC (Aldrich), 100 m*M* TBE, pH 8.2, 500 ng/mL YO-PRO-1; *temperature:* 25°C; *injection:* 5 s @ 1 kV; *separation:* 5 kV (185 V/cm); *sample:* 1:50 dilution of a HUMTH01 allelic ladder with deionized water + 150 and 300 bp internal standards.

10. **Figure 4** illustrates a typical result for the HUMTH01 locus. The allelic ladder is overlaid on an individual sample. With the 150 and 300 bp standards to correct for changes in migration time between runs, the migration times for the individual alleles can easily be related to the allelic ladder.

4. Notes

1. Although the overall throuput of DNA samples using the CE method described in this chapter is approximately equivalent to traditional gel methods, using multiple CE channels promises to greatly increase the throughput. Capillary-array electrophoresis systems have been demonstrated with five or more parallel capillaries *(8)*. Microfabricated CE devices may further increase the speed and throughput of DNA samples used in typing STRs *(14)*.

2. The length of the cellulosic polymer can play an important role in resolving DNA fragments *(13)*. For example, a 1% solution of HEC separates tetranucleotide STRs nicely (**Fig. 5**) but fails to resolve the larger D1S80 alleles.

3. The capillaries may be purchased in 10 m rolls and cut to size by the user. In this case, the polyimide outer coating of the fused silica capillary must be removed to allow on-column detection.

Table 1
A Set of Ten Population Samples Amplified
with PCR Primers Specific for the HUMTH01 Locus (3)[a]

	Alleles (bp)						
	5	6	7	8	9	9.3	11
	182.3	186.4	190.4	194.3	198.4	201.5	206.4
(1)			190.9				
(2)					198.7	201.8	
(3)						201.8	
(4)		186.5			198.6		
(5)			190.7	194.6			
(6)		186.7				201.9	
(7)			190.7			201.7	
(8)				194.7	198.7		
(9)			190.5	194.5			
(10)			190.6		198.6		

[a]Before the samples ran, an allelic-ladder standard was run to calibrate the CE system. The allele sizes are in base pairs. Polyacrylamide gel electrophoresis of the same samples yielded the following genotypes: (1) 7,7; (2) 9,9.3; (3) 9.3,9.3; (4) 6,9; (5) 7,8; (6) 6,9.3; (7) 7,9.3; (8) 8,9; (9) 7,8; (10) 7,9. In a blind study comparing the typing results of almost 100 samples, complete agreement was obtained between CE and traditional gel electrophoresis with silver-staining *(2)*.

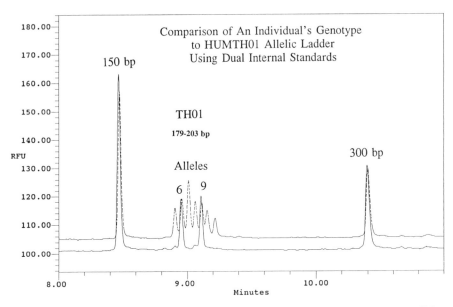

Fig. 4. Overlay of HUMTH01 allelic ladder and an individual sample. Conditions as in **Fig. 3**.

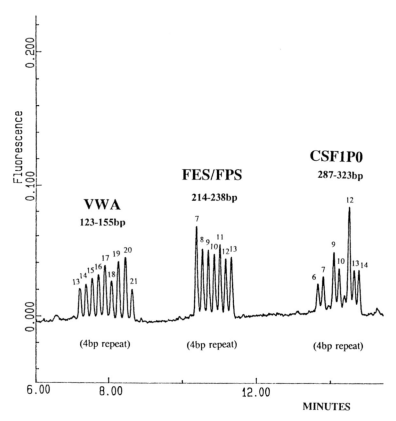

Fig. 5. Multiplex STR allelic ladders for the VWA, FES/FPS, and CSF1PO loci. The numbers above the peaks indicate the number of repeats in each allele. Conditions as in **Fig. 3** except separation voltage (0–5 min at 7.5 kV, 5–16 min at 1.5 kV, 16–18 min at 7.5 kV). The sample was a gift from Dr. Marcia Eisenberg of Laboratory Corporation of America.

4. Filling the capillary from a third vial, which is not involved in the electrophoresis process, aids the reproducibility of multiple CE separations. Amber vials were typically used to protect the light-sensitive intercalating dye.

5. A water dip between the buffer fill and the injection removes buffer salts from the edge of the capillary that can interfere in the injection process.

6. Changing the voltage during the run can benefit resolution and speed. By running at a higher voltage at the beginning of the run, the DNA fragments move rapidly through the column. Then, shortly before the fragments pass the detector, the voltage may be reduced in a single step in order to improve the resolution in a particular region of the separation. **Figure 6** illustrates the separation improvement with a HUMTH01 allelic ladder. Unfortunately, while separation speed is benefited, run-to-run precision is not as consistent as constant voltage separations *(7,13)*.

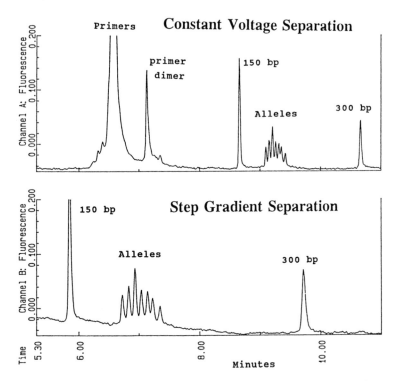

Fig. 6. A comparison of separation speed for the HUMTH01 alleles between a constant voltage and a step-gradient separation *(7)*. The time scales are the same. In the lower frame, the primers and primer dimer migrate faster than 5 min and are not seen. Alleles represented include 5 (179 bp), 6 (183 bp), 7 (187 bp), 8 (191 bp), 9 (195 bp), 9.3 (198 bp), and 11 (203 bp). Conditions as in **Fig. 3** except 37 cm DB-17 coated capillary, 50 ng/mL YO-PRO- 1, and separation: constant voltage = 10 kV; step gradient = 0–5 min at 15 kV, 5–11 min at 5 kV.

Acknowledgments

The development of this procedure involved the efforts of a number of individuals including Bruce McCord, David McClure, and Janet Jung Doyle at the FBI Laboratory's Forensic Science Research Unit. Names of commercial manufacturers are provided for identification purposes only and inclusion does not imply endorsement by the National Institute of Standards and Technology.

References

1. Oda, R. P. and Landers, J. P. (1994) Introduction to capillary electrophoresis, in *CRC Handbook of Capillary Electrophoresis: A Practical Approach* (Landers, J. P., ed.), CRC, Boca Raton, FL, pp. 9–42.

2. Butler, J. M., McCord, B. R., Jung, J. M., Lee, J. A., Budowle, B., and Allen, R. O. (1995) Application of dual internal standards for precise sizing of polymerase chain reaction products using capillary electrophoresis. *Electrophoresis* **16,** 974–980.

3. Edwards, A., Civitello, A., Hammond, H. A., and Caskey, C. T. (1991) DNA typing and genetic mapping with trimeric and tetrameric tandem repeats. *Am. J. Hum. Genet.* **49,** 746–756.

4. Puers, C., Hammond, H. A., Jin, L., Caskey, C. T., and Schumm, J. W. (1993) Identification of repeat sequence heterogeneity at the polymorphic short tandem repeat locus HUMTH01 [AATG]$_n$ and reassignment of alleles in population analysis by using a locus-specific allelic ladder. *Am. J. Hum. Genet.* **53,** 953–958.

5. McCord, B. R., McClure, D. M., and Jung, J. M. (1993) Capillary electrophoresis of PCR-amplified DNA using fluorescence detection with an intercalating dye. *J. Chromatogr.* **652,** 75–82.

6. Williams, P. E., Marino, M. A., Del Rio, S. A., Turni, L. A., and Devaney, J. M. (1994) Analysis of DNA restriction fragments and polymerase chain reaction products by capillary electrophoresis. *J. Chromatogr. A* **680,** 525–540.

7. Butler, J. M., McCord, B. R., Jung, J. M., and Allen, R. O. (1994) Rapid analysis of the short tandem repeat HUMTH01 by capillary electrophoresis. *Biotechniques* **17,** 1062–1070.

8. Wang, Y., Ju, J., Carpenter, B., Atherton, J. M., Sensabaugh, G. F., and Mathies, R. A. (1995) High-speed, high-throughput TH01 allelic sizing using energy transfer fluorescent primers and capillary array electrophoresis. *Anal. Chem.* **67,** 1197–1203.

9. McCord, B. R., Jung, J. M., and Holleran, E. A. (1993) High resolution capillary electrophoresis of forensic DNA using a non-gel sieving buffer. *J. Liq. Chromatogr.* **16,** 1963–1981.

10. Srinivasan, K., Morris, S. C., Girard, J. E., Kline, M. C., and Reeder, D. J. (1993) Use of TOTO and YOYO intercalating dyes to enhance detection of PCR products with laser induced fluorescence-capillary electrophoresis. *Appl. Theor. Electrophoresis* **3,** 235–239.

11. Srinivasan, K., Girard, J. E., Williams, P., Roby, R. K., Weedn, V. W., Morris, S. C., Kline, M. C., and Reeder, D. J. (1993) Electrophoretic separations of polymerase chain reaction-amplified DNA fragments in DNA typing using a capillary electrophoresis-laser induced fluorescence system. *J. Chromatogr.* **652,** 83–91.

12. Pearce, M. J. and Watson, N. D. (1993) Rapid analysis of PCR components and products by acidic non-gel capillary electrophoresis, in *DNA Fingerprinting: State of the Science* (Pena, S. D. J., Chakraborty, R., Epplen, J., and Jeffreys, A. J., eds.), Birkauser-Verlag, Basel, pp. 117–124.

13. Butler, J. M. (1995) Sizing and quantitation of polymerase chain reaction products by capillary electrophoresis for use in DNA typing. Ph. D. dissertation, Univ. Virginia, Charlottesville.

14. Woolley, A. T. and Mathies, R. A. (1994) Ultra-high-speed DNA fragment separations using microfabricated capillary array electrophoresis chips. *Proc. Natl. Acad. Sci. USA* **91,** 11,348–11,352.

15. Heiger, D. N., Cohen, A. S., and Karger, B. L. (1990) Separation of DNA restriction fragments by high performance capillary electrophoresis with low and zero crosslinked polyacrylamide using continuous and pulsed electric fields. *J. Chromatogr.* **516,** 33–48.

25

Solid-Phase Minisequencing
as a Tool to Detect DNA Polymorphism

Ann-Christine Syvänen

1. Introduction

Single-base substitutions, which give rise to biallelic sequence polymorphism, have been estimated to occur on the average at one out of a thousand nucleotides in the human genome *(1)*. Analysis of this allelic variation can be utilized in population genetic studies, in genetic-linkage analysis, for the discrimination between individuals and for tissue typing. Genotyping by analyzing single-nucleotide polymorphisms provides some clear advantages compared to genotyping by analyzing multiallelic microsatellite markers. The mutation rate of single nucleotides is lower, single-base substitutions can be detected by technically simple, automatable methods, and the computational interpretation of the results is simpler.

We have developed an efficient method, denoted solid-phase minisequencing, for the detection of known single-nucleotide variations or point mutations *(2)*. The method is routinely used in our laboratory in genetic linkage and association studies for analyzing biallelic markers as a complement to microsatellite markers. We have also applied the solid-phase minisequencing method to the identification of individuals in forensic analyses and paternity testing by analyzing a panel of 12 polymorphic nucleotides located on different chromosomes *(3)*. The power of discrimination between individuals using the selected marker panel is 0.99996 in the Finnish population, and the average probability of exclusion in paternity testing is 0.90 *(3)*. Because of the technically simple format of the method, more polymorphic nucleotides can easily be added to the marker panel to increase the discrimination power of the system.

The principle of the solid-phase minisequencing method is to identify a polymorphic nucleotide at a predetermined site in a DNA template by specific extension of a primer by a single nucleotide using a DNA polymerase. First, a

From: *Methods in Molecular Biology, Vol. 98: Forensic DNA Profiling Protocols*
Edited by: P. J. Lincoln and J. Thomson © Humana Press Inc., Totowa, NJ

DNA fragment spanning the polymorphic nucleotide position is amplified with one unbiotinylated and one biotinylated PCR primer. The amplified biotinylated fragment is then captured on an avidin- or streptavidin-coated solid support, the excess of PCR reagents are removed by washing the support, and the captured DNA template is rendered single-stranded by alkaline treatment. The nucleotide(s) at the polymorphic site are identified in the remaining immobilized single-stranded DNA fragment by "minisequencing" reactions, in which a detection-step primer designed to anneal immediately adjacent to the polymorphism is extended with a single labeled nucleoside triphosphate complementary to the nucleotide at the polymorphic site (**Fig. 1**).

The specificity of the solid-phase minisequencing method originates from the fidelity of the nucleotide incorporation catalyzed by the DNA polymerase, while the primer annealing reaction is carried out in nonstringent conditions. Therefore, all polymorphic nucleotides can be detected at the same reaction conditions, irrespectively of the flanking nucleotide sequence. The method can potentially be fully automated because it is carried out in a simple solid-phase format, and the results are obtained as objective numeric values that are easy to interpret and store. Furthermore, the solid-phase minisequencing method allows accurate quantitative PCR analysis of two sequences present as a mixture in a sample. This possibility has proven to be useful for rapid determination of the population frequencies of polymorphic nucleotides by quantitative analysis of large pooled DNA samples *(3)*.

In the protocol for the solid-phase minisequencing method presented below, [^{3}H]dNTPs are used as labels, and streptavidin-coated microtiter plates serve as the solid support. All reagents and equipment required are readily available and the protocol presented in **Subheading 3.4.** is generally applicable for detecting any polymorphic nucleotide.

2. Materials

2.1. Equipment

1. Access to oligonucleotide synthesis.
2. Programmable heat block and facilities for PCR.
3. Streptavidin-coated microtiter plates (e.g., Combiplate 8, Labsystems, Helsinki, Finland).
4. Incubator or water bath with shaker at 37°C.
5. Incubator or water bath at 50°C.
6. Liquid scintillation counter.
7. Multichannel pipets and/or microtiter plate washer (optional).

2.2. Reagents

1. Biotinyl phosphoramidite reagent (e.g., RPN 2012, Amersham) for biotinylation of one of the PCR primers.

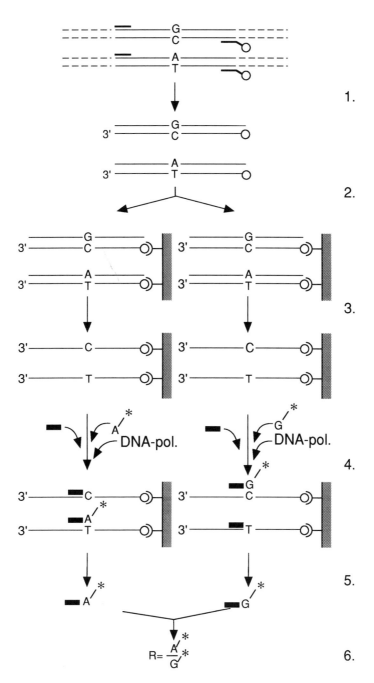

Fig. 1. Principle and steps of the solid-phase minisequencing method. 1) PCR with a biotinylated and an unbiotinylated primer. 2) Affinity-capture in streptavidin-coated microtiter plate wells. 3) Washing and denaturation. 4) Minisequencing primer extension reactions. 5) Measurement of the extended primers. 6) Interpretation of the result.

Fig. 2. Positions of the primers for solid-phase minisequencing. The variable nucleotide position is indicated with a vertical arrow. The minisequencing detection-step primer is marked MS, the PCR primers are marked PCR, and B denotes a biotin residue.

2. Thermostable DNA polymerase (e.g., *Taq* DNA polymerase, Promega Biotech or Dynazyme™ DNA polymerase, Finnzymes, Espoo, Finland) for PCR and the minisequencing reaction.
3. 10X concentrated DNA polymerase buffer: 500 mM Tris-HCl, pH 8.8, 150 mM $(NH_4)_2SO_4$, 15 mM $MgCl_2$, 1% (v/v) Triton X-100, 0.1% (w/v) gelatin. Store at –20°C, thaw completely before use.
4. dNTP mixture for PCR: 2 mM dATP, 2 mM dCTP, 2 mM dGTP, 2 mM dTTP. Store at –20°C.
5. Buffer for the capturing reaction (PBS-Tween): 20 mM sodium phosphate buffer, pH 7.5, 100 mM NaCl, 0.1% (v/v) Tween-20. Store at 4°C.
6. Washing buffer (TENT): 40 mM Tris-HCl, pH 8.8, 1 mM EDTA, 50 mM NaCl, 0.1% Tween-20. Store at 4°C.
7. Denaturing solution: 50 mM NaOH.
8. [^{3}H]-labeled deoxynucleoside triphosphates ([^{3}H]dATP, TRK 625; [^{3}H]dCTP, TRK 576; [^{3}H]dGTP, TRK 627; [^{3}H]dTTP, TRK 633, Amersham).
9. Scintillation fluid.

3. Methods

3.1. Primer Design

1. PCR primers that amplify a fragment, preferably between 50 and 200 base pairs in size, and contain the variable nucleotide position(s), are required. The PCR primers should be 20–23 nucleotides in size, have similar melting temperatures and noncomplementary 3'-ends *(4)*. One of the PCR primers is biotinylated in its 5'-end during the synthesis (*see* **Note 1**).
2. The minisequencing-detection-step primer is designed to be complementary to the biotinylated strand of the PCR product immediately 3' of the variable nucleotide position (**Fig. 2**). It should preferably be 20 nucleotides long and it should be at least five nucleotides nested in relation to the unbiotinylated PCR primer.

3.2. PCR Amplification

1. Various types of DNA samples, treated as is suitable for PCR amplification, can be analyzed *(5)*. PCR is performed according to a standard protocol *(6)*, except that the

concentration of the biotinylated primer should not exceed 0.2 μM (10 pmol/50 μL reaction) (*see* **Notes 2** and **3**).

3.3. Affinity-Capture

1. Transfer two 10 μL aliquots (or four 10 μL aliquots for parallel assays of both nucleotides) of each PCR product to streptavidin-coated microtiter plate wells. Include two negative controls containing DNA polymerase buffer only.
2. Add 40 μL of PBS-Tween buffer to each well and seal the wells with a sticker. Incubate the microtiter plate for 1.5 h at 37°C with gentle shaking. Discard the contents of the wells.
3. Wash the wells manually three times at about 20°C by adding 200 μL of TENT buffer. Empty the wells thoroughly between the washes by tapping the plates upside down against a tissue paper. The use of an automatic microtiter plate washer saves time and labor and improves the washing efficiency (*see* **Note 4**).
4. Denature the captured PCR product by adding 100 μL of 50 mM NaOH to each well, followed by incubation at 20°C for 2–5 min. Discard the NaOH and wash as in **step 3** above.

3.4. Solid-Phase Minisequencing

1. Prepare the reaction mixtures for detecting each polymorphic nucleotide by combining 5 μL of 10X DNA polymerase buffer, 2 μL of 5 μM detection step primer (10 pmol), 0.1 μCi (usually 0.1 μL) of a [^{3}H]dNTP complementary to the nucleotide to be detected, 0.05 U of DNA polymerase and distilled water to 50 μL per reaction (*see* **Note 5**). It is convenient to prepare master mixtures for the desired number of reactions during the capturing reaction. The mixtures can be stored at room temperature until use.
2. For each amplified sample, add 50 μL of reaction mixture to the microtiter plate wells that correspond to the nucleotide to be identified. Seal the wells with a sticker and incubate the plate for 10 min at 50°C. Discard the contents of the wells and wash as in **Subheading 3.3.**, **step 3**.
3. Release the primer by adding 60 μL of 50 mM NaOH to each well and incubate at 20°C for 2–5 min. Transfer the 50 mM NaOH solution containing the eluted primer to scintillation vials, add scintillation fluid, and measure the eluted ^{3}H in a liquid scintillation counter (*see* **Note 6**).

3.5. Interpretation of the Result

1. The result from the scintillation counter is a numeric cpm value corresponding to the amount of [^{3}H]dNTP incorporated in a minisequencing reaction. When the assay has been successful, the measured radioactivity will be 1000–5000 cpm when incorporation of a [^{3}H]dNTP has occurred, and the background will be below 100 cpm. **Table 1** shows a representative example of the result of the assay.
2. Calculate the ratio (R-value) between the cpm value obtained in the reaction corresponding to one of the variable nucleotides and the cpm value obtained in the reaction corresponding to the other nucleotide (*see* **Table 1**). Calculation of the

Table 1
Result From the Analysis of a Polymorphic Nucleotide (A or G)
in the LDL Receptor Gene by Solid Phase Minisequencing[a]

	[³H]dNTP incorporated (cpm)[b]		
Genotype of sample	A-allele	G-allele	R-value A_{cpm}/G_{cpm}
AA	3610	48	75
AG	2240	1610	1,4
GG	46	3560	0,013

[a]The data are from *(3)*, and the polymorphism is described in *(14)*.
[b]The specific activities of the [³H]dATP and [³H]dGTP were 40 and 31 Ci/mmol, respectively.

R-value eliminates variations in the amount of incorporated [³H]dNTPs caused by variation between samples in the efficiency of PCR. The R-values fall into three distinct categories that unequivocally define the genotype of the sample. The R-value will be >10 or <0.1 in samples from subjects homozygous for one nucleotide and between 0.5 and 2 in samples from heterozygous subjects, depending on the specific activities of the [³H]dNTPs used (*see* **Notes 7** and **8**).

4. Notes

1. An obvious prerequisite for complete capture of the biotinylated PCR product in the streptavidin-coated wells is that the primer has been efficiently biotinylated (80–90%) during the synthesis. Normally the biotinylated primer can be used without further purification, but if necessary, biotinylated oligonucleotides can be purified from unbiotinylated oligonucleotides by HPLC *(7)*, PAGE *(8)*, or with the aid of disposable ion-exchange chromatography columns (Perkin-Elmer/Applied Biosystems).

2. Because [³H]dNTPs that are of low specific activity are used as labels in the minisequencing reaction, it is important that the PCR amplification be efficient. Ten microliters of the PCR product should be clearly visible on an agarose gel stained with ethidium bromide.

3. The biotin binding capacity of the microtiter plate wells sets an upper limit to the amount of biotinylated PCR product (and excess of biotinylated primer) that can be present during the capturing reaction. The biotin binding capacity of the wells given in **Subheading 3.** (Combiplate 8, Labsystems, Helsinki, Finland) is 2–5 pmol of biotinylated oligonucleotide. Therefore, in our standard protocol, the biotinylated primer is used at 0.2 µ*M* (0.2 pmol/µL) concentration, and 10 µL of the PCR product is analyzed per well. If products of multiplex PCR reactions are to be analyzed, the amount of biotinylated primer during PCR and the volume of the analyzed aliquot should be reduced *(3)*. If a higher biotin binding capacity is required, another affinity matrix with higher biotin binding capacity, such as avidin-coated polystyrene microparticles (Fluoricon assay particles, 0.7–0.9 µ*M*, IDEXX Corp., Portland, ME) with

extremely high biotin binding capacity (>2 nmol/mg of particles) or streptavidin-coated magnetic polystyrene beads (Dynabeads M-280, Dynal, Oslo, Norway, binding capacity 150–300 pmol of biotinylated primer/mg of beads) can be used *(9)*.

4. It is important for the specificity of the minisequencing reaction that all dNTPs from the PCR be completely removed by the washing steps. The presence of other dNTPs than the intended [^{3}H]dNTP during the minisequencing reaction will cause unspecific extension of the detection-step primer.

5. Other than thermostable DNA polymerases (e.g., T7 DNA polymerase), and dNTPs or ddNTPs labeled with other isotopes, haptens, or fluorphores *(2,10,11)*, can also be used in the minisequencing reaction.

6. If a scintillation counter for microtiter plates is available, streptavidin-coated microtiter plates manufactured from scintillating plastic (ScintiStrips™, Wallac, Turku, Finland) can be used as the solid phase *(12)*. This will simplify the procedure in that the final washing and denaturing steps and the transfer of the eluted primer to scintillation vials can be omitted.

7. If, despite thorough washing, the R-value obtained in a sample does not fit into one of the three distinct categories defining the genotype of the sample, this indicates the presence of a mixed sample, e.g., contaminating DNA or a PCR contamination.

8. The R-value obtained in the solid-phase minisequencing method reflects the ratio between two sequences when they are present in a sample as a mixture in any other ratio than that in samples from homozygous subjects (allele ratio 2:0) or heterozygous subjects (allele ratio 1:1). Since the two sequences are identical, with the exception of a single nucleotide, they are amplified with equal efficiency during PCR. Thus the ratio between the sequences in the amplified sample measured by the minisequencing assay directly reflects the initial ratio between the sequences in the original sample *(3)*. The ratio between two sequences in a sample can be calculated from the R-value, taking into account the specific activities of the [^{3}H]dNTPs used in the minisequencing reactions. If the sequence contains one (or more) identical nucleotides immediately next to the nucleotide at the variable site, one (or more) additional [^{3}H]dNTP will become incorporated, which obviously affects the R-value. Both of these factors are known in advance and can easily be accounted for. Alternatively, the ratio between two sequences can be determined by comparing the obtained R-value with a standard curve prepared by analyzing mixtures of known amounts of the corresponding two sequences. The high specificity of the single nucleotide incorporation catalyzed by the DNA polymerase allows detection of one sequence present as a small minority (<1%) in a sample *(13)*. The use of a standard curve corrects for differences in specific activity and the number of [^{3}H]dNTPs incorporated, but also for a possible small misincorporation of a [^{3}H]dNTP by the DNA polymerase, which may affect the result, particularly when a sequence present as a small minority of a sample is to be quantified.

References

1. Bowcock, A. and Cavalli-Sforza, L. (1991) The study of variation in the human genome. *Genomics* **11**, 491–498.

2. Syvänen, A.-C., Aalto-Setälä, K., Harju, L., Kontula, K., and Söderlund, H. (1990) A primer-guided nucleotide incorporation assay in the genotyping of apolipoprotein E. *Genomics* **8,** 684–692.

3. Syvänen, A.-C., Sajantila, A., and Lukka, M. (1993) Identification of individuals by analysis of biallelic DNA markers using PCR and solid-phase minisequencing. *Am. J. Hum. Genet.* **52,** 46–59.

4. Dieffenbach, C. W., Lowe, T. M. J., and Dveksler, G. S. (1993) General concepts for PCR primer design. *PCR Meth. Applicat.* **3,** S30–S37.

5. Higuchi, R. (1989) Simple and rapid preparation of samples for PCR, in *PCR Technology. Principles and Applications* (Erlich, H. A., ed.), Stockton, New York, pp. 31–37.

6. Innis, M. A. and Gelfand, D. H. (1990) Optimization of PCRs, in *PCR Protocols: A Guide to Methods and Applications* (Innis, M. A., Gelfand, D. H., Sninsky, J. J., and White, T. J., eds.), Academic, San Diego, CA, pp. 3–12.

7. Bengtström, M., Jungell-Nortamo, A., and Syvänen, A.-C. (1990) Biotinylation of oligonucleotides using a water soluble biotin ester. *Nucleosides Nucleotides* **9,** 123–127.

8. Wu, R., Wu, N.-H., Georges, F., and Narang, S. (1984) Purification and sequence analysis of synthetic oligodeoxyribonucleotides, in *Oligonucleotide Synthesis: A Practical Approach* (Gait, M. J., ed.), IRL, Oxford, UK, pp. 135–151.

9. Syvänen, A.-C. and Söderlund, H. (1993) Quantification of polymerase chain reaction products by affinity-based collection, in *Methods in Enzymology, vol. 218: Recombinant DNA,* Part I (Wu, R., ed.), Academic, New York, pp. 474–490.

10. Harju, L., Weber, T., Alexandrova, L., Lukin, M., Ranki, M., and Jalanko, A., (1993) Colorimetric solid-phase minisequencing assay for the detection of alpha-1-antitrypsin Z mutation. *Clin. Chem.* **2,** 2282–2287.

11. Livak, K. J. and Hainer, J. W. (1994) A microtiter plate assay for determining apolipoprotein E genotype and discovery of a rare allele. *Human Mutation* **3,** 379–385.

12. Ihalainen, J., Siitari, H., Laine, S., Syvänen, A.-C., and Palotie, A. (1994) Towards automatic detection of point mutations: use of scintillating microplates in solid-phase minisequencing, *Biotechniques* **16,** 938–943.

13. Syvänen, A.-C., Söderlund, H., Laaksonen, E., Bengström, M., Turunen, M., and Palotie, A. (1992) N-ras gene mutations in acute myeloid leukemia: accurate detection by solid-phase minisequencing. *Int. J. Cancer* **50,** 713–718.

14. Leitersdorf, E. and Hobbs, H. H. (1988) Human LDL receptor gene: Hinc II polymorphism detected by gene amplification. *Nucleic Acids Res.* **16,** 7215.

Index